Historical Perspectives of Fisheries Exploitation in the Indo-Pacific

MARE Publication Series

Volume 12

Series Editors

Svein Jentoft, *University of Tromsø, Norway*
Svein.jentoft@uit.no
Maarten Bavinck, *University of Amsterdam, The Netherlands*
J.M.Bavinck@uva.nl

The MARE Publication Series is an initiative of the Centre for Maritime Research (MARE). MARE is an interdisciplinary social-science network devoted to studying the use and management of marine resources. It is based jointly at the University of Amsterdam and Wageningen University (www.marecentre.nl).

The MARE Publication Series addresses topics of contemporary relevance in the wide field of 'people and the sea'. It has a global scope and includes contributions from a wide range of social science disciplines as well as from applied sciences. Topics range from fisheries, to integrated management, coastal tourism, and environmental conservation. The series was previously hosted by Amsterdam University Press and joined Springer in 2011.

The MARE Publication Series is complemented by the Journal of Maritime Studies (MAST) and the biennial People and the Sea Conferences in Amsterdam.

For further volumes:
http://www.springer.com/series/10413

Joseph Christensen • Malcolm Tull

Editors

Historical Perspectives of Fisheries Exploitation in the Indo-Pacific

Editors
Joseph Christensen
Asia Research Centre
Murdoch University
Murdoch
West Australia
Australia

Malcolm Tull
School of Management and Governance
Murdoch University
Murdoch
West Australia
Australia

Cover picture: Tuna in the freezer of a Longline fishing vessel, Federated States of Micronesia, 2006. By permission of Alex Hofford (http://www.alexhoffordphotography.com).

ISSN 2212-6260   ISSN 2212-6279 (electronic)
ISBN 978-94-017-8726-0   ISBN 978-94-017-8727-7 (eBook)
DOI 10.1007/978-94-017-8727-7
Springer Dordrecht Heidelberg New York London

Library of Congress Control Number: 2014935154

Acknowledgements

This volume is an outcome of the HMAP Asia project, a sub-project of the History of Marine Animal Populations (HMAP) initiative (see hmapcoml.org/). A number of individuals have contributed to this project, and the Editors would especially like to thank the following: Maarten Bavinck, Lynnath Beckley, John Butcher, Shelton Harley, Poul Holm, Neil Klaer, Alison MacDiarmid, Rod Lenanton, Neil Loneragan, Seamus McElroy, Kira Pauli Pravato, Ian Poiner, Bo Poulsen, Robb Robinson, Peter and Noelene Reeves, Peter Rogers, Kevin Rowling, Natasha Stacey, Kathleen Schwerdtner Máñez, David Starkey, Simon Viera, and the participants at the two HMAP Asia workshops in 2009 and 2010.

Figure 2.2 is reproduced courtesy of the Art Gallery of South Australia; Figs. 8.1, 9.1 and 9.2 were supplied by the Australian Bureau of Agricultural and Resource Economics and Sciences; Fig. 8.2 uses images supplied by Robert Gillet; and Fig. 10.2 is reproduced courtesy of the State Library of New South Wales. Unless indicated otherwise, all other maps, charts and tables were produced with the assistance of Camille Vogel. Joao Vilas Boas made a valuable contributing to formatting and indexing the chapters.

James Francis Warren generously supported the production of the volume by providing Joseph Christensen with extended relief in his position at Murdoch University, and the Asia Research Centre at Murdoch University provided an institutional home for the HMAP Asia project.

Contents

Abbreviations

ABARES	Australian Bureau of Agricultural and Resource Economics and Sciences
ABS	Australian Bureau of Statistics
ADB	Asian Development Bank
AFZ	Australian Fishing Zone
ASC	Aquaculture Stewardship Council
BBS	Bangladesh Bureau of Statistics
CALM	Department of Conservation and Land Management (Western Australia)
CCSBT	Commission for the Conservation of Southern Bluefin Tuna
CITES	Convention on International Trade in Endangered Species of Wild Fauna and Flora
CoML	Census of Marine Life
CPUE	Catch Per Unit Effort
CTI-CFF	Coral Triangle Initiative on Coral Reefs, Fisheries and Food Security
DEC	Department of Environment and Conservation (Western Australia)
DEPM	Daily Egg Production Method
EEZ	Exclusive Economic Zone
ENSO	El Niño-Southern Oscillation
EPO	Eastern Pacific Ocean
EU	European Union
FAD	Fish Aggregating Device
FAO	Food and Agricultural Organisation of the United Nations
FCY	Fish Curing Yards
FFA	Pacific Islands Forum Fisheries Agency
FSM	Federated States of Micronesia
FSMA	Federated States of Micronesia Arrangement for Regional Fisheries Access
GRT	Gross register tonnage
GPS	Global Positioning System
HMAP	History of Marine Animal Populations
HYV	High Yield Varieties

ICZMS	Integrated Coastal Zone Management Strategy
ITQ	Individual Transferable Quota
IUCN	International Union for the Conservation of Nature
IUU	Illegal, Unreported and Unregulated fishing
IWC	International Whaling Commission
MFB	Madras Fisheries Bureau
MFD	Madras Fisheries Department
MOU	Memorandum of Understanding
MPA	Marine Protected Area
MSC	Marine Stewardship Council
MTI	Marine Trophic Index
MTL	Mean Trophic Levels
NGO	Non-Governmental Organisation
NSW	New South Wales
OFCF	Overseas Fisheries Cooperation Foundation (Japan)
OIOC	Oriental and India Office Collections (British Library)
OPEC	Organization of the Petroleum Exporting Countries
PNA	Parties to the Nauru Agreement Concerning Cooperation in the Management of Fisheries of Common Interest
PNG	Papua New Guinea
QLD	Queensland
RAN	Royal Australian Navy
RFAC	Recreational Fishing Advisory Committee (Western Australia)
RFMO	Regional Fisheries Management Organizations
SA	South Australia
SEF	South-East Continental Shelf (Australia)
SBT	Southern Bluefin Tuna
SPC	Secretariat for the Pacific Community
STI	State Trawling Industry (New South Wales)
TAC	Total Allowable Catch
TBOA	Tuna Boat Owners Association (Australia)
UK	United Kingdom
ULT	Ultra Low Temperature
UNESCO	United Nations Educational, Scientific and Cultural Organization
USA	United States of America
VDS	Vessel Day Scheme
VPA	Virtual Population Analysis
WA	Western Australia
WCPFC	Western and Central Pacific Fisheries Commission
WCPO	Western and Central Pacific Ocean
WTO	World Trade Organisation
WWF	World Wildlife Fund
WWI	First World War
WWII	Second World War

Contributors

Jo Marie V. Acebes Asia Research Centre, Murdoch University, Murdoch, WA, Australia

Sid Adams Asia Research Centre, Murdoch University, Murdoch, WA, Australia

Kate Barclay Faculty of Arts and Social Sciences, University of Technology Sydney, Broadway, NSW, Australia

Ta-Yuan Chen Asia Research Centre, Murdoch University, Murdoch, WA, Australia

Joseph Christensen Asia Research Centre, Murdoch University, Murdoch, WA, Australia

Andrea Gaynor History, The University of Western Australia, Crawley, WA, Australia

Brooke Halkyard Department of Parks and Wildlife, Exmouth, WA, Australia

Poul Holm Trinity Long Room Hub, Trinity College Dublin, Dublin 2, Ireland

Gary Jackson Western Australian Fisheries and Marine Research Laboratories, Department of Fisheries, Western Australia, North Beach, WA, Australia

Anne Lif Lund Jacobsen Rigsarkivet, the Danish State Archives, København K, Denmark

John McGuire South Asia Research Unit, School of Media Culture & Creative Arts, Curtin University, Perth, WA, Australia

Bob Pokrant Department of Social Sciences & International Studies, School of Media, Culture & Creative Arts, Curtin University, Perth, WA, Australia

Peter Reeves South Asia Research Unit, School of Media Culture & Creative Arts, Curtin University, Perth, WA, Australia

Malcolm Tull School of Management and Governance, Murdoch University, Murdoch, WA, Australia

About the authors

Dr. Jo Marie Acebes   has a biology degree from the Ateneo de Manila University and a Doctor of Veterinary Medicine from the University of the Philippines. Jom worked for WWF-Philippines as the project manager of the cetacean and humpback whale research and conservation projects. She obtained a Masters degree in Biodiversity, Conservation and Management at the University of Oxford in 2005. In 2007, she became involved with the History of Marine Animal Populations—Asia. She continues to work in the field of cetacean and marine research and conservation as the principal investigator of a non-profit group called Balyena.org which she co-founded in 2009. From 2009 to 2013 she was a postgraduate student at the Asia Research Centre, Murdoch University.

Dr. Sid Adams   is a Fellow of the Asia Research Centre, and a former Lecturer in Politics, at Murdoch University. He now lives in Melbourne.

Dr. Kate Barclay   researches the international political economy of food, focusing particularly on tuna fisheries in the Asia Pacific Region. The main themes of her work include the development potential of tuna resources for Pacific Island countries and governing for sustainability along global tuna commodity chains. Kate has acted as researcher for several international organizations, including the WWF and the United Nations Development Programme. Her major publications include *A Japanese Joint Venture in the Pacific* (Routledge 2008) and *Capturing Wealth From Tuna* (ANU ePress 2007). Kate teaches in the International Studies Program at the University of Technology Sydney.

Dr. Ta-Yuan (Henry T) Chen   is a Fellow and former postgraduate student of the Asia Research Centre, and a former Postdoctoral Fellow at the University of Technology Sydney. He now lives in Sydney.

Dr. Joseph Christensen   is a Postdoctoral Fellow at the Asia Research Centre, Murdoch University. His research focusses on the environmental and maritime history of Western Australia and the Indian Ocean, and his past publications include contributions to the *Historical Encylopedia of Western Australia* (UWA Press, 2008) and the Centenary History of The University of Western Australia (2013). He was formerly Secretary of the Australian Association for Maritime History and is a past recipient of the Frank Broeze Postgraduate Award, Faculty of Arts and Humanities, The University of Western Australia. He teaches Southeast Asian and Australian history at Murdoch University.

Prof. Andrea Gaynor is Professor of History at The University of Western Australia. Primarily an environmental historian, she pursues various questions relating to historical relationships between the human and non-human. Her publications include articles on topics as diverse as landscape art and feral cats, and the book *Harvest of the Suburbs: An environmental history of growing food in Australian cities* (UWA Press, 2006).

Brooke Halkyard has been affiliated with the Department of Parks and Wildlife (Western Australia) since 2005. In her role, she has been fortunate to work with a variety of threatened species, including marine turtles. Brooke's experience includes several years training volunteers as part of the Ningaloo Turtle Program and satellite tagging of nesting loggerhead turtles at Ningaloo Marine Park. She has also spent several seasons volunteering with the flatback tagging program at Barrow Island. Brooke's love of the Ningaloo Coast, her fondness of marine turtles and her interest in history, provided the impetus for this project.

Poul Holm earned a Dr.Phil. at the University of Aarhus, Denmark, with a study of Scandinavian early modern maritime history. He is currently President of the European Consortium of Humanities Institutes and Centres and Director of the Irish Structured PhD Programme in Digital Arts and Humanities. He is a former President of the European Society for Environmental History, and he serves as global chair of the research project History of Marine Animal Populations. Holm is the author of ten books and more than one hundred research papers, including 'World War II and the "Great Acceleration" of North Atlantic fisheries', *Global Environment10*, 2013, and Holm et al., 'Marine Animal Populations: A New Look Back in Time', in Alasdair D. McIntyre (ed), *Life in the World's Oceans: Diversity, Distribution, and Abundance*, Oxford, Blackwell, 2010.

Dr. Anne Lif Lund Jacobsen is an environmental historian with a doctoral degree from University of Tasmania. Her research interests are history of science, marine environment and resource management. Her approach is cross disciplinary, drawing on methods from both science and economy applied to historic data. Between 2001 and 2010 she participated in the international research programme, History of Marine Animal Population (HMAP) under the Census of Marine Life (CoML). She is currently a postdoc at Center for Science Studies, Aarhus University.

Dr. Gary Jackson PhD (Murdoch), is a Principal Research Scientist with the Research Division of the Western Australian Department of Fisheries. He is based at the WA Fisheries and Marine Research Laboratories in Hillarys Boat Harbour to the north of Perth. Gary has been involved with research on Shark Bay's pink snapper stocks and the recreational fishery since 1997. He considers himself to have been very fortunate to have seen some of the bay's many natural splendors and met some of the unique characters involved in the local fishing industry.

Prof. John McGuire   PhD (SOAS), is Adjunct Professor of Modern History and a former Director of the South Asia Research Unit at Curtin University. His research focuses on South Asian history and politics, and he has published widely on Bengal, aspects of the Indian economy under colonial rule, and the Bharatiya Janata Party (BJP). He was co-editor with Ian Copeland of *Hindu nationalism and governance* (Oxford University Press, 2007).

Prof. Bob Pokrant   BA, MA (Northwestern), PhD (Cambridge), is Professor of Anthropology in the Department of Social Sciences and International Studies and a member of the management board of the Australia-Asia-Pacific Institute at Curtin University, Australia. His research interests include aquaculture and fisheries in South Asia and community adaptation to climate change in Bangladesh. He has published widely on the history of South Asian fisheries, export-oriented shrimp sector in Bangladesh and the anthropology of adaptation to climate change. He was co-editor with Michael Gillan of *Trade, Labour and the Transformation of Community in Asia*, (Palgrave-Macmillan, 2009).

Peter Reeves   BA, MA (Tasmania), PhD (ANU), is Emeritus Professor of South Asian History, Curtin University in Perth, Western Australia and Fellow of the Academy of the Humanities in Australia. He was Head of the South Asian Studies Programme at the National University of Singapore from 1999 to 2005 and has also taught at Sussex, Western Australia and Michigan. His research interests include: the political and agrarian history of northern India; the maritime history of the Indian Ocean since 1800; the history of fisheries in colonial and postcolonial South Asia; the history of Singapore's inshore and culture fisheries; and the history of the Indian diaspora. He was Executive Editor of *The Encyclopedia of the Indian Diaspora* (Singapore: Editions Didier Millet, 2006).

Prof. Malcolm Tull   is a Professor in the School of Management and Governance at Murdoch University in Perth, Western Australia. His main research field is maritime economics and maritime economic history. His publications include *A Community Enterprise: the History of the Port of Fremantle, 1897–1997* (1997), and the co-edited *Port Privatisation: The Asia-Pacific Experience* (2008). Malcolm was joint editor of the *International Journal of Maritime History* from 2000 to 2008 and led the Asian component of the History of Marine Animal Populations Project. He is a life member of the Australian Association for Maritime History and currently Vice-President of the International Maritime Economic History Association.

Chapter 1
Introduction: Historical Perspectives of Fisheries Exploitation in the Indo-Pacific

Joseph Christensen and Malcolm Tull

Abstract Historical knowledge has an important role in addressing the problems facing marine capture fisheries today. The growing awareness of the value of historical perspectives underpinned the History of Marine Animal Populations (HMAP) project, a 10-year global research collaboration concerned with the long-term interaction of humans and the marine environment. The chapters presented in this volume developed out of HMAP Asia, one of HMAP's 12 regional case-studies, and a sub-project designed specifically to address a lack of knowledge about the history of fishing and the historic impact of human activity on marine environments in Asia and Oceania. At a time when overfishing and declining fish stocks remain pressing problems for marine scientists and fisheries managers, the task of establishing baselines that expose the full extent of ecological change is as important as ever; understanding the scale and extent of historic change is a necessary first step towards achieving sustainability in marine capture fisheries. *Historical Perspectives of Fisheries Exploitation in the Indo-Pacific* represents an important step in what we hope will be ongoing international research on the marine environmental history of Asian and Pacific seas.

Keywords Asia fisheries history · Oceania fisheries history · Indo-Pacific fishing history · HMAP Asia · Marine environmental history

Overfishing of the Earth's oceans and seas has become one of the most pressing environmental problems of the twenty-first century. The world's wild marine harvest is widely believed to have peaked in the late 1980s, and the consensus amongst fisheries scientists and marine ecologists is that too much of the global catch is now being taken from stocks that are overexploited and continuing to decline (Worm et al. 2009; Mora et al. 2009; Pikitch 2012; Watson et al. 2012; Worm and Branch 2012). A lot is at stake in this crisis confronting global fisheries. Many millions of

J. Christensen (✉)
Asia Research Centre, Murdoch University, 90 South Street,
6150, Murdoch, WA, Australia
e-mail: J.Christensen@murdoch.edu.au

M. Tull
School of Management and Governance, Murdoch University, 90 South Street,
6150, Murdoch, WA, Australia
e-mail: M.Tull@murdoch.edu.au

J. Christensen, M. Tull (eds.), *Historical Perspectives of Fisheries Exploitation in the Indo-Pacific*, MARE Publication Series 12, DOI 10.1007/978-94-017-8727-7_1,

people worldwide depend on marine capture fisheries for their daily protein requirements, and fishing industries are the mainstay of coastal communities in many parts of the world. These challenges of food security and economic livelihood are particularly acute in the developing world, including the nations of Asia, Africa, and Latin America (Kooiman and Bavinck 2005). Yet unsustainable fishing is only one element in a larger environmental crisis that also involves widespread habitat destruction, the loss of biodiversity, increasing marine pollution, and the onset of rapid and unpredictable climate change (Roberts 2012). It seems that the human relationship with the sea has never been more precarious.

A number of strategies have been developed to address overfishing, although the results are mixed at best. The Code of Conduct for Responsible Fisheries, developed in 1995 by the Food and Agricultural Organisation (FAO) of the United Nations, established international standards for the sustainable exploitation of marine resources and encouraged the adoption of ecosystem-based approaches to fisheries management, where the management of exploited stocks is incorporated into a broader strategy of upholding ecosystem health. But compliance with the code has been poor, with clear differences between the relatively positive performance of Europe, North America and Oceania and the comparatively poor performance of Asia, Africa and Latin America (Pitcher et al. 2009). A similar pattern can be observed in relation to the targets for recovery and sustainable exploitation of marine fisheries established at the World Summit on Sustainable Development in 2002 (Kompass 2013). During the 2000s momentum built globally for the creation of marine protected areas (MPAs), where human activity is curtailed in order to preserve the natural environment, but when the development of MPAs is included as a benchmark for assessing marine resource management only a small number of nations from the developed world emerge with successful ratings(Alder and Pauly 2008). In western countries, public education programs actively seek to promote the consumption of seafood sourced from sustainable fisheries such as those certified by the Marine Stewardship Council (MSC). However, there has been no comparable impact on the eating habits of consumers in other parts of the world, particularly Asian consumers, whom account for nearly two-thirds of all seafood consumed globally (Roberts 2012).

These differences reflect a growing trend in which the status and prospects of global fisheries is increasingly unevenly divided around the world. In 2010 the FAO reported that 15 % of assessed fish stocks were low or moderately exploited, 53 % were fully exploited, 28 % were overexploited and 3 % were depleted (FAO 2010). However, when data from fisheries not traditionally included in FAO assessments of global fisheries are also considered, a more worrying picture emerges. As much of 80 % of the global catch now comes from stocks that are not formally monitored and where biomass is likely to be either fully exploited or declining (Costello et al. 2012).Such stocks are located disproportionately in the waters of Asia and Africa, where the challenges of food security and employment in fishing communities are already at their most acute (Pikitch 2012; Watson and Pauly 2013; Mora et. al. 2009). This division between countries with well-assessed and sustainably-managed fisheries, and countries with data-poor and poorly-managed fisheries, is of increasing concern to fisheries managers and scientists. The global limits to marine fisheries exploitation have now been reached, and, by necessity, promoting recovery of depleted stocks has become the basis of fisheries management strategies in many parts of the world. The need to direct science, management and conservation efforts to the poorly-performing nations and regions is, arguably, the most pressing problem in global marine fisheries today (Worm and Branch 2012).

What has changed during the quarter of a century since the peak in marine capture harvests is the role of historical knowledge in addressing the challenges of sustainability in world fisheries. History became an important factor in debates over fisheries resource management through the development of the concept of the 'shifting baseline syndrome', which exposed the tendency for marine scientists to adopt steady-state models of the marine environment and ignore historic changes in marine animal populations. As a consequence of the shifting baseline syndrome, ideas of what constitutes 'natural' have shifted over time towards more degraded environments, and the past impact of human harvesting activity on the marine environment has tended to be overlooked (Pauly 1995). Shifting baselines has been a groundbreaking concept, encouraging the integration of historical, archaeological and palaeoecological evidence into long-term assessments of human impacts on marine ecosystems (Jackson et al. 2001; Lotze and Worm 2009; Jackson and Alexander 2011). Much can come from an enhanced understanding of past patterns of diversity and abundance of life in the world's oceans. Awareness of the former numbers and geographical range of marine animals helps to clarify both the causes and the extent of long-term environmental changes and reveals the impact of human harvesting activity across timeframes that reach further, often many hundreds of years further, than the period for which reliable statistical data or detailed scientific assessments are available. History can also can also carry a powerful argument in favour of conservation, highlighting the full magnitude of anthropogenic changes in marine ecosystems and exposing the human values and behaviours responsible for destructive actions in the past (Bolster 2006; Bolster et al. 2011; Roberts 2007, 2012).

The growing awareness of the value of historical perspectives underpinned the History of Marine Animal Populations (HMAP) project, a 10-year global research collaboration concerned with the long-term interaction of humans and the marine environment. Conceived as the historical component of the inaugural Census of Marine Life (CoML), HMAP was a unique initiative that incorporated historical, archaeological, biological and ecological investigations of historic changes in marine animal populations across time (Holm 2003). Researchers involved in HMAP's 12 regional and 3 species-based projects uncovered many new sources of historical information, and developed new ways to analyse historical data, including methods for the calculating biodiversity and biogeographic range, techniques for the standardization of catch per unit of effort (CPUE), and approaches to modelling statistical data from historic and long-term data sets. This research has in turn contributed to the rise of two new and inter-related academic disciplines, marine environmental history and historical marine ecology, which have continued to develop beyond the formal end to HMAP's decade-long research program in 2010 (Holm et al. 2010).

This volume has developed out of HMAP Asia, one of HMAP's 12 regional case-studies, and a sub-project designed specifically to address a lack of knowledge about the history of fishing and the past impact of human activity on marine environments in Asia. One work dominates the literature in this area. John Butcher's *The Closing of the Frontier: A History of the Marine Fisheries of Southeast Asia, c.1850–2000* (2004), the first general history of marine capture fisheries across the vast Southeast Asian region, provides an overview of the changes that have taken place during a period that witnessed the rise and often devastating ecological impact of modern industrial fishing. The chapters that make up this volume cover a broader area of the eastern Indian and western Pacific oceans, or the Indo-Pacific, comprising of case-studies from South Asia, Southeast Asia, the Southwest Pacific, and Australia. This area covers large parts of Asia and Oceania, the statistical divisions used by the FAO and in other assessments of the status and prospects of global fisheries. It is also one of the world's principal marine biogeographic provinces (Spalding et al. 2007). The challenges that confront marine fisheries in Asia and Oceania, and the deficiencies in knowledge that exist in terms of the past, present, and future of marine fisheries in these parts of world, provides the backdrop for the studies in marine environmental history and historical ecology presented in *Historical Perspectives of Fisheries Exploitation in the Indo-Pacific*.

The following chapter by Joseph Christensen deals with the marine environmental history of the Central Indo-Pacific, a body of water that neatly encompasses the geographical range of the volume's other case-studies. The first part of Chap. 2 elaborates on the challenges that confront historians studying human impacts on marine environments in what is one of the most environmentally and socially diverse regions of the globe. In the second part of the chapter, Christensen extends Butcher's original (2004) framework for the expansion of marine capture fisheries between the second half of the nineteenth century and the present day. Their chapter shows how understandings of the history of marine animal populations is tied up inextricably in changing drivers of fishing intensity, the role of governments and their fisheries agencies in the development and management of fishing industries, the potential for different species to withstand fishing pressure and, finally, in the nature and availability of source materials and of techniques for analysing such sources. In this way, Christensen outlines a contextual framework for the chapters that follow. Each chapter, in turn, offers new insights into the historic impact of fishing industries and of past changes in the abundance and diversity of Indo-Pacific marine animal populations.

In Chap. 4, Jo Marie Acebes reconstructs the origins and impacts of both indigenous and foreign whaling activities. Confronted with limited archival and scientific records, but aware that Filipino peoples have extensive traditions of catching whales and other large marine mammals, Acebes draws extensively on oral histories to construct historical baselines for key species. This is a valuable exercise in ethnohistory that highlights the potential for such approaches to yield insight into historic change in the marine environment. In Chap. 7, Ta-Yuan Chen draws extensively on contemporary fishery reports to chart the origins, development and expansion of the island's distant-water fishing fleet. Such material rarely comes to light in analyses of the global expansion of marine capture fisheries. Chen's study of Taiwan's rise to global dominance in tuna fisheries, and Acebes exploration of the origins and impacts of indigenous and foreign whaling in the Philippines, are of particular significance in this collection. Historical studies written in English but based on the extensive use of Asian-language source materials remain uncommon, and books such as Micah Muscolino's *Fishing Wars and Environmental Change in Late Imperial and Modern China* (2009) are welcome additions to the literature. Along with other works by Chen (2009) and Acebes (2009), such multi-lingual research ultimately holds the key to unlocking the full potential for marine environmental history in the Asian region.

The role of central governments and their fisheries agencies is another important theme in Chen's chapter. Tuna fishing became crucial to Taiwan's economy and society during the 1920s and 1930s as the government encouraged investment in boats and equipment and the migration of skilled Japanese fishers to the island. In the 1950s and 1960s, as overfishing around Taiwan led to large decreases in tuna stocks, the island's government supported the expansion of fishing operations in more distant fishing grounds in Southeast Asia and the Indian and Pacific oceans, leading in time to the nation's distant-water tuna catch being surpassed only by that of Japan and the United States. Yet the relationship between the state and industry was often complex and, when viewed in retrospect, could occasionally be contradictory or misguided. Collapsing fish stocks are one example of state failures in policy-making, but as Peter Reeves, Bob Pokrant and John McGuire demonstrate in Chap. 3, fisheries agencies can also fail to meet government objectives even in situations where depletion of fish stocks is not an issue. Their study of the origins and development of fish curing yards in the Madras Presidency shows how reforms designed to benefit small-scale fishers by providing access to cheap salt for preserving catches had the effect of benefitting merchants and traders instead. Fishers typically lacked the capital to construct and equip curing sheds, providing an opportunity for wealthy merchants to enter the industry. Such investors rapidly gained a high degree of control by becoming boat owners and tying fishers to them by the granting of credit, so that merchants and financiers emerged as the main beneficiaries of the creation of the fish curing yards, while fishers remained a marginalised social group.

The plight of small-scale fishers is also central to Malcolm Tull's chapter on the history of Indonesia's shark fishery. Chapter 5 examines the rapid expansion of industrial shark fishing from the 1970s, driven mainly by the demand for shark fins in China. By the early twenty first century Indonesia had the largest harvest of elasmobranchs (sharks and rays) in the world. The main beneficiaries of this growth, however, were also boat owners and traders rather than fishers and their families. Traditionally, Indonesia's small scale fishers utilised most of the sharks, including the skin for leather, liver oil for medical purposes and the flesh for food. But the fishery was effectively open access, and a large amount of shark came to be caught as by-catch in industrial fisheries for high value species such as tuna, leading to increasing frequency of the cruel and wasteful practice of 'finning'. Declining elasmobranch populations emerged as serious threat to the livelihood of millions of coastal people during the 1990s and early 2000s, creating new obstacles to the implementation of effective management policies and driving the growth of illegal fishing in Australian territorial waters. A National Plan of Action is needed but complicated by fiscal constraints and the division of powers between the national, *Kabupaten* (district/regency) and provincial governments. Similar governance failures are a widespread problem in the Indian Ocean and Southeast Asian regions (Rumley Y; Chuenpagdee et al. 2005; Butcher 2004). Yet whereas the imperative to develop resource management strategies that are both ecologically sustainable and socially just is one of the greatest challenges facing fisheries governance in the twenty-first century, the case-studies presented here serve as reminders that such challenges are invariably bound up in the intertwined histories of fishing communities, state policymaking, and the impact of fishing activities on targeted populations.

As marine capture fisheries have declined, aquaculture has expanded to supply demand for seafood. Between 1995 and 2010, the contribution of aquaculture to total global fish production increased from 21 to 40 %, driving the ongoing expansion of fish production during this period (FAO 2012). But fish farming raises its own set of environmental and social challenges. Some of these are revealed by Bob Pokrant in Chap. 6 through his study of the development of shrimp farming in Bangladesh. The Bengali people have a long history of finfish and shrimp farming for domestic consumption, and during the 1970s the Bangladeshi government began to promote export-orientated development of aquaculture in order to generate foreign exchange. Farmed shrimp eventually accounted for almost 90 % of total seafood exports, and was the nation's second largest earner of foreign exchange. However, the development of aquaculture on a large scale led to the displacement of landless labourers from public lands which have been taken over by shrimp farming, and reduced opportunities for share cropping as farmers have switched from rice to shrimp production. There was also a severe environmental toll. Large areas of mangrove forest have been reclaimed in the course of aquaculture development, leading to a loss of nursery habitats and thus reducing the productivity of coastal fisheries. Some areas were also affected by increasing salinity of ground water, which reduces the productivity of surrounding rice growing and grazing lands. Disease also began to spread through shrimp farms. As has been the case in many developing nations, the growth of successful global export fisheries masks a range of negative impacts on marginal coastal and agricultural populations, although Pokrant gives due attention to acts of local resistance to the spread of aquaculture in Bangladesh.

Tuna fisheries, among the largest and most lucrative commercial fisheries in the world, helped to sustain the rapid global expansion of marine capture fisheries in the second half of the twentieth century. Perhaps surprisingly, these fisheries have not always attracted a corresponding level of historical attention (Robinson 2011). This volumes covers tuna fishing in three separate chapters: by Chen on Taiwanese distant-water tuna fishing; by Kate Barclay in Chap. 8 on tuna fishing in the Pacific Islands; and by Sid Adams in Chap. 9 on the Southern Bluefin tuna fishery. Barclay's focus is on the development of industrial tuna fishing amongst the islands of the Western and Central Pacific Ocean (WCPO). The role of Japan in the development of Indo-Pacific fisheries is an important theme in this chapter. Japanese pole-and-line and longline fishers supplying canned tuna for their own domestic market helped to open the sparsely-populated Pacific Islands to commercial fishing, before the development of Ultra Low Temperature (ULT) freezing technologies and the growing affluence of Japanese consumers led to growing catches of sashimi tunas such as bigeye and yellowfin during the late 1960s and 1970s. Following the declaration of Exclusive Economic Zones (EEZs) in the late-1970s many Japanese companies established joint ventures with Pacific Island states based mainly on medium-scale pole-and-line and longline operations, whilst during this same period foreign distant-water fleets employing large purse seines and longlines also developed a strong presence in these waters. Fishing pressure has not seriously impacted on faster-breeding skipjack and albacore populations, but yellowfin and bigeye tunas have been seriously depleted. The development of sustainable fishing practises for highly migratory species such as tunas necessitates the effective implementation of regional governance arrangements, although as Barclay shows, the co-operation of many Pacific Island nations to restrict catches to sustainable levels does not yet include the activities of distant-water foreign longliners nor ensure adequate protection in the adjoining seas of Indonesia and the Philippines.

The complexities that surround multinational cooperation in the management of migratory stocks provides a starting point for Sid Adams in his analysis of the Southern Bluefin Tuna (SBT) fishery. Although a truly international industry, this fishery has historically been dominated by Australian pole-and-line and purse seine vessels and distant-water Japanese longliners. Following decades of expansion after the Second World War (WWII) and in the face of declining catches, the first major restrictions on these operations were introduced in the 1980s, before the Commission for the Conservation of Southern Bluefin Tuna (CCSBT) was established in 1994 to promote sustainable harvests levels amongst the major fishing nations. However, ongoing declines in catches and the corresponding reduction in official quotas allocated to each nation underlined the difficulties confronted by the CCSBT. A review of market-based data in 2006 revealed significant Japanese under-reporting of Southern Bluefin Tuna catches, one of the most infamous cases of Illegal, Unreported and Unregulated (IUU) fishing of recent times (Polacheck 2012). Adams argues that conflicting priorities in the international forums for governance explain such failures. Whereas Australia's position in relation to stock management has been underpinned by the precautionary principle with preservation of stock the main goal, Japanese scientists has been caught up in a corporatist alliance with the fishing industry and so tend to interpret scientific data in ways that benefit the industry's interest, at least in the short-term. Politics is never far removed from the management of fisheries resources; a point that has been made elsewhere in relation to Australian and Asian fisheries by Meryl Williams, a former Director of the World Fish Centre. She describes Australian and Southeast Asian fisheries as 'enmeshed' through trade in fish products, shared fish stocks and bilateral management arrangements, and argues that the pathway to resolving current and potential problems requires a more concerted effort to engage and understand the priorities and positions of foreign countries (Williams 2007). The same can be said for the Indo-Pacific at large, as the chapters by Barclay and Adams indicate.

The remaining chapters deal exclusively with the history of Australian fisheries. Chapter 10, by Anne Lif Lund Jacobsen, examines the south-east Australian trawl fishery. This fishery represents a relatively well-documented example of the impact modern industrial fishing on a previously untouched ecosystem, and for this reason, it is of particular interest to historical ecologists and environmental historians (Klaer 2001; Gowers 2008). Jacobsen's analysis covers the political, economic, and ecological aspects of this industry's rise and decline; rarely has a single fishery anywhere in the Indo-Pacific region received such comprehensive historical treatment. Established by the New South Wales (NSW) government in 1915 in order to boost the commercial fishing industry and provide a new and cheap supply of fish, the State Trawling Industry (STI) was soon sold into private ownership and, despite a period of success in the 1920s, due to the combined impact of the Great Depression and the collapse of the richest trawling grounds near Sydney, a protracted decline in the financial position of the main trawling companies set in by the end of the decade. A revival occurred during WWII, before a second and ultimately final phase of decline took place between 1954 and 1961 in the context of changing species composition on the main trawling grounds caused by sustained trawling over the preceding decades. This chapter shows how the commercial position of the industry was tied closely to the biomass of the Southeast Australian continental shelf; though gear changes and the spatial expansion of trawling compensated for a time for declining CPUE, a long-term decline in abundance of the main target species, tiger flathead, combined with fleet overcapacity and changed marketing arrangements, eventually made the industry unviable. Research has shown that the tiger flathead population was fished down to a level of approximately 20 % of its unfished size between 1915 and the 1950s, and although a recovery to around 40 % by the mid-2000s was an encouraging sign, the South-East Australian trawl fishery nonetheless stands as a reminder of the devastating and long-term impacts of trawl fisheries on marine ecosystems.

Few fisheries yield such stark insights into the impact of fishing on marine animal populations– the nature of historical records often makes it impossible to reach precise conclusions, as Christensen discusses in Chap. 2. This is a problem that Andrea Gaynor confronts in her chapter on fishing and environmental change in the south-west Capes Region of Western Australia. The Capes Region is characterized by low productivity, high biodiversity, and variations in the influence of currents, which can impact significantly on the regional abundance and distribution of the main species of fish targeted in these waters. Fishing effort has been multi-method and multi-species, involving both the commercial and recreational sector. Gaynor disentangles this complex history by examining sources that include commercial catch data, angling club records, newspaper reports, oral histories, and biological and oceanographic surveys. The pattern that emerges is one of persistent anxieties over stock declines and limited evidence of actual depletions within a broader context of cyclical and long-term environmental change. Since 2011 this region has been part of the Ngari Capes Marine Park, one of an increasing number of MPAs in Australian waters. Historical knowledge that predates scientific surveys and provides insights into long-term changes in marine animal populations, of the kind that

Gaynor presents here, provides a basis for future monitoring of such populations in areas protected by marine parks. Similair studies, undertaken in Western Australia's Ningaloo Reef marine park (Fowles and Gaynor 2011), further demonstrate the role of historical knowledge in setting benchmarks for marine conservation efforts, provided of course that methodological challenges can be managed.

Catch records and oral histories also provide the basis of Brooke Halkard's study of the turtle fishery in the Ningaloo Reef region. Six of the world's seven species of marine turtles are found in Western Australian waters, and two species, green and hawksbill turtles, are the main subject of attention in Halkyard's chapter. She shows that although there is evidence of commercial exploitation dating back to the late 1860s, only sporadic harvesting occurred for the next 90 years, before a commercial industry utilising small catcher boats and large freezer craft was eventually established at the start of the 1960s. Over the next 10 years approximately 69,000 green turtles and 20,400 hawksbill turtles were taken from the Northwest Australian coast. In 1973 the fishery was closed by the Western Australian government in response to evidence of overfishing and in a context of rising public opposition, influenced by international concerns, to the commercial exploitation of marine turtles. Green and hawksbill turtle populations now show signs of recovery, and are today the largest in the Indo-Pacific region.

The closure of Western Australia's marine turtle fishery is an example of the role of turtle species as flagships for marine conservation (Frazier 2005). Large marine mammals such as the humpback whale, once hunted to verge of extinction in many parts of the world, are another flagship species that act as symbols of endangered species, attracting international interest in population recovery and preservation. In their respective chapters, Halkyard and Acebes underline the vulnerability of these animals to human harvesting and outline pathways for constructing baselines for populations subjected to harvesting pressure in the past. Each author combines archival research, oral histories and insights from modern population studies in their quest to document the impact of human harvesting practises. For Acebes this task is complicated by evidence that whaling has continued in parts of the Philippines subsequent to the introduction of a ban introduced by the Philippine government in 1992; it may be continuing today, in a scenario that reflects the wider challenges faced by scientists, managers and conservationists alike in poorly performing jurisdictions of the Indo-Pacific. Knowing what has been lost provides a valuable target for conservation efforts, and while recoveries in long-lived and slow-breeding species of turtles and whales can be measured only over a period of decades, the chapters by Halkyard and Acebes are valuable contributions to the global effort to protect such iconic marine animals.

Recreational fisheries, once overlooked, now attract increasing attention from scientists and managers. Historians, too, have begun to show an interest in recreational fisheries (McClenachan 2013; Cooke and Cowx 2006). In Chap. 13 Joseph Christensen and Gary Jackson chart the decline and recovery of the recreational fishery for Pink Snapper at Shark Bay on the Western Australian coast. The story told here is one that is common to many developed countries, where recreational fishers, utilising small boats and fish-finding technologies such as GPS and sonar, have developed the potential to deplete targeted stocks that is in some instances comparable to the impact of commercial fisheries. Shark Bay is somewhat exceptional, however, in that the level of recreational catch and effort is comparatively well studied over a period of several years. The collapse of snapper stocks in the Bay's inner gulf led the state's fisheries department to trial an array of management strategies that were entirely novel in the context of recreational fisheries management, including closed seasons, total fishing bans, and in 2003, the first Total Allowable Catch (TAC) for a non-commercial fishery anywhere in Australia. Extensive biological research underpinned these management strategies. Today, the recovery of Shark Bay's inner gulf snapper stocks, and the widespread public support for the strong management controls that apply to recreational fishers, stands as a fine example of the potential for scientists, managers and fishers to work cooperatively to reverse population declines.

The potential for recoveries in exploited marine animal populations, highlighted in several chapters, emerges as one of the main themes of this volume as a whole. Such positive messages reinforce the importance of historical perspectives. At a time when overfishing and declining fish stocks remain pressing problems for marine scientists and fisheries managers, the task of establishing baselines that expose the full extent of ecological change is as important as ever–understanding the scale and extent of historic change is a necessary first step towards achieving sustainability in marine capture fisheries. This is particularly the case in Indo-Pacific seas, which contain some of the world's most heavily exploited fisheries, yet some of least well understood fisheries from both scientific and historical perspectives. As Meryl Williams has argued in relation to Southeast Asia and Australia, and as we posit for the Indo-Pacific at large, the fisheries of the developing and the developed world are fundamentally enmeshed as parts of a single global ocean. A truly international collaboration, uniting researchers in marine environmental history and historical marine ecology, will ultimately be required to place our understanding of marine animal populations in Asia and Oceania on a par with knowledge of oceans past in other parts of the world. *Historical Perspectives of Fisheries Exploitation in the Indo-Pacific* represents an important step in what we hope will be ongoing international research on the marine environmental history of Asian and Pacific seas.

References

Acebes JMV (2009) Historical whaling in the Philippines: origins of 'indigenous subsistence whaling', mapping whaling grounds and comparisons with current known distribution: A HMAP Asia project paper. Working paper No. 161. Murdoch University, Perth, Western Australia

Alder J, Pauly D (2008) Aggregate performance of countries in managing their EEZs. In: Alder J, Pauly D (eds) Fisheries Centre Research Reports, 16(7)

Bolster WJ (2006) Opportunities in marine environmental history. Environmental History 11(3):567–598

Bolster WJ, Alexander KE, Leavenworth WB (2011) Historical abundance of Cod on the Nova Scotia Shelf. In: Jackson JBC, Alexander KE, Sala E (eds) Shifting baselines: the past and the future of ocean fisheries. Island Press, Washington

Butcher JG (2004) The closing of the frontier: a history of the marine fisheries of Southeast Asia, c.1850–2000. ISEAS Publications, Singapore

Chen HT (2009) Taiwanese distant-water fisheries in Southeast Asia, 1936–1977. International Maritime Economic History Association, St. John's Newfoundland

Cooke S, Cowx IG (2006) Contrasting recreational and commercial fishing: searching for common issues to promote unified conservation of fisheries resources and aquatic environments. Biol Conserv 128:93–108

Costello C, Ovando D, Hilborn R, Gaines SD, Deschenes O, Lester SE (2012) Status and solutions for the world's unassessed fisheries. Science 338:517–520

Chuenpagdee R, P Degnbol, M Bavinck, S Jentoft, D Johnson, R Pullin, S Williams (2005) Challenges and concerns in capture fisheries and aquaculture. In: Kooiman J, Bavinck M, S Jentoft S, Pullin R (eds) Fish for life: interactive governance for fisheries. Amsterdam University Press, Amsterdam, MARE Publication Series No. 3

FAO Fisheries and Aquaculture Department (2010) The state of world fisheries and aquaculture. FAO, Rome

FAO Fisheries and Aquaculture Department (2012) The state of world fisheries and aquaculture. FAO, Rome

Fowles B, Gaynor A (2011) The challenge of creating a scientifically-robust historical description of changing finfish populations in the Ningaloo Marine Park. In: Gaynor A, Davis J (eds) Environmental exchanges: studies in Western Australian history (27):99–124

Frazier J (2005) Marine turtles: the role of flagship species in interactions between people and the sea. MAST 3/4:241–271 (Maritime Studies)

Gowers R (2008) Selling the 'untold wealth' in the seas: a social and cultural history of the Southeast Australian shelf trawling industry, 1916–1961. Environ Hist 14:265–287

Holm P (2003) History of marine animal populations: a global research program of the census of marine life. Oceanol Acta 25:201–211

Holm P, Marboe AH, Poulsen B, Mackenzie BR (2010) Marine animal populations: a new look back in time. In: McIntyre AD (ed) Life in the world's oceans: diversity, distribution and abundance. Wiley Blackwell, Chichester

Jackson JBC, Alexander K (2011) Introduction: the importance of shifting baselines. In: Jackson JBC, Alexander KE, Sala E (eds) Shifting baselines: the past and the future of ocean fisheries. Island Press, Washington

Jackson JBC, Kirby MX, Berger WH, Bjorndal KA, Botsford LW, Bourque BJ, Bradbury RH, Cooke R, Erlandson J, Estes JA, Hughes TP, Kidwell S, Lange CB, Lenihan HS, Pan-dolfi JM, Peterson CH, Steneck RS, Tegner MJ, Warner RR (2001) Historical overfishing and the recent collapse of coastal ecosystems. Science 293(5530):629–637

Klaer NL (2001) Steam trawl catches from South-Eastern Australia from 1918 to 1957: trends in catch rates and species composition. Mar Freshw Res 52:399–410

Kompass T (2013) Saving the fish in Asia and the Pacific. East Asia Forum 30 January 2013

Kooiman J, Bavinck M (2005) The governance perspective. In: Kooiman J, Bavinck M, Jentoft S, Pullin R (eds) Fish for life: interactive governance for fisheries, Mare Publication Series, vol 3. Amsterdam University Press, Amsterdam

Lotze H, Worm B (2009) Historical baselines for large marine animals. Trends Ecol Evol 24(5):254–262

McClenachan L (2013) Recreation and the "Right to Fish" movement: anglers and ecological degradation in the Florida keys. Environ Hist 18(1):76–87

Mora C, Myers RA, Coll M, Libralato S, Pitcher TJ et al (2009) Management effectiveness of the world's marine fisheries. PLoS Biol 7(6):e1000131. doi:10.1371/journal.pbio.1000131

Muscolino M (2009) Fishing wars and environmental change in late imperial and modern China. Harvard East Asian monographs. Harvard University Press, Cambridge

Pauly D (1995) Anecdotes and the shifting baseline syndrome of fisheries. Trends Ecol Evol 10(10):430

Pikitch EK (2012) The risks of overfishing. Science 388:474–475

Pitcher T, Kalikoski D, Pramod G, Short K (2009) Not honouring the code. Nature 475(5):658–659

Polacheck T (2012) Assessment of IUU fishing for Southern Bluefin Tuna. Mar Policy 36(5):1150–1165

Roberts C (2007) The unnatural history of the sea. Island Press, Washington

Roberts C (2012) Ocean of life: how our seas are changing. Penguin Books, London

Robinson R (2011) Hook, line and sinker: fishing history—where we have been, where we are now and where we are going? Mar Mirror 97(1):167–179

Rumley D (2009) A policy framework for fisheries conflicts in the Indian Ocean. In: Rumley D, Chaturvedi S, Sakhuja V (eds) Fisheries exploitation in the Indian ocean: threats and opportunities. ISEAS Publications, Singapore

Spalding MD, Fox HE, Allen GR, Davidson N, Ferda ZA, Finlayson M, Halpern BS, Martin K D, Mcmanus E, Molnar J, Recchia CA, Robertson J (2007) Marine ecoregions of the world: a bioregionalization of coastal and shelf areas. Bioscience 57(7): 573–583

Watson RA, Pauly D (2013) The changing face of global fisheries—The 1950s vs. the 2000s. Mar Policy 42:1–4. http://dx.doi.org/10.1016/j.marpol.2013.01.022

Watson RA, Cheung WWL, Anticamara JA, Sumaila RU, Zeller D, Pauly D (2012) Global marine yield halved as fishing intensity redoubles. Fish Fish 14:493–503 doi:10.1111/j.1467-2979.2012.00483.x

Williams M (2007) Enmeshed: Australia and Southeast Asia's fisheries. Lowy Institute for Public Policy, Double Bay

Worm B, Branch TA (2012) The future of fish. Trends Ecol Evol 27(11):594–599

Worm B, Hilborn R, Baum JK, Branch TA, Collie JS, Costello C, Fogarty MJ, Fulton EA, Hutchings JA, Jennings S, Jensen OP, Lotze HK, Mace PM, McClanahan TR, Minto C, Palumbi SR, Parma AM, Ricard D, Rosenberg AA, Watson R, Zeller D (2009) Rebuilding global fisheries. Science 325(5940):578–585

Chapter 2
Unsettled Seas: Towards a History of Marine Animal Populations in the Central Indo-Pacific

Joseph Christensen

Abstract A central theme of this book concerns the importance of historical perspectives for understanding the challenges that confront marine capture fisheries in the twenty-first century. This chapter explores this theme in relation to the Central Indo-Pacific, a body of water that lies at the geographic and geopolitical heart of the different case studies brought together in this volume. The Central Indo-Pacific is one of the world's principal marine biogeographic realms. It is made up of the eastern Indian and western Pacific oceans, and the seas linking the two—the South China Sea, the seas and straits of Southeast Asia, the Coral Sea, the waters separating Australia from Indonesia and Papua New Guinea, and Australia's northern continental shelf. Here, I cover a period similar to the timeframe of the book's other chapters, which extends from the late nineteenth to the early twenty-first century. This was a period of profound transformation in the marine fisheries of the Central Indo-Pacific, brought about by the intensification of established fisheries and the advent of new industrial fishing practices. My aims are two-fold: to discuss some of the challenges that confront marine environmental historians working in this region; and to describe the major patterns to the transformation of fishing during the period under review, which propelled the Central Indo-Pacific to the centre of the global expansion of marine capture fisheries.

Keywords Asia fisheries history · Oceania fisheries history · Indo-Pacific fishing history · HMAP Asia · Marine environmental history

A central theme of this book concerns the importance of historical perspectives for understanding the challenges that confront marine capture fisheries in the twenty-first century. This chapter explores this theme in relation to the Central Indo-Pacific, a body of water that lies at the geographic and geopolitical heart of the different case studies brought together in this volume. The Central Indo-Pacific is one of the world's principal marine biogeographic realms. It is made up of the eastern Indian and western Pacific oceans, and the seas linking the two—the South China Sea, the seas and straits of Southeast Asia, the Coral Sea, the waters separating Australia from Indonesia and Papua New Guinea, and Australia's northern continental shelf.

J. Christensen (✉)
Asia Research Centre, Murdoch University, 90 South Street,
6150, Murdoch, WA, Australia
e-mail: J.Christensen@murdoch.edu.au

J. Christensen, M. Tull (eds.), *Historical Perspectives of Fisheries Exploitation in the Indo-Pacific*, MARE Publication Series 12, DOI 10.1007/978-94-017-8727-7_2,
© Springer Science+Business Media Dordrecht 2014

Here, I cover a period similar to the timeframe of the book's other chapters, which extends from the late nineteenth to the early twenty-first century. This was a period of profound transformation in the marine fisheries of the Central Indo-Pacific, brought about by the intensification of established fisheries and the advent of new industrial fishing practices. My aims are two-fold: to discuss some of the challenges that confront marine environmental historians working in this region; and to describe the major patterns to the transformation of fishing during the period under review, which propelled the Central Indo-Pacific to the centre of the global expansion of marine capture fisheries.

Catch statistics for most Central Indo-Pacific fisheries are available only from 1950 onwards, when the FAO began to collect information on fisheries production globally. These statistics point to the truly massive expansion in harvests of wild fish and other marine animals in the second half of the twentieth century. Between 1950 and 2000 the fishing fleets of Asia and Oceania, the two continental regions that border the Central Indo-Pacific, increased their recorded marine catches by 422 and 1,218 % respectively, against a global average of 344 % (Watson and Pauly 2013). This growth was driven by a series of innovations in fishing vessels, gears, and fish-finding technologies, which in the case of the Asia produced a 25-fold increase in effective fishing power, and by a relentless expansion of the geographic frontiers of fishing activity that moved at an average rate of 1° of latitude per year during the period 1950 to 2000 (Swartz et al. 2010). There is, however, more to the history of this expansion than is revealed by analysis of catch and effort data. The maritime peoples of South Asia, Southeast Asia and the South Pacific have strong indigenous fishing traditions, and the intensification of existing and predominantly small-scale fishing practises in response to economic and demographic growth was a major factor in the expansion in capture fisheries. The second major factor involves the advent of industrial fishing, or fishing powered by fossil fuels, which facilitated the remarkable increase in fishing power. A third and related factor lies in the increasing presence of foreign fishing fleets, particularly the Japanese, in Central Indo-Pacific seas. Together, these developments help to explain the dramatic transformation of marine capture fisheries that took place between the late nineteenth and early twentieth century (Fig. 2.1).

The Marine Environment

The hallmark of the Central Indo-Pacific is its high level of biodiversity, with the Coral Triangle, the 'global centre of marine biodiversity', at the very heart of region (Allen 2007). This rich biological diversity is a factor of the complexity of the marine environment on broad physical and ecological scales. In their global system for the regional classification of coastal and shelf areas, Spalding and his colleagues delineate 12 separate provinces and 41 distinct ecoregions within the Central Indo-Pacific biogeographic realm (Spalding et al. 2007). As this scheme suggests, the outstanding feature of this realm is the extent of its shallow coastal

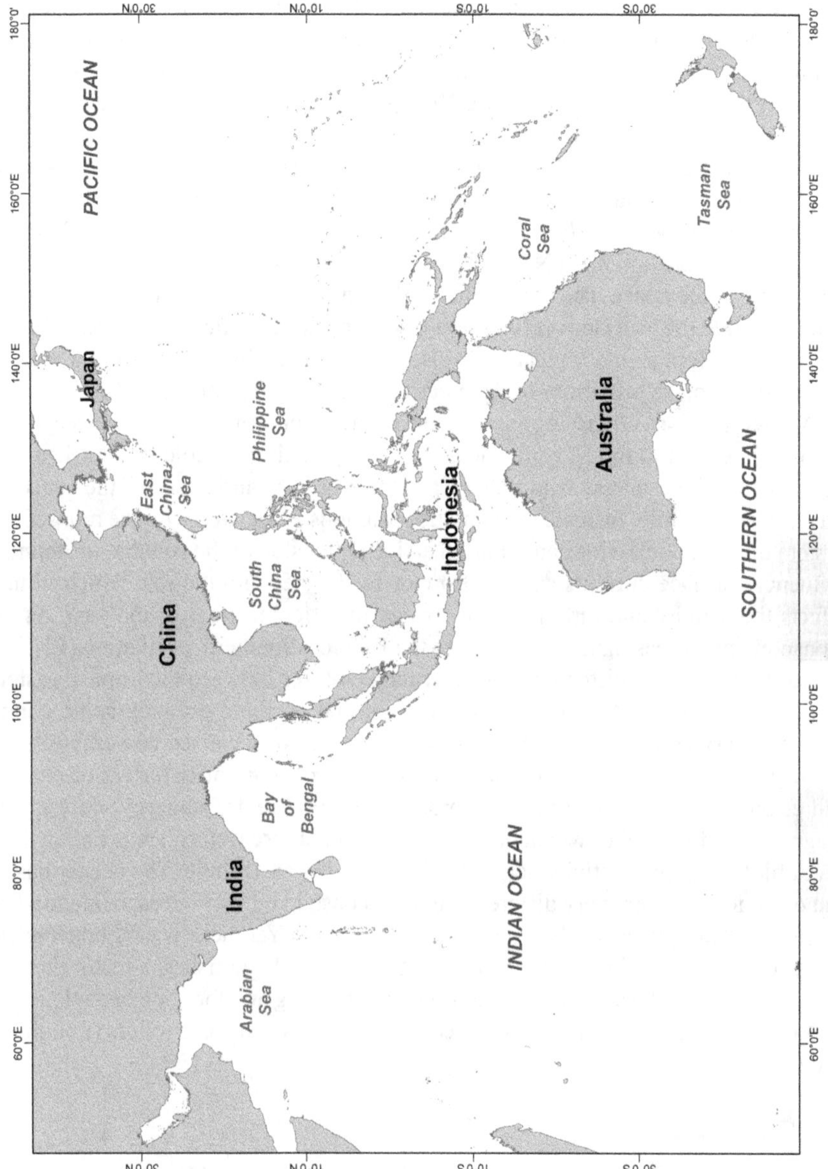

Fig. 2.1 Map of the Central Indo-Pacific

and shelf, as opposed to oceanic, waters. Two large shelf areas, which together form around one-fifth of the shelf area of the entire globe, lie at the centre of the Central Indo-Pacific. The Sunda Shelf encompasses the coastal waters of Burma, the Malacca Straits, the Gulfs of Tonkin and Thailand, the southern part of the South China

Sea, the Java Sea, and parts of the Philippines. The Sahul or Arafura Shelf is located to the south and comprises of the Timor and Arafura Seas separating Australian and Indonesia, the Gulf of Carpentaria, and Torres Strait. All of these waters are less than 200 m deep, and many parts are 50 m or less. Deeper basins are formed by the eastern parts of the Bay of Bengal and the Andaman Sea, the northern half of the South China Sea, the Sulawesi, Flores and Banda Seas, the Coral Sea, the Arafura Sea between Australia and Papua New Guinea, and the Timor Sea and Northwest Australian shelf between Australia and Indonesia. South of Java, west of Sumatra, and north and east of Papua New Guinea are narrow shelves and deep seas in close proximity to the coast. The island chains that fall within the Central Indo-Pacific, which include the Marianas Islands, the Caroline Islands, Micronesia, the Solomon Islands, Fiji, Tonga, and Vanuatu, are also characterized by narrow shelf areas and the relatively close proximity of deep oceanic waters (Longhurst 2007).

Across these seas there is considerable variation in terms of tides, currents, and salinity. The Central Indo-Pacific lacks the vast upwellings of cold and nutrient-rich waters of the kind that sustain fisheries in the North Atlantic and off the coasts of California and Chile. Instead, the climatic variations associated with the monsoon, a system of alternating winds and rainfall active either side of the equator, profoundly influences surface currents and the flow of freshwater into the sea, which in turn affects the salinity and turbidity of the water at different times of the year. As the monsoon varies seasonally according to the El Niño-Southern Oscillation (ENSO), caused by prolonged differences in water temperatures between the tropical eastern and tropical western Pacific Ocean, localised and regional oceanographic conditions can vary considerably from year to year. Areas of localised upwellings, such as the eastern part of the Banda Sea, the existence of nutrient-rich river discharges and estuarine systems, and the high biomass of coral reefs, mangrove forests and seagrass meadows, all of which occur extensively in these waters, are the main drivers of biological productivity across the Indo-Pacific as a whole. The deeper basins and oceanic waters are very different from the coastal and shelf areas, being marked by significantly lower levels of primary productivity. Yet these waters nonetheless support large populations of tuna, billfish and other pelagic fishes, a factor that has been attributed to the simplified conditions for breeding and foraging behaviour that result from the highly stratified water column that exists in these waters (Longhurst 2007).

Many States of the Sea: Marine Environmental History in the Central Indo-Pacific

The History of Marine Animal Populations (HMAP) initiative proposes two general principles in relation to the study of past patterns of diversity and abundance in the marine environment and the historic impact of humans on marine populations. The first concerns data sources. Human fishing activity has historically been concentrated in coastal and near-shore environments or, in other words, in that part of the sea that is most accessible to people. Such areas constitute the 'the human edges' of the ocean, and represent the areas where historical records are most abundant and hence where knowledge of past interactions with the marine environment are most likely to be recoverable (Holm et al. 2010). The second principle concerns methodology. Research in marine environmental history and historical marine ecology required a fundamentally interdisciplinary approach that draws on history, archaeology, marine science, and ecology. These principles underpin some of HMAP's main insights, which are to clarify both the potential and pathways for recovering knowledge of oceans past and to sharpen understanding of what is likely to always remain beyond the reach of human understanding (Holm 2003; Holm et al. 2010). HMAP's principles are also universal, applying to all the world's seas across time. Yet the Central Indo-Pacific region is not as straightforward a prospect for marine environmental historians as the seas of Europe and North America, a situation that helps to explain why Asia and Oceania have yet to attract the same level of attention as the extensively studied and better understood marine animal populations of the northern hemisphere. In this section I discuss some of the reasons why marine environmental history presents such a challenge in the Central Indo-Pacific.

In the first instance, an important distinction needs to be made between histories of general sea fishing, or fishing that yields food for human consumption, and high-value fisheries linked to long-distance trade networks, such as whaling, pearling, trepang or beche-de-mer fisheries, and fisheries for tortoiseshell, trochus, cowries, and chank. Whereas the former have tended to be neglected, the latter fisheries have been the subject of a number of detailed studies including several studies undertaken from a historical-ecological approach (e.g. Schwerdtner Mánez 2010; Schwerdtner Mánez and Ferse 2010; Anderson et al. 2011). Whaling, pearling and trepang fishing in particular share similar historical characteristics. In essence these relate to a comparatively early entry into global trading networks and the methodical exploitation of marine resources by large and mobile fishing fleets, linked to the demand for high-value commodities in markets often located at considerable distance from the seas where such resources are exploited: the market for whale oil and other whale products in Europe and North America; for trepang and other marine exotica in China; for pearls in Persia, India, China and Europe, and for mother of pearl in Europe and North America. By virtue of their link to global trading networks these industries produce more extensive historical records, often accessible in European archives, which helps to explain why a more extensive historiography exists (Boomgaard 2005). By contrast, there is little historical literature relating to the history of general sea fishing. John Butcher's *The Closing of the Frontier* (2004), a history of the marine fisheries of Southeast Asia between the mid-nineteenth and early twenty-first centuries, is the outstanding exception to this general neglect. Three factors in particular combine to make the marine fisheries of the Central Indo-Pacific such a challenging prospect: one relates to the nature of the marine environment; a second relates to the nature of human society; and a third relates to the broader history of human society in this part of the world.

The rich biodiversity of the Central Indo-Pacific is reflected in the variety of species that have been harvested across time for human consumption. These warm tropical and subtropical waters provide a habitat for many tens of thousands of species of worms, corals, crustaceans, molluscs, echinoderms, fish, reptiles and marine mammals, which along with phytoplankton, algae, seagrasses and detritus form complex ecosystems that are, even today, at best only partially understood (Butcher 2005). This means that we are dealing with many hundreds of, if not several thousand, exploited species, ranging from whales and dugongs, to turtles, to fish of every size, to shellfish collected by hand along the shoreline. There are two important consequences that arise from this high level of diversity. The first is that most fisheries tend to be multi-species rather than single-species, where fishers readily switch target species, location and even gears in response to weather conditions or the vagaries of fish behaviour in order to obtain an adequate catch on any given day. This is a feature common to most tropical fisheries, and it makes it difficult or even impossible to determine the composition of catches across time (Allsopp 1977). The second consequence of high biodiversity relates to the problem of naming. A wealth of indigenous names for the same or similar species once existed across this extended region, and although many of these names are still used, a great many are only poorly recorded in the present and many others are now rarely used and may not have been recorded at all. The converse of this problem is the tendency for Europeans and speakers of European languages to refer to many different species by simple generic names. 'Snapper', a name that applies to over a 100 species in the Lutjanidae family of fishes, is perhaps the extreme example of this tendency, although other common names such as 'shark', 'mackerel' and 'tuna' have also been applied in the past to a range of species that may not even occupy similar ecological strata nor bear more than a superficial resemblance to the Northern Hemisphere fishes after which they are named (Fig. 2.2).

Secondly, in relation to human society, it must be noted that the coastal peoples of South Asia, Southeast Asia, the South Pacific Islands, and northern Australia possess a wide diversity of social and political systems, cultures, and languages. This has certainly been the case in the past, and remains true today. In terms of fishing, this translates into an almost bewildering array of fishing techniques and technologies across time. Of equal significance is the characteristic that most fishing communities share in common across this region. Fisherfolk are traditionally amongst the most socially marginal members of society. The vast majority of traditional fishers across Asia and the Pacific have historically been illiterate peoples, and part of communities that lie at the very margins or even beyond the authority of centralised states, at least on a practical basis (Firth 1966; Reeves et al. 1988; Pearson 2003). This creates a problem in terms of historical records, insofar as they simply don't exist for many fisheries across the region before the second half of the twentieth century (Butcher 2004). Furthermore, extensive areas of the Central Indo-Pacific are highly vulnerable to natural disasters such as typhoons or tsunamis or to the activities of pirates and slave raiders, which may cause discontinuities or silences to appear in the historical records that do exist.

The third challenge relates to history, or rather, to the periodization and chronology of Central Indo-Pacific fisheries. Models that explain changes in fisheries across time propose three principal phases of development, moving from 'Aboriginal'

Fig. 2.2 JW Lewin, *Fish Catch at Dawes Point* (Sydney Harbour, 1813). Reproduced by permission of the Art Gallery of South Australia

or subsistence exploitation of coastal and near-shore environments using simple technologies, to 'Colonial' fisheries involving systematic resource exploitation within the frameworks of developing market economies, to the 'Global' era where marine resources exploitation is incorporated into global patterns of consumption and trade (Jackson et al. 2001). Another approach involves looking at the nature of data sources. The documentary sources for marine environmental history can be divided amongst three chronological categories: a 'historical' period that covers an extended era when archival and other records are mainly unscientific or otherwise unsophisticated; a 'proto-statistical' period, which covers the period from the mid-nineteenth century when port, customs and other similar records become abundant; and a 'statistical' period, which begins around 1900 with the collection of the first national or regional data on catches (Holm et al. 2001). Neither of these approaches fit comfortably with the pattern of historical change in Central Indo-Pacific fisheries, where the development of general fisheries cannot be so easily divided into neat chronological divisions. This is because the transition from indigenous to colonial to global fisheries in Southeast Asia, South Asia, and Pacific is strongly confounded across space and time, or in other words, it tends to takes place in different places at different times (Jackson et al. 2001). Instead, one of the defining features of the twentieth century is the transformation and intensification of artisanal or small-scale fisheries alongside, and sometimes in competition with, the development of new kinds of industrial and globalised fishing, a situation that continues in some places even today.

Within this context a few simple conclusions can be reached. Broadly speaking, the size and structure of any marine animal population in the Central Indo-Pacific at any given moment is an outcome of three factors: 'natural' factors, or long-term and cyclical environmental changes; the nature and intensity of human predation; and human modification of the environment, including pollution and habitat destruction. As a general rule, the second and third of these factors become significantly more important causes of change in population size and structure during the era of 'industrial fishing', or fishing that is powered by fossil fuels, an era that dawned with the introduction of steam trawlers to the North Sea in the 1880s and the subsequent spread of new kinds of fishing vessels and gears to other parts of the world (Cushing 1988; Roberts 2007). During the 'Great Acceleration' of industrial fishing after the Second World War (WWII) this nexus involving human fishing pressure and habitat modification as drivers of ecosystem change intensifies (Holm 2013), although the relative impact of different causal factors and the links or synergies between them remain poorly understood, and even in the late twentieth century it was generally only the most important commercial species that tended to attract any significant level of scientific attention (Cowen et al. 2007; Roberts 2012). The implications are straightforward. Although in an ideal scenario historians would write about the environment using the same analytical categories and concepts as ecosystem modellers so as to yield the fullest insights that a historical perspective can offer, this is often impossible for the vast majority of Central Indo-Pacific fisheries.

Instead, it is possible only to speak in terms of generalisations and basic principles. What can be said with confidence is that, as time has gone by, more and more marine animals have come within the reach of humans trying to catch them, and that humans have developed the power to catch particular animals over more and more of their range. This has meant that wild populations have ever-smaller refuges that are free of fishing pressure, or as Callum Roberts has written, that increasingly fish have 'no place left to hide' (Roberts 2007). This in turn creates the potential for more and more populations collapsing, but collapse has not been inevitable, since the impact of fishing is also related to the life history characteristics of a particular species. How vulnerable animals are to fishing pressure depends on a range of factors, although the species most vulnerable to overfishing typically possess one or more of the following characteristics: they are commercially valuable; have small population sizes; are easily caught due to their size, habits of feeding or breeding or regular movement through particular areas; attain reproductive capacity at a relatively late age; produce relatively few offspring; are confined to a narrow ecological nice; or are not highly mobile (Butcher 2005). The selectivity of fishing gears is also crucial. Gears such as harpoons, handlines, and small nets such as beach seines are highly selective and unlikely to directly impact on non-targeted species in any significant way. On the other hand, advanced techniques have the potential to be highly destructive. Industrial fishing has been estimated to reduce ecosystem biomass by an estimated average of 80% within the first 15 years of exploitation (Myers and Worm 2003). Modern purse seines are capable of taking entire shoals of schooling pelagic fishes, and are implicated in the by-catch of species such as dolphins and turtles. Longlines can devastate long-lived populations of large ocean predators and also record high levels of by-catch for non-target species including seabirds. Trawling is perhaps the most destructive practise of all through the indiscriminate capture of non-target species and the often devastating impact on benthic species of corals and sponges wherever demersal trawls are used (Butcher 2004). In other words, the impact of human harvesting activity on marine animal population should not be considered solely in relation to targeted species. Rather, the full impact of fishing practises must be taken into account, in terms of by-catch and habitat destruction, wherever the nature of historical records allows.

The Transformation of Marine Fisheries in the Central Indo-Pacific

Fish has always been a staple foodstuff of the island and coastal populations of South and Southeast Asia, northern Australia and the Pacific, but despite this, there are relatively few examples of long-distance or bulk trade in fish for human consumption in the period before the mid-nineteenth century. This can be explained by the difficulties in preserving fish in large quantities in tropical climates, the limitations on fast and efficient transport in the age of sail, and most importantly of all, by the ability of traditional fisheries to meet local needs from resources located in coastal and near-shore waters. The trade in dried and salted fish from the Persian Gulf and other parts of the western Indian Ocean to India, and trade in pastes and sauces made from fermented and salted fishes like shrimp paste or *belacan* in Southeast Asia, are among the main examples that can be highlighted (Reeves et al. 1988; Butcher 2004). Yet even these fisheries relied on simple methods such as stakes or hand-drawn nets, and employed small open vessels such as canoes or skiffs. Technological changes did occur during the nineteenth century, as with the adoption of metals for making fish hooks in the South Pacific Islands and the diffusion of beach seining from India to Sri Lanka (O'Meara et al. 2011; Paulin 2011). Most marine fisheries, however, remained characteristically small-scale and directed principally at meeting subsistence needs or localised demand, and across the Central Indo-Pacific as a whole, there remained a great diversity of practises and techniques employed by indigenous and artisanal fishers. As a result fishing pressure tended to affect the marine environment only on a localised scale, and there are few cases of systematic depletions of entire species through the demands of human consumption in the Central Indo-Pacific. The decline of dugong populations in Southeast Asian waters is one of the few examples that have been documented for this period (Butcher 2004).

John Butcher's *The Closing of the Frontier* (2004) describes the processes by which this situation was fundamentally altered in the Southeast Asian context by the expansion of marine capture fisheries that began in the second half of the nineteenth century. There are three broad and overlapping phases to this expansion, made up of an initial phase that witnessed the advent of industrial fishing practises and an intensification of established fishing practises to meet growing demand for fish as a staple foodstuff; a second phase, the 'great fish race', which begins after the WWII and involves the rapid expansion of industrial fisheries across the region; and a third period, the 'closing of the frontier', which coincides with the creation of exclusive economic zones (EEZs) by Southeast Asian nations, the exploitation of the last remaining commercially viable fish populations, and stagnating or declining harvests as the limits to growth are reached. Butcher's main themes are the geographic and bathymetric extensions of the 'frontiers' of fishing that results from the intensification of established fisheries and the introduction of new and more powerful fishing gears, and the typical boom-bust pattern of fishing industries that sees the opening of a new fishing grounds leading inevitably to the decline of exploited stocks and a subsequent movement to more distant or deeper waters in search of untapped populations (Butcher 2004). To a large extent, these same processes played out across the Central Indo-Pacific as a whole during the late nineteenth and twentieth centuries. The remainder of this chapter outlines a brief history of marine fisheries, using a similar three-phase framework to make sense of the profound change in the nature, scale and extent of the human impact on the marine environment that has taken place during the last 150 years.

State, Economy, and Technology: The Origins of Intensive and Industrial Fisheries

In the 1800s it was common to find explorers and naturalists expressing their astonishment at the rich and varied marine life of the Central Indo-Pacific seas. For example, Alfred Russel Wallace wrote of Ambon harbour in the Indonesian archipelago that 'there is perhaps no place in the world richer in marine productions, corals, shell and fishes' (Wallace 1869). As the nineteenth century progresses, the literature on the sea and its products begins to take on a different character, where the prospects for the commercial development of fisheries is increasingly the focus of discussion. Francis Day, an ichthyologist who served as Inspector General of Fisheries in India and Burma during the 1870s, was one of the pioneers of this style of writing about the marine environment. His landmark *Report on the sea fish and fisheries of India and Burma* (1873), a work that also contains numerous references to the diversity and abundance of Indo-Pacific marine life, dealt extensively with the existing artisanal fisheries of British India and outlined a range of measures by which production might be expanded and new fisheries developed (Day 1873). More fisheries experts soon followed in other parts of the region. An early report on the fisheries of the Dutch East Indies was completed by colonial authorities in 1882 (Butcher 2004). In the late 1880s and early 1890s the English marine biologist William Saville-Kent travelled extensively through the Australian colonies as an advisor on fisheries development, including extended visits to two colonies with waters extending north of the Tropic of Capricorn, Queensland (1889–1892) and Western Australia (1893–1895); the marine fisheries of northwest Australia, he suggested in 1896, 'presents an inexhaustible field for future enterprise' (Saville-Kent 1896; Harrison 1997). The United States government commissioned a survey by the fisheries expert A.B. Alexander in the South Pacific Islands in 1899–1900 (Alexander 1902). Underlying each of these assessments, there lay a growing conviction that fish had the potential to become a major item of capitalist production and trade.

The basis for these optimistic appraisals lay in the advance of European colonial states. During the second half of the nineteenth century imperial powers strengthened their control across an increasingly wide area, beginning with the formation of the British Raj in 1857. The British in the Straits Settlements and Australia, the Dutch in the East Indies, the French in Indo-China, the Spanish and later the Americans in the Philippines, and the Germans in New Guinea and the Pacific, all extended or consolidated their overseas empires, so that by the early twentieth century most of the region fell within the boundaries of one or another colonial power. State expansion and the consolidation of political power was accompanied by, if not propelled by, rapid economic and demographic growth. This led to increasing demand for fish as a staple food, especially in the growing urban markets of cities and towns and amongst the sizeable labouring populations of the mines and plantations located across the different colonies. Technology offered solutions to the problem of meeting this growing demand. Steamships, faster and with larger cargo capacities than sailing vessels, were able to supply salt cheaply to fishing grounds to allow catches to be preserved and transport catches to major markets at economical rates. Railways offered a means to transport fish in bulk quantities to inland cities and towns and cities inland. And the advent of steam trawling promised to greatly expand catches, both by increasing harvest in established fishing grounds and opening up potentially rich fishing grounds in deeper waters offshore (Cushing 1988; Reeves et al. 1996).

These factors combined to lead many colonial governments to support experimental fisheries surveys designed to locate new grounds and test new fishing gears. In British India steam trawlers were trialled on several occasions, including the *Golden Crown* in the Bay of Bengal in 1908–1909, the *William Carrick* off Bombay in 1921–1922, and the *Lady Goschen* along the Madras coast in 1927–1930 (Reeves et al. 1996). The United States government followed Alexander's survey of pelagic resources in the South Pacific in the steamer *Albatross* during 1899–1900 with a survey of demersal fisheries in the Philippines in the same vessel during 1907–1909. In the Dutch East Indies, the steamer *Gier* was engaged for trawl surveys in the Java Sea in 1910–1911. Other surveys included the French vessel *de Lanessan* off the southeast coast of Vietnam in 1925, and the steamer *Tongkol* in the Straits of Malacca and the South China Sea on behalf of the government of British Malaya in 1926–1927 (Butcher 2004). The Australian government supported a number of surveys by the custom-built trawler *Endeavour* during the early 1900s, although these did not extend into northern waters nor were they continued after the loss of this trawler in 1914 (Roughley 1966).

From a historical perspective, these trawl surveys provide valuable baselines for marine environments that would later be transformed by fishing. At the time, however, little came in terms of new fishing enterprises. This was partly because trawlers, which had relatively high capital costs, could not yet compete with established fisheries based in shallower coastal and nearshore waters, where abundant fish stocks were located, and partly because knowledge of the location of suitable trawling grounds and of the best gear and techniques to employ had not yet built up amongst fishers (Reeves et al. 1996). There was, however, one important exception:

the Japanese. During the first four decades of the twentieth century, the Japanese led the way in pioneering new fishing technologies in the regions. Japanese beam trawlers, powered initially by sail, began operating in the Philippines during the late 1890s, encouraged by a state policy that provided training and subsidies to the commercial fishing sector in order to boost the supply of fish to the rapidly growing working class in the nation's new industrial centres. Diesel-powered vessels, capable of towing much larger nets, were introduced to these waters in the 1920s, and by the early 1930s a fleet of at least 70 was operating in the Philippines (Butcher 2004). Another major catalyst for expansion occurred after the First World War, when Japan secured control over Germany's Pacific Island possessions north of the equator. During the 1920s, following a systematic survey of marine resources in these waters, pole-and-line fisheries targeting tuna and skipjack and exported as a canned product directly to Japan were established in the Celebes Sea, at Ambon, and at Palau in Micronesia. Additional surveys of tuna stocks extending through Southeast Asia and into the Indian Ocean were undertaken during the 1930s (Butcher 2004; Barclay, this volume).

Other fisheries spread from waters close to Japan into the Central Indo-Pacific during this period. Motorized pair trawlers moved into the waters off Taiwan and from here into the South China Sea and the Tonkin Gulf in the 1920s, and by the 1930s large otter trawlers were also fishing in these waters. One trawler, the 473-t *Shinko Maru*, was based at Singapore in the 1930s and, equipped with a freezer, was able to range as far afield as the Northwest Australian shelf (Butcher 2004). The Japanese practise of netting reef fishes like fusiliers using a long encircling net called the *muro ami* also spread widely during these years. *Muro ami* fishers, employing large and well organised teams of divers equipped with carrier boats and smaller fishing boats, drove fish into traps formed by these nets at the edges of coral reefs, working from reef to reef as each was depleted in turn. They spread southward through the Philippines and the South China Sea in the 1920s and 1930s, eventually basing several operations in Singapore and Batavia and operating widely in the Indonesian archipelago. Singapore was also a principal base for motorized driftnetting and trolling operations, two fisheries that, much like trawling and *muro ami* fishing, were established by Japanese fishers during the 1920s and spread out in 1930s across the waters of Southeast Asia (Butcher 2004; Morgan and Staples 2006).

Alongside these initiatives there are comparatively few examples of major technological change in Central Indo-Pacific fisheries. Chinese fishers based at Singapore introduced purse seines to the Straits of Malacca in the early 1900s, employing these nets initially from sail-powered junks before transitioning to motored vessels in the latter 1930s. At Batavia the quantity of fish landed by motorized vessels doubled between 1935 and 1938, mainly through the adoption of engines aboard boats employing the *paying*, a large sack-like net, in conjunction with *rumpon*, a form of fish aggregating device consisting of palm fronds suspended from a float, in the Java Sea. In the 1930s, Filipino fishers developed an operation known as a *lawag*, which involved the use of a powered boat and a number of canoes or smaller vessels equipped with lamps to increase the power of the *sapyaw*, or round haul seine, by attracting fish with the lights and trapping them with the net (Butcher 2004). For the most part, however, marine fisheries remained characteristically small-scale and basic in method even into the 1920s and 1930s. What did occur was simply an intensification of such fisheries. Examples of this include the proliferation of fish stakes and other fixed gears in Southeast Asian waters, greater use of simple beach seines, gillnets and purse seines from vast fleets of small canoes and other open boats in India, and higher catches of demersal fishes through the use of hook-and-line across the Central Indo-Pacific (Butcher 2004; Hornell 2004). Much of this catch was for the trade in salted fish. Access to cheap salt, to means of transport to markets, and to established trading networks helps to explain regional variations in the intensity of fishing. It also meant that expansion was by no means uniform. For example, whereas marine fisheries in Madras expanded due to the creation of fish curing yards and fishing industries based at Singapore also grew through the influence of efficient transport and marketing networks, other places, such as Burma, actually witnessed a local decline in marine fisheries due to the availability of cheap imported fish (Butcher 2004; Reeves et al., this volume).

In this way, demand for fish as a staple food was met during the first half of the twentieth century. Per capita consumption of fish probably did not increase dramatically, as it was to after 1950, but because of population growth across South and Southeast Asia and the growing quantities of fish exported to places like Japan a much larger quantity of fish was being taken from the Central Indo-Pacific as a whole. In the absence of statistical data for most fisheries the size of the increase cannot be calculated precisely, but a tripling of the marine catch between 1900 and the late 1930s is a reasonable estimate (see Butcher 2005). Just as significantly, fishers had pioneered new types of fisheries such as trawling and purse seining, and as a consequence, were reaching into new ecological strata and into more remote waters. Vast sections of sea were as yet hardly touched by fishing, including most of Bay of Bengal outside of coastal fisheries, much of the Banda, Arafura, and Andaman Seas outside of indigenous fisheries, and much of the northern Australian coastline apart from small and sporadic fisheries for turtle and dugong and limited indigenous and subsistence fisheries in coastal waters. This was however set to change after 1950, through an expansion in capture fisheries that eventually left few parts of the Central Indo-Pacific untouched by commercial fishing pressure.

The Great Acceleration: Post-WWII Intensification and Expansion

The Second World War devastated fishing industries across the Central Indo-Pacific. From the Southwest Pacific to Burma the war resulted in the destruction of boats and other equipment, brought a halt to the import of materials such as twine, nets, hooks and salt necessary to sustain fishing operations, disrupted transport and marketing networks, and reduced demand for fish. As a result, many important stocks had a brief reprieve from fishing pressure. But as soon as the war ended most marine fisheries were re-established, the first stage in a process that led not only to the restoration of pre-war harvests but also to a massive increase in catches across the

Central Indo-Pacific.[1] Across the region people looked to the sea with a similar set of objectives: as a source of food and a guarantee of food security for fast growing populations; as a source of employment and means of improving the welfare for coastal communities; and as a potential source of income through the development of export industries. The great expansion in marine capture harvests that took place during the post-war decades largely came about through an intensification of processes initiated earlier in the twentieth century, involving state commitment to fisheries development, a broader context of substantial and sustained demographic growth, economic modernisation, and a series of technological innovations that enabled the intensification of established fisheries, the development of new fisheries, and the movement of fishing activity into remoter and deeper seas. Compared to the pre-WWII period, however, developments between the late 1940s and 1970s were on an entirely different scale.

Southeast Asia was once again at the forefront of these changes. Beginning with the Philippines, and then followed by British Malaya (later Malaysia), Thailand, and Indonesia, fishing industries revived and quickly reached pre-war rates of production, before the massive and unprecedented surge in catches that took place during the 1950s and 1960s. This expansion was supported through a range of national programmes designed to boost fisheries production, many of which relied upon foreign aid in the form of education and training, technical assistance, and funding or loans to build ports and facilities for storing and processing catches. In South Asia a boom in fisheries exploitation also began in the 1950s, and in time the results as expressed through national production statistics were equally spectacular. India implemented the first in what became a succession of national 5-year plans designed to boost capture fisheries in 1951, the start of a process that had been dubbed the 'Blue Revolution' because its ideals, aims and outcomes all resemble the 'Green Revolution' in agricultural of the post-war era (Bavinck 2001). The focus of the 'Blue Revolution' shifted over time, from modernising fishing enterprises and supporting the development of ports and other infrastructure, to supporting the development of export fisheries and promoting deep-sea fishing. Sri Lanka's government followed a similar programme in the 1960s and 1970s (Bathal 2005; O'Meara et al. 2011). Pacific Island nations also moved to develop their fishery resources after the war, often with the support of the United States or other foreign governments, or in conjunction with the Japanese fishing companies that returned to these seas to fish for tuna in increasing numbers after the early 1950s (Gillet 2007; Barclay, this volume).

The return of Japanese fishers to the Pacific Islands highlights the important contribution of foreign nations to the expansion of marine fisheries across the Central Indo-Pacific. This contribution took a number of forms. The post-war decades witnessed ever-growing demand for the region's seafood in developed nations, principally Japan, the United States, and parts of western Europe, which encouraged investment in boats and gear across the region. Transfer of technology from developed to developing nations in the region was also crucial. Fishing vessels of

---

[1] Butcher labels this post-WW11 period as 'the Great Fish Race'; Holm, writing on the North Atlantic, uses the phrase 'the Great Acceleration' (Butcher 2004; Holm 2013).

all kinds were increasingly likely to be powered by engines, and thus able to reach fishing grounds quickly and spend more time actually fishing, and to travel more readily to remoter or deeper waters to exploit stocks that had previously attracted little or no fishing pressure. Fish-finding technologies, led by the echo-sounder, were adopted for the first time during these decades. Ice began to be used more frequently both at sea and on land, enabling catches to be easily preserved and providing a more marketable commodity to consumers. Nylon and other synthetic fibres replaced cotton and hemp in fishing nets and lines, making fishing gears stronger, more durable and less visible to fish, and allowing for the use of nets that were much larger, and lines that were much longer, than those widely used in earlier times (Butcher 2004).

Differences in the priorities of state development and the timing and magnitude of export market growth help to explain regional variations in capture production increases during the 1950s, 1960s, and 1970s. Civil conflicts and foreign wars held back the development of fisheries in some areas, as in Vietnam and Cambodia in the 1960s and 1970s, Bangladesh in the 1970s, and Sri Lanka in the 1980s. Across the Central Indo-Pacific as whole, however, similar processes of intensification of established fisheries and the extension of these fisheries into new areas, in the emergence and expansion of new industrial fisheries, and in levels of foreign fishing can all be observed during this period. Four developments in particular highlight these changes that swept widely across Central Indo-Pacific fisheries in the post-WWII period: a massive expansion in trawling for small demersal fishes and shrimp; the rapid growth in the use of different types of purse seines to catch tunas as well as smaller pelagic species; the spread of longlining; and the mechanization and modernization of small-scale fisheries. Each was characterized by a broadly similar pattern of movement, spreading outwards from waters close to major ports and markets to increasingly remote seas in the search for new or less heavily exploited stocks of fish.

In the Philippines, beam trawlers began operating shortly after WWII ended, along similar lines to the operations undertaken by the Japanese in the 1930s. More powerful otter trawls and the larger boast required to pull such gear through the water appeared by the late 1940s and became increasingly common in the 1950s, initially in Manila Bay, and later in San Miguel Bay and the Visayan and Samar Seas, as new grounds were sought to replace depleted grounds close to ports and supply ever-growing domestic markets with fresh fish. The most spectacular expansion in trawling, however, took place in Thailand. Aided by foreign aid programmes designed to introduce inshore trawling, the number of registered Thai trawlers rose from 99 to 2,700 between 1960 and 1966. Most of the catch was used for domestic consumption, although 'trash fish', which represented as much as 40 % of an average catch, came to be processed as fish meal and animal feed, thus making the industry profitable (Butcher 2004). An outward spread of trawling brought about by sharply falling catch rates in the more heavily fished waters occurred, and in the late 1960s Thai trawlers moved increasingly from the inshore waters of the west and north of the Gulf of Thailand towards the west coast of Vietnam, into the Mergui Archipelago on the coast of Burma, and into the waters of the Indonesian archipelago.

By the late 1970s more than half of Thailand's recorded catch came from beyond its own territorial waters. In Malaysia and Indonesia, trawling operations spread through the Straits of Malacca and thence along the coasts of Sumatra, Java and Borneo between the late 1960s and early 1970s, eventually reaching as far afield as Irian Jaya and the Arafura Sea (Butcher 2002, 2004).

Trawling, particularly for paneid shrimps, developed elsewhere as a major export industry during this period. In Australia, lucrative trawl fisheries for shrimp, scallops and demersal fishes were established during the late 1950s and 1960s at Shark Bay, Exmouth Gulf and Nikol Bay in Western Australia, at Moreton Bay and along the northern New South Wales and southern and central Queensland coasts, and in the Gulf of Carpentaria in the continent's north, alongside the expansion of lobster, trawl and net fisheries in the continent's temperate southern waters (Williams and Stewart 1993). Australian companies contributed to the development of shrimp trawling in Papua New Guinea during the 1970s. Off Australia's Northwest, Japanese and Taiwanese trawlers fished for snappers and other demersal species during the 1960s and 1970s (Gillet 2007). In India, trawling for shrimp and small demersals began at Kerala in the early 1960s, and, as in Southeast Asia, it spread rapidly in the late 1960s and 1970s as the profitability of trawl fisheries attracted increasing investment and government support. As in other parts of the Central Indo-Pacific, the main market was the United States, followed by Europe. Trawling eventually emerged as the single most important sector of the Indian fisheries, accounting for around 50% of the nation's total catch (Bathal 2005; Bavinck 2001).

The seemingly ubiquitous spread of trawling in the Central Indo-Pacific has a counterpart in the growth of tuna fisheries after WWII. Tuna fishing was an industry particularly affected by the war, and although the industry revived somewhat belatedly owing to the restrictions placed on the movement of Japanese vessels in the immediate post-war years, the lifting of these restrictions in 1952 sparked the first stages of an expansion that carried on through throughout the second half of the twentieth century. Purse seining was the mainstay of this expansion. Large nylon nets and power blocks to haul such nets, initially introduced by American fishers operating in the eastern Pacific, were rapidly adopted by Japanese vessels operating in Southeast Asian and Pacific waters in the 1960s (Gillet 2007). Other innovations followed. The development of ultra-low freezing, which maintained catches in better condition, assisted in the expansion of the distant-water fleets of Japan and Taiwan. From about 1975 tuna fishers in the Philippines and Indonesia began to fish with the aid of floating lures, or *payaw*, a type of FAD that was particularly effective in lifting catch rates for skipjack and yellowfin tunas. Pole-and-line and trolling, more suited to smaller-scale operators, also expanded in these years, helping to maintain the increasing marine capture harvests of Indonesia, the Philippines and Thailand in the 1970s and 1980s (Butcher 2004). During the 1970s Japanese fishing companies took to the extensive employment of longlines, including deepwater longlines that reached down into parts of the water columns inhabited by species such as bigeye tuna. This fishing method, which typically employed a number of catching boats attached to a large and well-equipped mothership, was directed at supplying the high-value sashimi tuna increasingly favoured by affluent Japanese consumers. Southern and Pacific Bluefin tuna were also exploited heavily by long-line vessels in this period. Longlining operations on a similar industrial scale were also developed by Taiwanese and Filipino fishing companies during the 1970s. The Japanese were also influential in supporting the expansion of land-based pole-and-line fishing in Southeast Asia and Pacific Islands in the 1960s and 1970s, often in partnership with local fishers and fishing companies based in Indonesia and the Pacific Islands, who generally provided labour both at sea and in the canneries that processed the catch for export overseas (Morgan and Staples 2006; Butcher 2004).

Purse seines are also effective in capturing smaller pelagic fishes such as mack-erels and sardines, and with abundant stocks of such fishes occurring across the Central Indo-Pacific, this too developed as an important part of the post-WWII expansion of marine fisheries. Declining demersal stocks caused by the excesses of trawling contributed to the rise of this fishery. Trawl vessels were readily adaptable to the use of smaller seines and other towed nets, and with purse seining requiring less engine power and hence fuel than trawling, numbers of vessels in Thailand, In-donesia and the Philippines made this transition during the second half of the 1960s. This process resumed after the oil price shock of 1973. According to one source, the total pelagic catch in the Gulf of Thailand increased 63,000–480,000 t between 1971 and 1977 (Butcher 2004). Purse seining for small pelagic fishes began along the west coast of India in the late 1970s (Bathal 2005).

To a large extent the expansion of small pelagic fisheries reflected a wider mod-ernisation of the small-scale fishing sector. This was also the case in relation to the expansion of driftnet, gillnet, and dropline (bottom longline) and handline fisher-ies. Replacement of traditional vessels with relatively inexpensive fibreglass and other kinds of small vessels, the adoption of inboard engines and the installation of ice-boxes or freezers, and the use of synthetic nets in place of cotton or hemp all contributed to the expansion in net and line fisheries across the region. Driftnetting in the Straits of Malacca and along the east coast of the Malay Peninsula was one of the first fisheries to benefit from engines and synthetic nets; one report from 1958 credited the introduction of nylon drift nets in the Straits of Malacca with a doubling of catches (Butcher 2004). In time small-scale fishers in the Philippines, Vietnam and Indonesia also began to adopt these technologies, although the process was slower in the beginning and the uptake less rapid than, for example, fisheries based at Singapore. Other developments were however embraced, such as the spread of motorized trolling in Southeast Asian waters in the 1950s and 1960s, the greater use of powered boats to support a massive expansion of *muro ami* fishing, and greater numbers of powered boats and electric lights employed in *lawag* and *basnigan* (or bag-net) fishing operations (Butcher 2004). Nylon nets and lines began to be used in India and Sri Lanka's small-scale fisheries from around the late-1950s, and al-though motorization in the small-scale sector did not begin on a large scale until the early 1980s, the use of gillnets in particular became increasingly important in place of beach seining amongst artisanal fishers (Devaraj and Vivekanandan 1999; O'Meara et al. 2011). Powered vessels and synthetic nets and lines also became more common as well amongst small-scale operators in Australia's north-western and north-eastern waters in the 1950s and 1960s, along with other techniques such as the use of steel traps to catch schooling snappers and other demersal fishes (Haysom 2001).

The significance of this kind of modernisation, which took a multitude of forms, cannot be underestimated. It meant that small-scale fishers across the Central Indo-Pacific were able to increase their fishing power and reach further beyond the more heavily fished coastal waters and, by so doing, maintain or increase catches in the face of competition from trawling and other industrial fisheries and declining yields from traditional inshore stocks. In this way the small-scale sector, by virtue of the sheer numbers of fishers that fell into this category, was able to remain the dominant fishing sector in terms of its contribution to total national catch, especially in heavily populated countries with large coastal populations such as the Philippines and Indonesia.

How to measure the impact of this expansion? FAO statistics show that between 1950 and 1980 nominal marine fish landings increased by factors of three in Japan and India, four in Australia, Malaysia, and Bangladesh, five in Sri Lanka, six in the Philippines and Papua New Guinea, more than ten in Thailand and Burma, and several Pacific Island nations, and, most spectacularly, 20 in Indonesia. The vital point is that this boom took place across virtually the entire Central Indo-Pacific, excluding waters more than 200 m deep. By 1980 few areas remained untouched by industrial fisheries, and large stretches of coastal and nearshore waters were now subjected to intensive fishing pressure from industrial and small-scale fishing alike. Stock declines were now a common occurrence. Purse seines are capable of capturing an entire school of fish, and while the sheer profusion of pelagic species meant that stocks could sustain even enormous increases in fishing pressure, in places like the Java Sea sharp declines in catch rates for small pelagic fishes occurred during the 1980s (McElroy 1991). Analyses of longline hooking rates show sharp declines for larger predator species after only a few years of fishing effort (Myers and Worm 2003). The same can be said in relation to driftnets and gillnets in more intensively-fished areas. Some practises could be particularly destructive on a localised scale, such as *muro ami* fishing or the use of explosives to catch demersal reef fishes, a technique that had been common in Philippines and other parts of Southeast Asia and the Pacific in the immediate post-WWII years (Butcher 2004).

Nothing, however, compared to the destructive impact of trawling. In San Miguel Bay in the Philippines, the trawlable biomass or quantity of fish and shrimps accessible to trawlers fell at a rate of around 5% a year from 8,900 t in 1948 to 1,600 t in 1980 (Butcher 2004). On the Northwest Australian shelf, intensive trawling by Taiwanese pair trawlers resulted by-catch of sponges falling from around 500 kg an hour to only a few kg per hour between 1972 and the mid-1980s, evidence of the devastating impact of trawling on the benthic environment (Sainsbury et al. 1992). In the Gulf of Thailand a trawler could catch about 230 kg of fish per hour in 1963, but by 1967 catch rates had fallen to around half that figure; it is estimated that the Gulf lost 60% of its large finfish, shark and skate populations in the first 5 years of industrial trawling, although catches of squids, shrimps and other smaller species rose during the 1960s as a result of the removal of predator species. In fact, the Gulf of Thailand has come to be recognised as a prime example of 'fishing down the food web', or the systematic depletion of high trophic level species and the progressive shift of fishing effort to lower tropic levels (Butcher 2004; Christensen 1998). Similar patterns have been observed in other areas subjected to intensive industrial fishing (e.g. Bathal and Pauly 2008), although few areas have yet rivalled the Gulf of Thailand in terms of the extent to which industrial and intensive fishing in the 1960s and 1970s was responsible for fundamental change in the marine environment.

Closing the Frontier: Towards a New Ocean, 1980s–2010s

The principle of the 'Tragedy of the Commons' posits that individuals acting out of self-interest will inevitably deplete a finite resource held in common ownership due to the lack of any inherent mechanism to encourage conservation for the common good. It is a powerful tool for explaining depletion of marine fisheries, where nation-states take the place of individuals and the world's oceans represent a global commons. In the late 1960s and 1970s there were increasing signs that such a scenario was playing out across the Central Indo-Pacific. Clashes between small-scale fishers and trawlers, arising from competition over dwindling inshore stocks, the destructive impact of trawling, and lax enforcement of regulations designed to protect small-scale fishers, occurred on several occasions: in the Straits of Malacca, along the shores of Thailand and Burma facing the Andaman Sea, along the east coast of Sumatra and off both the north and south coasts of Java, in the Indian states of Kerala, Madras, and Tamil Nadu; and in many other places in between (Butcher 2004; Bavinck 2001; Rumley 2009) After years of conflict and illegal trawling, the Indonesian government banned trawlers from the waters surrounding Java and Bali, before extending the ban to include Sumatra in 1981 (Butcher 2004). Such conflicts were a sign that a turning point had been reached in the history of the region's fisheries. The spectacular spatial expansion of fishing effort and the prolonged growth of marine capture harvests that had marked the post-WWII decades were drawing to a close, and fishing nations were becoming more assertive in their claims to jurisdiction over marine resources. By the early 1980s the commons was effectively closed across the Central Indo-Pacific, the cycle of boom-bust that had propelled the spatial expansion of fishing effort was drawing to a close, and a new phase in the history of marine capture fisheries was beginning (Fig. 2.3).

Claims to ownership of marine animal populations in offshore waters were progressively strengthened during the post-WWII period. Until 1950 the widely accepted principle was that states could legitimately claim exclusive ownership over waters extending up to three nautical miles from the coast. In 1952 the Australian government unilaterally rejected this principle by claiming ownership of benthic resources to edge of the continental shelf, an act designed to prevent Japanese pearling from re-establishing operations off the northern and northwestern coasts after WWII. In 1957 Indonesia declared itself to be an archipelagic state and lay claim to all waters existing within 12 miles of baselines drawn around the entire archipelago, and this act that was repeated by the Philippines in 1961. Twelve-mile ter-

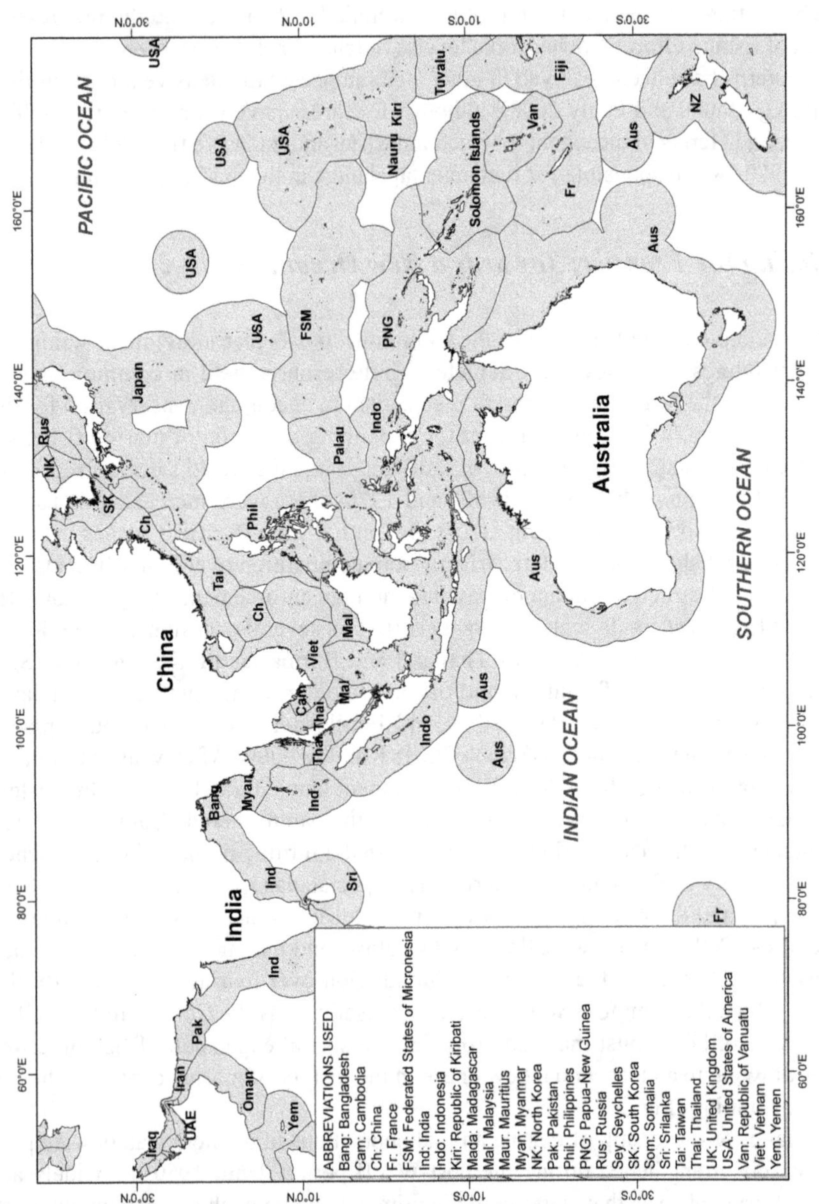

Fig. 2.3 Map of the Central Indo-Pacific showing EEZs

ritorial seas were declared by most of other coastal states in South Asia, Southeast Asia, the Pacific and Australia during the late 1960s (Butcher 2004; Campbell and Wilson 1993). New maritime boundaries also began to be negotiated. Australia and Indonesia agreed to a division spanning the Timor Sea in 1973. This was followed by the attempted expulsion of Indonesian fishers from Australian territorial waters and, in 1979, to an Australian claim to a 200 mile exclusive fishing zone. These claims were formalised by the Third United Nations Conference on the Law of the Sea (UNCLOS III) in 1982 (Campbell and Wilson 1993). Much of the Central Indo-Pacific thereafter fell within the EEZ of one coastal state or another, leaving only the open Indian and Pacific oceans outside of any national jurisdiction.

By the mid-1980s the Philippines, Thailand, Indonesia, Malaysia and India were among the top 12 fishing nations in the world, in terms of size of the reported marine catches, whilst Japan, Taiwan and China were also ranked amongst the world's top fishing nations and taking a large share of their catch from the Central Indo-Pacific. Catches continued to expand for a time, driven by rising catches of tuna and in the few other remaining fisheries that could sustain large-scale expansions, and by the drive to attain maximum exploitation rates in established fishing sectors; these developments are described in more detail below. Amongst most countries, however, marine catches began to stagnate and even decline. In some cases, such as Australia's tropical trawl and demersal line and net fisheries, expansion was curtailed through management strategies such as quota restrictions, closed seasons, and license limitations designed to restrict catches to levels deemed to be sustainable in the long-term (Williams and Stewart 1993). However, across much of the Central Indo-Pacific, the end to growth was the inevitable result of the decreasing number and size of new and untouched fish stocks to exploit. To put it another way, growth began to push up against the biological productivity of the sea itself, a situation that was most pronounced in the coastal and near-shore waters that had been the focus of the post-WWII expansion in capture fisheries. Yet across the region nations remained committed to earning revenue and supporting coastal communities from the marine environment, and demographic and economic growth continued across Asia and the Pacific. These factors combined to create challenges of reducing over-capacity and protecting the livelihoods of small-scale fishers which emerged, alongside the problems of declining stocks, habitat loss and marine pollution, as some of the most pressing problems in Central Indo-Pacific fisheries in the final decades of the twentieth century. The other major problem from a governance perspective was the spread of Illegal, Unreported and Unregulated (IUU) fishing, particularly in relation to high-value species, an outcome of weak enforcement of regulations, high market prices, and the growing competition for scarce fishery resources (Agnew et al. 2009)

Tuna held out the most potential for maintaining growth at a similar rate to the decades prior to the 1980s. Initially, market factors including high fuel prices and lower demand for canned and sashimi tuna in the major markets of Japan, the United States and Europe brought about a decline in investment in tuna fishing and therefore in fishing effort, but generally favourable conditions beginning in the early-1980s drove expansion across the Central Indo-Pacific. Indonesia and the Philippines together accounted for the bulk of the Southeast Asian tuna catch, and Japan and Taiwan continued to fish for tuna in the waters of these and other Southeast Asian countries, but during the 1980s stocks in the main tuna fishing grounds began to decline through the widespread employment of *payaw* and other FADs and the increasing use of more efficient purse seines in competition with the more traditional pole-and-line, trolling, and smaller scale purse seine and longline methods (Butcher 2004; Morgan and Staples 2006). Declining stocks encouraged movement eastward into the tropical and subtropical waters of the Pacific, where foreign distant-water fleets and fledgling joint-venture operations between Pacific Island states and developed nations also began to increase during the 1980s and 1990s. Chinese and Taiwanese longliners became more active in these waters, alongside the large and modern purse seine fleets of Japan and the United States, and the smaller pole-and-line fleets acquired by nations such as Samoa, Fiji, New Caledonia and the Solomon Islands. Catches of the main species (skipjack, yellowfin, bigeye, albacore) tripled in the Western and Central Pacific in the 20 years after 1980 (see chapters by Chen and Barclay, this volume). Taiwanese and Japanese longliners also pushed deeper into the Indian Ocean in the 1980s and 1990s in search of bigeye and Southern Bluefin tuna for the lucrative sashimi market, increasingly using the deep-water lines that reached down to depths that had once provided a haven for these species. Another major development, and one that pointed towards the fishing practises of the future, was the capture of juvenile tunas in the wild to be raised in large mobile sea cages for sale at a later stage (Gillet 2007).

Trawling offered less scope for expansion. Shrimp fisheries had benefitted from rising prices during the late 1970s, but declining yields from wild stocks and the development of shrimp farming on a large scale (see Pokrant, this volume) curbed the industry's expansion during the 1980s. The incorporation of coastal and near-shore waters within EEZs also hampered trawling, particularly the activities of the sizeable Thai trawling fleet. During the 1980s and 1990s illegal fishing was carried out extensively in the surrounding waters of Malaysia, Indonesia and Burma, but the development of more effective surveillance and regulatory systems led increasingly to the implementation of joint-venture and licensing arrangements that allowed foreign fishing vessels to operate within the zones of these countries and as far afield as Vietnam, Bangladesh and Australia during the late 1980s and 1990s (Butcher 2004; Williams 2007). Deep water trawling at depths between 50 and 100 m, where abundant stocks of demersal fishes such as snappers could still be obtained, became more common across the region in the 1990s and 2000s, encouraged in cases like India and Sri Lanka by government support for such deep-sea fishing ventures. But trawling remained a controversial practise, strongly opposed by small-scale fishers and subject to increasingly strict management conditions and high license fees. Bans on foreign trawling were imposed, not only in Indonesia, but also in India, Bangladesh, Burma, and Australia in the late 1980s and 1990s, although illegal fishing continued in some areas such as Indonesia's Arafura Sea (Butcher 2004; Bathal 2005). During the 2000s Thai trawlers were fishing as far afield as the western Indian Ocean in the waters of Oman, Somalia and Madagascar in the western region of the Indian Ocean.

In many respects the situation with tuna and trawl fisheries reflects the wider development of industrial fisheries after 1980. Large and mobile fleets were becoming more common, operating under license in a particular country's EEZ, often on a short-term basis as a quota was filled, a stock declined to the point where it was no longer viable to fish it, or market factors such as fuel prices and demand for seafood made a particular fishery uneconomical. Technological innovations continued to be deployed in an effort to increase fishing efficiency, not just in terms of gears such as deep-water trawls and longlines, but also through fish-finding aids such as side-scanning sonar and GPS (Anticamara et al. 2011). Overfishing was a constant threat, partly as a result of weak governance arrangements that were in turn a legacy of the largely unrestrained expansion of earlier decades, and partly due to overcapacity in industrial fleets. The common theme across the Central Indo-Pacific was the full exploitation of marine resources wherever a commercial profit could be made and fishing rights could be obtained. One indication of this tendency was an expansion in invertebrate fisheries targeting squids, crabs and other crustaceans. Another was the push into waters that lay beyond the Central Indo-Pacific proper. The open seas in the Pacific and Indian oceans, outside of EEZs and hitherto avoided due to remoteness and low productivity, began to attract more fishing effort during the 1980s. For larger factory fleets, the final frontier for expansion lay in the Southern Ocean, which also began attracting increasing fishing activity in the 1980s and 1990s (Anticamara et al. 2011; Watson et al. 2012; Swartz et al. 2010). The other major development was the growth of aquaculture in coastal areas.

Small-scale fisheries also moved inexorably towards full exploitation of available resources. This was the case across of a multitude of sectors. Bans on trawling benefitted small-scale fishers operating in coastal and near-shore waters by removing industrial-scale competition, and stocks of demersal fishes recovered in many areas following these bans, only to be exploited by gillnets, dropnets, droplines, handlines and other relatively basic methods employed by such small-scale fishers supplying fish to ever-growing regional populations. Small pelagic fisheries also continued to expand, where scope for expansion existed, in concert with rising domestic demand. Coral reefs offered such scope, which was met in a number of ways. Across Southeast Asia demersal reef fish continued to be targeted by *muro ami* fishing and its variants, and by net and line fisheries, including by new fleets of deep-water demersal longlines operating in waters unsuitable for trawling, and small vessels that made use of GPS and sonar to target spawning aggregations. Cyanide began to be used more widely to stun and capture reef fishes. Important new markets developed in the form of the live fish trade favoured by increasingly wealthy consumers in cities like Honk Kong and Singapore, and by the live aquarium trade in small reef fishes (Butcher 2004). The rise of a middle class in Hong Kong and mainland China also contributed to the growing demand for shark fin that led to a boom in elasmobranch fisheries in Indonesia and other countries and, in time, to the decimation of shark populations (see Tull, this volume), and to a revival in trepang fishing in Southeast Asian and parts of the Pacific. Specialised or niche fisheries also developed and spread. Landings of cephalods expanded rapidly in many countries in the 1980s and 1990s as small-scale operators made greater use of electric lights and small casting or lifting nets to capture squids, which were actually prolific in many areas due to the overfishing of larger predator species. Other fishers targeted giant clams, lobsters, crabs, or different kinds of mollusc or crustaceans, often for consumption in distant markets (Butcher 2004; Gillet 2007). In this way the global trade in fishery products, a major factor in the expansion of tuna fishing and trawling, reached down into the small-scale sector. Towards the end of the twentieth century a third sector of fisheries also appeared in the Indo-Pacific in the form of recreational fishing, an increasingly important factor in fisheries exploitation of parts of the northwest and northeast Australian coastline.

It is important to point out that population declines in response to fishing pressure are not irreversible, and indeed, many important commercial stocks stabilised and even began to recover during the 2000s in response to improved management arrangements. This reflected a growing commitment to rebuilding regional fisheries expressed in international forums such as an Association of Southeast Asian Nations special meeting on fisheries in 2001 and the World Summit on Sustainable Development in 2002 (Butcher 2004). The nations of Oceania, which include Australia and the Pacific Island countries, have tended to perform better in measures of management effectiveness than the nations of Asia, although the benefits of restoring biomass in depleted ecosystems in order to sustain long-term exploitation is a widely-shared goal, and poorly-performing nations now receive greater international assistance to achieve sustainability in fisheries (Worm et al. 2009; Mora et al. 2009; Worm and Branch 2012). However, despite these positive signs, the outlook for marine animal populations in the Central Indo-Pacific is a vision of life in a very different ocean. Even in sustainably-managed fisheries, the impact of intensive fishing pressure is such that targeted populations now possess life history characteristics and population dynamics that are far removed from 'pristine' or unfished populations, and in this sense, can be viewed as fundamentally 'new' species (Longhurst 2007). An increasing proportion of wild capture fisheries is now processed and used as feed in ocean fish farms. The synergistic effects of fishing pressure also changes trophic structure within ecosystems, where higher-order predators are less abundant and short-lived species such as shrimps, squids and jellyfish are more prolific as both predators and competitors for food sources are removed through human harvesting (Butcher 2004). Marine pollution, habitat loss through the destruction of mangroves, seagrass beds and coral reefs, the spread of invasive species, and the onset of climate change caused by global warming combine to exacerbate the impact of fishing, reducing the productivity of marine environments and posing serious threats to the maintenance of biodiversity in the twenty-first century (Roberts 2012).

Conclusion: Challenges and Opportunities

The Asian tsunami of 2004 was the biggest single shock to the marine fisheries of the Central Indo-Pacific since WWII. Small-scale fishers in India, Sri Lanka, Thailand and Indonesia were devastated. Much like the restoration of fishing capacity in the late 1940s, fleets were quickly rebuilt in the wake of this disaster. In the 2000s, however, there was no longer the vast and largely untouched wealth of marine resources that had existed in the 1950s and 1960s; the frontier for expansion of industrial and intensive fisheries had closed. Viewed from a long-term perspective, the expansion of marine capture fisheries in the Central Indo-Pacific is remarkable both on account of its rapidity, developing over the course of just over a century and concentrated in the three decades between 1950 and 1980, and its pervasiveness, extending across virtually the entire body of water bordered by the Asian subcontinent, northern Australia, the western Pacific, and the East China Sea, and reaching into the open Indian and Pacific oceans. Within this area marine animal populations had been fundamentally transformed by fishing pressure, perhaps for all time. The period between the late nineteenth and the early twentieth century was truly an era of unsettled seas.

Acknowledgments   I would like to thank John Butcher for sharing the notes of his presentations at the HMAP Asia I (2009) and II (2010) workshops, and for commenting on an earlier version of this this paper. I also thank Robb Robinson for his comments on an earlier draft. All errors and omissions are the responsibility of the author.

References

Agnew DJ, Pearce J, Pramod G, Peatman T, Watson R, Beddington JR, Pitcher TJ (2009) Estimating the worldwide extent of illegal fishing. PLoS ONE 4(2):e4570. doi:10.1371/journal.pone.0004570

Alexander A (1902) Notes on the boats, apparatus, and fishing methods employed by the natives of the south sea islands, and results of fishing trials by the Albatross. Report of the Commissioner, Part XXVII, USA Commission of Fish and Fisheries, Washington, pp 741–829

Allen GR (2007) Conservation hotspots of biodiversity and endemism for Indo-Pacific coral reef fishes. Aquat Conserv Mar Freshw Ecosyst 18(5):541–556

Allsopp WHL (1977) Problems and perspectives in tropical fisheries. In: Winslow JH (ed) The Melanesian environment. Australian National University Press, Canberra

Anderson SC, Flemming JM, Watson R, Lotze HK (2011) Serial exploitation of global sea cucumber fisheries. Fish Fish 12:317–339

Anticamara JA, Watson R, Gelchu A, Pauly D (2011) Global fishing effort (1950–2010): Trends, gaps, and implications. Fish Res 107 (1–3):131–136. http://dx.doi.org/10.1016/j.fishres.2010.10.016

Bathal B (2005) Historical reconstruction of Indian marine fisheries catches, 1950–2002, as a basis for testing the 'Marine Trophic Index'. Fisheries Centre Research Reports 13(5). The Fisheries Centre, University of British Columbia

Bathal B, Pauly D (2008) 'Fishing down marine food webs' and spatial expansion of coastal fisheries in India, 1950–2008. Fish Res 91:26–34

Bavinck M (2001) Marine resource management: conflict and regulation in the fisheries of the Coromandel Coast. Sage Publications, New Delhi

Boomgaard P (2005) Resources and people of the sea in and around the Indonesian archipelago, 900–1900. In: Boomgaard P, Henley D, Osseweijer M (eds) Muddied waters: historical and contemporary perspectives on the management of forests and fisheries in island Southeast Asia. KITLV Press, Leiden

Butcher JG (2002) Getting into trouble: the diaspora of Thai trawlers, 1965–2002. Int J Marit Hist 14:85–121

Butcher JG (2004) The closing of the frontier: a history of the marine fisheries of Southeast Asia, c.1850–2000. ISEAS Publications, Singapore

Butcher JG (2005) The marine animals of Southeast Asia: towards a demographic history, 1850–2000. In: Boomgaard P, Henley D, Osseweijer M (eds) Muddied waters: historical and contemporary perspectives on the management of forests and fisheries in island Southeast Asia. KITLV Press, Leiden

Campbell BC, Wilson BVE (1993) The politics of exclusion: Indonesian fishing in the Australian fishing zone. Indian Ocean Centre for Peace Studies, Perth

Christensen V (1998) Fishery-induced changes in a marine ecosystem: insight from models of the Gulf of Thailand. J Fish Biol 53(Suppl A):128–142

Cowen RK, Gawarkiewicz G, Pineda J, Thorrold SR, Werner FE (2007) Population connectivity in marine systems: an overview. Oceanography 20(3):14–21 http://dx.doi.org/10.5670/oceanog.2007.26

Cushing DH (1988) The provident sea. Cambridge University Press, Cambridge

Day F (1873) Report on the sea fish and fisheries of India and Burma. Superintendent of Government Printing, Calcutta

Devaraj M, Vivekanandan E (1999) Marine capture fisheries of India: challenges and opportunities. Curr Sci 76(3):314–332

Firth R (1966) Malay fishermen: their peasant economy. Routledge & Kegan Paul, London

Gillet R (2007) A short history of industrial fishing in the Pacific Islands. Food and Agricultural Organisation of the United Nations, Regional Office for Asia and the Pacific, Bangkok

Harrison AJ (1997) Savant of the Australian Seas: William Saville-Kent (1845–1908) and Australian fisheries. Tasmanian Historical Research Association, Hobart

Haysom N (2001) Trawlers, trollers and trepangers: the story of the Queensland commercial fishing industry pre-1988. Department of Primary Industries, Queensland Government, Brisbane

Holm P (2003) History of marine animal populations: a global research program of the census of marine life. Oceanol Acta 25:201–211

Holm P (2013) World War II and the 'Great Acceleration' of North Atlantic Fisheries. Global Environ 10:66–91

Holm P, Starkey D, Smith T (2001) Introduction. In: Holm P, Smith T, Starkey D (eds) The exploited seas: new directions for marine environmental history. Research in Maritime History 21. International Maritime Economic History Association, St. John's, Newfoundland

Holm P, Marboe AH, Poulsen B, Mackenzie BR (2010) Marine animal populations: a new look back in time. In: McIntyre AD (ed) Life in the world's oceans: diversity, distribution and abundance. Wiley Blackwell, Chichester

Hornell J (2004) [1914] Marine fisheries. In: Playne S (ed) Southern India: its history, people, commerce and industrial resources. Asian Educational Services, New Delhi

Jackson JBC, Kirby MX, Berger WH, Bjorndal KA, Botsford LW, Bourque BJ, Bradbury RH, Cooke R, Erlandson J, Estes JA, Hughes TP, Kidwell S, Lange CB, Lenihan HS, Pandolfi JM, Peterson CH, Steneck RS, Tegner MJ, Warner RR (2001) Historical overfishing and the recent collapse of coastal ecosystems. Science 293(5530):629–637

Longhurst A (2007) Ecological geography of the sea. Elsevier, Burlington

McElroy JK (1991) The Java Sea purse seine fishery: a modern-day 'tragedy of the commons'. Mar Policy 15:255–271

Mora C, Myers RA, Coll M, Libralato S, Pitcher TJ et al (2009) Management effectiveness of the world's marine fisheries. PLoS Biol 7(6):e1000131. doi:10.1371/journal.pbio.1000131

Morgan G, Staples D (2006) The History of Industrial Marine Fisheries in Southeast Asia, Asia-Pacific Fishery Commission, FAO Regional Office for Asia and the Pacific, Bangkok

Myers R, Worm B (2003) Rapid worldwide depletion of predatory fish communities. Nature 423:280–283

O'Meara D, Harper S, Perera N, Zeller D (2011) Reconstruction of Sri Lanka's fisheries catches: 1950–2008. In: Harper S, Zeller D (eds) Fisheries catches reconstruction. Islands, Part II. Fish Cent Res Rep 19(4):85–96

Paulin C (2011) The Maori fish hook: traditional materials, innovative design. Mem Connect 1(1):475–486

Pearson M (2003) The Indian Ocean. Routledge, London

Reeves P, Broeze F, McPherson K (1988) The maritime peoples of the Indian Ocean region since 1800. Mar Mirror 74(3):241–254

Reeves P, Pope A, McGuire J, Pokrant B (1996) Mapping India's marine resources: colonial state experiments, c.1908-1930. South Asia 19(1):13–35

Roberts C (2007) The unnatural history of the sea. Island Press, Washington

Roberts C (2012) Ocean of life: how our seas are changing. Penguin Books, London

Roughley TC (1966) Fish and fisheries of Australia. Angus and Robertson, Sydney

Rumley D (2009) A policy framework for fisheries conflicts in the Indian Ocean. In: Rumley D, Chaturvedi S, Sakhuja V (eds) Fisheries exploitation in the Indian Ocean: threats and opportunities. ISEAS Publications, Singapore

Sainsbury KJ, Campbell RA, Whitelaw AW (1992) Effects of trawling on the marine habitat on the north west shelf of Australia and implications for sustainable management. In: Hancock DA (ed) Sustainable fisheries through sustaining fish habitat. Department of Primary Industries and Energy, Bureau of Resource Sciences, Australian Government Publishing Service, Canberra

Saville-Kent W (1896) Fish and fisheries of Western Australia. Western Australian Yearbook, 1894–1895. Government Printer, Perth

Schwerdtner Mánez K (2010) Java's forgotten pearls: history and disappearance of pearl-fishing in the Segara Anakan lagoon, South Java, Indonesia. J Hist Geogr 36:367–376

Schwerdtner Mánez K, Ferse SCA (2010) The History of Makassan Trepang fishing and trade. PloS ONE 5(6):e11346. doi:10.1371/journal.pone.0011346

Spalding MD, Fox HE, Allen GR, Davidson N, Ferda ZA, Finlayson M, Halpern BS, Martin KD, Mcmanus E, Molnar J, Recchia CA, Robertson J (2007) Marine ecoregions of the world: a bioregionalization of coastal and shelf areas. Bioscience 57(7):573–583

Swartz W, Sala E, Tracey S, Watson R, Pauly D (2010) The spatial expansion and ecological footprint of fisheries (1950 to present). PLoS ONE 5(12):e15143. doi:10.1371/journal.pone.0015143

Wallace AR (1869) The Malay Archipelago. McMillan, London

Watson RA, Cheung WWL, Anticamara JA, Sumaila RU, Zeller D, Pauly D (2012) Global marine yield halved as fishing intensity redoubles. Fish Fish 14:493–503. doi:10.1111/j.1467-2979.2012.00483.x

Watson RA, Pauly D (2013) The changing face of global fisheries – The 1950s vs. the 2000s. Mari Policy 42:1–4. http://dx.doi.org/10.1016/j.marpol.2013.01.022

Williams M (2007) Enmeshed: Australia and Southeast Asia's fisheries. Lowy Institute for Public Policy, Double Bay

Williams M, Stewart P (1993) Australia's fisheries. In: Kailola PJ (ed) Australian fisheries resources. Bureau of Resource Sciences, Department of Primary Industries and Energy, Canberra

Worm B, Hilborn R, Baum JK, Branch TA, Collie JS, Costello C, Fogarty MJ, Fulton EA, Hutchings JA, Jennings S, Jensen OP, Lotze HK, Mace PM, McClanahan TR, Minto C, Pa-lumbi SR, Parma AM, Ricard D, Rosenberg AA, Watson R, Zeller D (2009) Rebuilding global fisheries. Science 325(5940):578–585

Worm B, Branch TA (2012) The Future of fish. Trends Ecol Evol. 27(11):594–599

Chapter 3
Changing Practice in the Madras Marine Fisheries: Legacies of the Fish Curing Yards

Peter Reeves, Bob Pokrant and John McGuire

Abstract This chapter is concerned with the ways in which institutional change in the structure of artisanal fisheries in colonial South Asia affected the position of those engaged in the industry. It examines the 'experiment' undertaken in the coastal districts of Madras Presidency in specially-instituted Fish-Curing Yards (FCY). These yards were promoted as a solution to the problems caused by the colonial government's Salt Tax, which increased greatly the price of salt needed to cure catches. The FCY soon became a central part of the structure of coastal fisheries in the Presidency. Government regulations, which officials claimed were established for the benefit of the fishers, meant that the yards brought fundamental changes to the structures and operations on the relations of production within the fishing industry, although it was curers, rather than the fishers, who became the real beneficiaries of the FCY. Curers emerged as the key players in the yards as they gained control of the entire curing process, from catching to curing to the sale and export to markets. The corollary of the strengthening of the curers' position was the marginalization of the members of the traditional fishing communities and the breakdown of their traditional role in sustaining the community. As a result, over the first 40 years of the introduction and development of the FCYs, traditional small-scale fishers came to be increasingly thought of as the 'problem' in Indian fisheries, and in time, officials saw the need for programmes of 'reform' and 'improvement' to change Indian fisheries.

Keywords Fish curing yards · Madras fisheries · Salt tax · India · History

---

Professor John McGuire passed away in July 2013. This chapter is dedicated to his memory.

B. Pokrant (✉)
Department of Social Sciences & International Studies, School of Media, Culture & Creative Arts, Curtin University, GPO Box U1987, Perth, WA 6845, Australia
e-mail: B.Pokrant@curtin.edu.au

P. Reeves · J. McGuire
South Asia Research Unit, School of Media, Culture & Creative Arts, Curtin University, GPO Box U1987, Perth, WA 6845, Australia

J. Christensen, M. Tull (eds.), *Historical Perspectives of Fisheries Exploitation in the Indo-Pacific*, MARE Publication Series 12, DOI 10.1007/978-94-017-8727-7_3,
© Springer Science+Business Media Dordrecht 2014

This chapter is concerned with the ways in which institutional change in the structure of artisanal fisheries in colonial South Asia affected the position of those engaged in the industry, particularly the fishers. It discusses the background to the 'experiment' undertaken in the coastal districts of Madras Presidency in specially-instituted Fish-Curing Yards (FCY). These yards were promoted from the mid-1870s as a solution to the problems caused by the colonial government's Salt Tax, which, by the 1860s, increased greatly the price of salt which fishers who needed to cure some part of their catch had to pay. The Government of India authorised the creation of FCY from the mid-1870s and the Madras Presidency took up the development of such yards along the full extent of its coastal districts on both its west and east coasts.

The effects of the new yards' structures and operations on the relations of production within the fishing industry are then examined. Although some fishers initially opposed and/or attempted to disregard the new institutions, the FCY became a central part of the structure of coastal fisheries in the Presidency, fisheries in which some 50–70% of the catch was cured. Moreover, since the FCY were introduced in all regions of the Presidency, fishers could not afford to ignore the yards if they wanted to access the cheaper salt that the FCY made available for curing fish.

Government regulations, which officials claimed were established for the benefit of the fishers, meant that the yards brought fundamental changes to the working of the fisheries and the trade in cured fish. Before the FCY were introduced, the operation of the fisheries took place on the beach: the fishers landed their catches and the catch was disposed of in a variety of ways: (a) to the women for sale as fresh fish in neighbouring villages within half a day's walking distance; (b) to the women for curing and subsequent sale; (c) to agents ('middlemen') for sale to merchants in major regional or local markets; and (d) to larger merchants or their agents for sale as dried or cured fish in larger inland or foreign export markets. With the FCY in place the situation was changed by the creation of a new set of players—the curers (or 'ticket-holders', as they became known)—who were those who were able to purchase a 'ticket' which gave them the right to buy the salt available for curing purposes in the FCY (and *only* in the FCY).

The 'curers' needed to have access to still more capital resources because while the Presidency government built the fences and administrative buildings for the yards, the 'curers' were responsible for the construction and equipping of the curing sheds within the yards. The result was that those in the locality who either had capital or could be secured on the basis of wealthy men's patronage—whether they had formerly dealt with fish or not—were able to disproportionally take the curers' tickets and so place themselves (or their agents) in the very centre of the curing process. Not only did they control the curing in the FCY, the wealthier merchants also learned very quickly the advantage of becoming boat owners so that they obtained fish at a favourable rate from fishers, who were—essentially—'their men'. Moreover, the wealthier merchants and financiers rapidly found that they could gain a hold over other independent catches by making credit available to independent fishers who needed such credit to sustain their fishing and who could be talked into accepting that, in return for credit, the curer gained a lien on the catch at a price well below the going rate in the market. A reporter from Tellicherry in Malabar wrote:

> An ordinary pair of fishing boats with a complete set of gear together with advances given to labourer-fishermen will cost Rs. 2,000 to Rs. 4,000. After investing so much money the owner of boats and nets may, owing to failure of fishing season or other causes be obliged to borrow a few hundreds from the fish merchants or other capitalists. In such cases all the fish caught by the borrower must be placed at the disposal of the lender, the latter has liberty to buy it for himself or sell it to others. The price in such cases is much less than that realised by other fishermen who have not borrowed money and who are free to sell their catches to anybody they like. Most of the boats of this place are thus controlled by a few capitalists. (Govindan 1916)

It became clear that although the FCYs were seen initially by the officials as institutions for the benefit of the fishers, it was the curers who became the real beneficiaries of the FCY. The curers, in fact, became the key players in the FCY as they gained control of the entire curing process in a vertically-integrated fishing process: from catching to curing, to the sale and export to the most significant markets.

The corollary of the strengthening of the curers' position was the marginalization of the members of the traditional fishing communities and the breakdown of their traditional role in sustaining the community. As a result, over the first 40 years of the introduction and development of the FCYs, traditional small-scale fishers came to be increasingly thought of as the 'problem' in Indian fisheries. Consequently, officials saw the need for programmes of 'reform' and 'improvement' to change Indian fisheries.

Historical Background

From the beginning of the nineteenth century, there was a growing level of interest in Indian fish and fisheries by colonial scientists and those with an eye for the commercial opportunities presented by Indian fisheries (Russell and Bulmer 1803).[1] This led also to (admittedly intermittent) attention to the ways in which British colonial rule affected fisheries. It was not until the 1860s, however, that there was sustained commentary on the problems affecting Indian fisheries. These interventions touched on a widening range of issues: the colonial government's neglect of fisheries; the lack of provision for fish to proceed upstream for the purpose of

---

[1] For example, Buchanan-Hamilton investigated the fisheries of Bengal in his 1807–1813 survey of the Bengal; the results were incorporated by Francis Day (1876b) to his account of *The Fisheries of Bengal*. Hamilton (1822), went on from this survey to publish his *An account of the fishes found in the River Ganges and its branches*. See also Sykes (1839) which was reported to the Company in 1831 but only published when communicated to the Zoological Society of London. Royle (1842), p. 88, gives a bibliography of works on Indian fishes up to 1842 and Day (1865b), discusses the literature to the early 1860s. Day (1873b), in his *Report on the sea fish and fisheries of India and Burma*, p. 1, cites Dr Jerdon's "Ichthyological gleanings from Madras", Dr Cantor's "Remarks on the sea-fisheries of the Bay of Bengal", Dr McLelland's "Observations on useful species" and Dr Helfer's report on fisheries in Mergui (Burma) from this early period.

breeding in the upper reaches of the major rivers in new irrigation works;[2] and the depletion ('destruction' in the eyes of some commentators) of the fisheries by careless and wasteful methods of fishing—as well as the failure to eliminate 'vermin' such as crocodiles (Day 1873a). It was in this context that questions were raised for the first time about the neglect of trade—especially the trade in salt fish into the inland districts—and of the problems created by the Salt Tax for fishers acting as curers of fish for that trade.

These issues were taken up very strongly by Dr Francis Day, a Surgeon-Major in the colonial administration of Madras Presidency, who became a key commentator on fisheries in the India of the Raj. Francis Day began with a scientific interest in the classification of Indian fishes, which was to lead ultimately to the publication of his *magnum opus*, the two-volume classic, *The Fishes of India* (Day 1876a).[3] But he was also closely interested in the society around him, including the position of the fisheries and the fishers (although he had a fairly traditional view of their 'failings'). He was aware of the debate about colonial neglect and his own early books—*Fishes of Cochin* and *Fishes of Malabar*—both raised general questions along with their ichthyological concerns (Day 1865a). These books presage, in fact, his more 'political' involvement in these matters. In 1867 Day sought a position as a 'naturalist' with the Government of India and in 1868 these overtures paid off when he was invited to undertake investigations in southern India on questions to do with the influence of irrigation works on fisheries. Then, in the late 1860s and early 1870s, as the first (and only) Inspector-General of Fisheries for the Government of India, he investigated fish and fisheries in a range of regions across India and Burma: Madras, Orissa, Bengal, Burma, Assam, the Andamans, the North-Western Provinces, Punjab and Sind. These investigations led to two major reports presented to the Government of India in 1872–1873: a *Report on the Freshwater Fish and Fisheries of India and Burma*; and a *Report on the Sea Fish and Fisheries of India and Burma*, which not only underpinned the major debates in Indian fisheries to the end of the century but also formed, in their catalogues of fish species, the basis of his *The Fishes of India* (Day 1876a).

From the outset of his work in the 1860s, Day recognised two broad sets of problems facing Indian fisheries. Firstly, there was a need for conservation given the nature of current fishing practices which raised questions regarding mesh sizes of nets, fixed engines, fishing with 'poisonous' substances and the lack of regulation and restraint in the management of fisheries. Secondly, attention had to be given to the difficulties that fishers and fish curers faced in obtaining salt at economically-

---

[2] In 1867 the Secretary of State for India had drawn the attention of Govt of Madras to Sir Arthur Cotton's claim that irrigation works on east coast rivers would harm fisheries; see Day (1873a), pp. 7–14. The Secretary of State's attention to these problems had been secured by Colonel George Haly who wrote to him in 1866 and drew his attention to Sir Arthur Cotton's concerns; see Whitehead and Talwar (1976), p. 42, 47.

[3] An abridgement was later made by Oates and Blanford (1889). It should be noted that both of the 1873 reports—and the earlier work on Malabar and Cochin fishes—had extensive taxonomic lists, which became the basis of the great work. For an account of the completion of this work and its definitive status see Whitehead and Talwar (1976), pp. 47–57.

viable rates to carry on the trade in dried and cured fish. He saw also the difficulties with which trading in processed fish was faced from 'vexatious and arbitrary practices' that worked to discourage trade. It was, however, salt that he identified as the key; 'it is', he declared, 'only the moneyed man who can engage in the curing of fish' (Day 1873a).[4]

Day used his position as Inspector-General to investigate and give prominence to these issues from his India-wide investigations and his subsequent reports. His time as Inspector-General, however, also provided him with a position from which to argue and the opportunity to assemble evidence across the Presidencies, through questionnaires which he could send with the authority of the British. Indian governments to Commissioners, Collectors and *Tehsildars*.[5] It was on this basis, therefore, that he could argue for corrective measures in the form of legislation and proper regulation of the fisheries; and more enlightened management of the salt required for the fish-drying and fish-curing.

The *Report on the Freshwater Fish and Fisheries of India and Burma* concentrated on the question of the 'destruction' of the freshwater fisheries and the need for a Fisheries Act for the whole of India (Day 1873b). The *Report on the Sea Fish and Fisheries of India and Burma*, which canvassed the salt issue, argued that 'Government or monopoly salt' was from twelve to sixteen times more expensive than untaxed salt.[6] The *Report* argued that a diminution of the price of salt was needed to allow fish-curing to continue and that this would be best carried on in 'enclosures' in which salt could be issued at cost price, without duty, for curing purposes.

Handling the Salt Question

The use of salt for curing and drying marine fish in India was long-standing (Achaya 1998). While sun-drying without salt had always been a major practice, both drying with salt and curing with salt (and tamarind) were also significant (Achaya 1994). These methods of processing were important for the trade in fish inland from the coast and for export to markets elsewhere in South Asia. As much as 50–70% of the marine catch on both the east and west coasts of India (of which sardine and mackerel were a large component (Achaya 1998) was regularly processed.

---

[4] Day (1865a), pp. xv–xvi, notes that the price per maund in British Malabar had risen from Rs. 1-2-6 to Rs. 1-8-0 from 1858–1859 to 1862–1863 and that, as a result of an agreement with the Madras Government on the price of salt in the Cochin State rose there over a decade ending in 1863–1864 from half the average cost of salt in British territory (Re. 0-9-2 in the mid-1850s) to the equivalent of the British price, 'equivalent to an increase of about 100%'.

[5] In December 1871, the Indian Civil Service's 'club' newspaper, *The Pioneer* of Allahabad, dubbed him, 'the Inspector-General of Stickelbacks'; see the article pasted into Day's cutting book, Day Library of Natural History, Cheltenham Public Library, vol. Q658. See also Reeves (1995); and Whitehead and Talwar (1976), p. 67.

[6] Day (1873b), pp. 49–50, indicates that the cost differential for salt for those using foreign, untaxed salt is 2 annas per maund compared to Rs. 12.00 per maund.

The basic difficulty concerning the use of salt in fish-curing in the mid-nineteenth century was the cost to the fish-curer of salt on which the duty was levied in the different Presidencies. Most Indian salt was produced from sea water by solar evaporation in coastal districts, which Madras Presidency had on both the east and west coasts. The cost of production was low but, by the mid-nineteenth century, the duty charged under the Salt Tax was considerable. As a result, despite low production costs, the price of salt to the public was high. In 1861, shortly after the assumption of the government of India by the Crown, the price of a maund (82.5 lbs.) of salt, which cost between 1 and 3 annas to produce was much higher in Bengal and north India than in Bombay or Madras: a *maund* of salt cost Rs. 3 and 4 annas in the Lower Provinces of Bengal and Rs. 3 in the Upper Provinces of Bengal, as against Re. 1 and 4 annas per *maund* in Madras and Bombay (Oriental and India Office Collections (OIOC) 1878).

The cost, moreover, began to rise from the mid-1860s as the Government of India decided to equalise rates across the country. The control of the manufacture, trade and taxation of salt was, from the mid-eighteenth century, a central concern of British colonial administration in India. Until the mid-nineteenth century, however, control over its manufacture and sale was organised at the Presidency level. The result of this piecemeal development was that the rates of taxation and the systems of administration varied between the Presidencies (Strachey and Strachey 1882; Watt 1908; Dodwell and Wheeler 1968).[7] Until the 1860s, the duty, and therefore the total cost, of salt varied between the Presidencies, with lower rates in the south (Madras) and the west (Bombay and Sind) and higher rates in the north (Punjab, the NWP, Oudh and the Central Provinces) and the east (Bengal, Bihar and Orissa). The move to equalise the duty and hence the price of salt, was partly to increase revenue—because the Salt Tax was a valuable contributor to the Imperial budget. But it was also intended as a means to undermine the rampant smuggling (from south and west to north and east) which the different Presidency prices made profitable. The policing of salt smuggling required a substantial establishment and an 800-mile 'hedge' which formed a manned 'Customs Line' from Rajasthan to Orissa (Moxham 2001).

The imperial government moved in the 1860s to increase the rates in Bombay and Madras. In 1866 and 1869 they took the first steps by increasing the duty by 3 annas and 5 annas per maund respectively (OIOC 1878).[8] In 1871, financial decentralisation made 'Salt' a Government of India revenue 'head' and this strength-

---

[7] Bengal, which tried a number of expedients in the early period of the Diwani, put in place in 1793 a system of monopoly of manufacture and trade, which lasted, to the 1840s and, with some modifications, to the 1890s. Madras created a government monopoly by Regulation I of 1805 under which the control of manufacture and sale were controlled by the government. Bombay established, as soon as it gained a foothold in Gujarat, control over the sources of salt there and in Saurashtra; and it moved, in 1837, to establish an excise throughout the Presidency, the Maharashtran 'Continental' territories, as well as Gujarat. See also, Achaya (1994) and Barui (1985).

[8] The result was that prices in Madras and Bombay increased to Re. 1-13-0 per maund as compared to Rs. 3 per maund in the Northwestern Provinces and Oudh, Punjab and the Central Provinces and Rs. 3-4-0 in the Lower Provinces of Bengal. See also, Strachey and Strachey (1882).

ened the Government of India's ability to move on the issue of the equalisation of the Salt Tax (Strachey and Strachey 1882). In 1874 it abolished the 800 mile 'Customs Line' which had been created to regulate the smuggling of salt (OIOC 1878; Strachey 1911; Strachey and Strachey 1882). Then, to move towards the full equalisation of rates, Act XVIII of 1877 increased Madras and Bombay rates to Rs. 2-8-0 per maund, while lowering the North-Western Provinces and Oudh (the 'Upper Provinces' of Bengal) to Rs. 2-12-0 per maund and the Lower Provinces to Rs. 3-2-0 per maund(OIOC 1878; Strachey and Strachey 1882). Act XII of 1882 completed the process by providing that the determination of the rate of the Salt Tax would be determined for the whole of India by the Government of India. The rate was to be Rs. 2-6-0 by 1882 and Rs. 2-8-0 by 1886–1887. Then, between 1903 and 1907, it was to be progressively reduced to Re 1.[9]

The problems for fishers arising from the high price of salt for marine fisheries were identified, as we have seen, from the mid-1860s by Francis Day. He argued that fishers would be unable to continue to commercially produce dried and cured fish in British Indian territory and that they would either go out of business—which would undermine the extensive trade in fish to the interior—or they would move into the Portuguese territories of Goa or Diu, or even further afield to the Persian Gulf, where they would be able to obtain untaxed salt.

Because Day was able to show that cured fish was a trade product—which could, of course, be taxed and which, given the growth of a cured fish trade, linked neatly with railway developments—the Government of India accepted that Day had pinpointed a basic problem. As a result, when consideration of Day's *Report of Sea Fish and Fisheries* began there was a preparedness to move on the question of how to relieve the problem of the cost of salt for fish-curing. The Government of India were prepared to listen to the enclosure idea and indicated that the Presidencies should 'experiment' with this scheme.

The Government of India declined to accept the full force of Day's argument on the score of the 'injurious effects of the Salt Tax on the salt-fish trade of India' (OIOC 1874a).[10] However, it was sufficiently concerned at the possible loss of (taxable) inland trade in dried fish to look at Day's suggestion that a means should be found to supply fishers with cheaper salt (OIOC 1874b, c). In forwarding copies of Day's *Sea Fish and Fisheries* to the Presidency governments, the Governor-General-in-Council sanctioned the establishment of fish-curing yards 'where salt will be supplied at a little above cost to fishermen and others willing to cure fish within the enclosure' (OIOC 1874a). It added that it would be open to further suggestions for the improvement of the salt-fish trade, 'a careful watch over which will, it is hoped, be maintained by the local maritime Governments and Administrations'(OIOC 1874a). The Government, however, was not prepared to forego the Salt Tax in general usage:

---

[9] The rate was Rs. 2 per maund in 1882; Rs. 2-8-0 in 1888; Rs. 2 in 1903; Re. 1-8-0 in 1905 and Re. 1 in 1907.

[10] The salt tax 'doubtless had a share in preventing that development of the trade which might have been looked for … but other causes such as the general poverty and want of enterprise of the fishermen, and their inability to purchase boats and deep-sea nets have also unquestionably contributed to dwarf this industry'.

> Should it be ultimately found, however, that such remedial measures as have been applied, and may hereafter be devised, are ineffectual to stimulate the trade, and that the existing rate of duty on salt is really incompatible with its full development, the Government of India must accept this evil as a small set-off against the advantages of a tax which produces a revenue of seven million sterling, with the least possible sensible pressure, and consequently at the cost of a minimum of discontent. (OIOC 1874a)

In this context, the Government of India accepted the idea of the fish curing yards on an experimental basis and each of the Presidencies was asked to run trials of the yards. The Government also looked to a 10 % import duty on salted fish coming into India from foreign sources as a means of protecting Indian cured fish by preventing untaxed salt giving outsiders an advantage in the Indian markets and each of the Presidencies was asked to establish such an import duty.

Of the three Presidencies, only Madras acted expeditiously.[11] It was reported that not all district officers were 'sanguine' about the outcome but, since the processing industry was important on the Malabar coast and most of Madras' salt was produced on the Coromandel coast, Madras clearly had an interest in the proposals. Moreover, since the price of salt was going up very markedly in Madras, there was an added incentive to accept the proposals. It was also a fact that in Madras, Day, together with his friend and active colleague, Henry Sullivan Thomas, the Collector of the important fishery district of South Kanara, who was also a fisheries enthusiast (Reeves 1995), were able to combine to gain credence in the debate. Thomas fed material on his negotiations with regard to a design for such a yard and Day utilised this material in making the case for the experiment. Madras also strengthened its case by referring the proposal to the Sanitary Commissioner, who argued that salt would preserve fish better and so reduce disease because it would help to eliminate 'putrid' fish. Consequently, Madras moved in 1873–1874 to set up experiments in Ganjam, Madura, Kistna and Malabar districts and it signalled further moves on Tinnevelly district. Significantly, Madras recognised that, because they traditionally used salt earth, fishers were likely to be reluctant to join the enclosures; but the government expressed the view that it was fully prepared to see 'European firms and local capitalists' (OIOC 1874c)move into the yards in the first instance, if that was necessary to get the yards started.

The Madras Yards and their Operation

Madras Presidency had a long coastline on the western coast of the Indian peninsula (the Malabar coast and a portion of the Canara coast) and the eastern coast of the Bay of Bengal (the Coromandel Coast). It also had a great diversity of fisher communities and, as the Fish-Curing Yards were created along the length of these lit-

---

[11] Bombay began grudgingly but, over time, the Konkan coast (Ratnagiri and North Canara districts) saw more successful establishments and a steady growth to 33 yards by the 1930s. Bengal played no effective part in the system.

toral areas, there were differences in their handling in different regions of the Presidency, of which James Hornell, who joined the Madras Fisheries Bureau (MFB) in 1908 and who rose to become its Director in the 1920s, provided a sketch in 1915 (Hornell 1915).

Hornell outlined four regions within the maritime space of the Presidency: on the west coast he saw the fisheries of the South Canara and Malabar districts—in which Mogarveera and Mukkuva fishers respectively had traditionally dominated—as key regions:

> In these districts sea-fishing attains the dignity of a national industry, employing many thousands of hands, a great fleet of boats and a capital of no mean magnitude. Here the industry is frequently well organised by wealthy and often enterprising merchants, particularly in Mangalore, Malpe and Cannanore, with an extensive export trade to Bombay, Ceylon and Burmah.

On the east coast—where local conditions are less uniform than on the west—he distinguished the northern reaches of the Coromandel coast from Ganjam, in the north, to Point Calimere, in the south, as the 'catarmaran coast', and the area southward to Point Calimere taking in Palk Bay and the pearl-fishing regions in the Gulf of Manaar, as the distinctive regions. These regions were marked by different fishing vessels with large and small canoes in Malabar, these, together with some built-up boats, in South Canara, catamarams on the northern and central stretches of the Coromandel coast, and a mixture of canoes, catamarams and built-up boats in the southern region. Hornell described catamarams in the following way:

> large catamarams truly so-called, consisting usually of three main logs with the aid of special key pieces every time they are used and subsequently taken apart to dry and regain full buoyancy at the end of each day's work.

On the South Canara coast, long beach seine nets (*rampani*) and drift nets along with bag seine nets using two canoes were deployed. In Malabar, and in parts of the catamaran coast, nets and large prawn fishing stakes were important; while in the southern regions of the east coast long-lining and beach nets were to be found. In this southeastern coastal region, whose waters were marked by strong surf, Hornell was able to identify 'true' boats:

> In Palk Strait true boats prevail, capable of standing heavy seas and going long distances. Further south, in the pearl fishing region, of which Tuticoron is the centre, two local, types have evolved, which may be termed respectively canoe-boats and boat-catamarans. Both sail well…and are able live in a sea that no Malabar canoe would ever dream of facing.

The fish-curing yards, which Hornell called an 'outstanding feature' of Madras fisheries since they provided 'the encouragement given by Government to the efficient curing of salt fish', were fenced and guarded physical structures with administrative offices, storage for salt and space for the construction of curing sheds and tubs (Achaya 1994). The government provided the security fence to protect the valuable salt stored in the yard together with the administrative and storage facilities, whilst the registered fish-curers ('ticket holders') had to provide the sheds and curing tubs themselves as a condition of being registered. Until the 1920s, when control passed to the Fisheries Department, the yards were controlled by an establishment

Table 3.1 Fish-Curing yards Madras presidency, 1885–1914: Number of fish curing yards. (Source: Govindan (comp.) 1916, pp. 7–9, 65–71)

Region	1885–1889	1890–1894	1895–1899	1900–1904	1905–1909	1910–1914	Notes
S. Canara	7	9	12	15	15	17	1 private
Malabar	20	24	26	32	32	32	
E. Coast North	26	32	49	55	65	64	
E. Coast South	10	10	13	15	17	16	
Total	63	75	100	117	129	129	

of Salt Department officials and their retainers—sub-inspectors, *duffadars* and peons—who provided the administrative and security services for the yard. Within the yards, registered fish curers were able to purchase salt at an officially designated price, solely for the purpose of curing fish. Those curing activities had, moreover, to be conducted totally within the yard itself. Any attempt to take salt purchased in the yard outside its confines was 'smuggling' and was an offence under the Salt Law. The price of the salt, although not at 'cost price', was well below the normal taxed price outside the yard, which in Madras was Rs. 2.00 per maund. Prices in the Madras yards varied from 2 annas per maund to Re 1 per maund, although the general price was about 8–10 annas per maund. Once a yard was established in an area, 'salt-earth', which was not taxed, could not be used for curing fish either inside or outside the yard because this would reduce the amount of Government salt which could be sold.

In addition to the Salt Department officers and the registered fish curers, there were three other groups of participants working in the yards. Firstly, there were the fishers who brought the fish to the yard, which in sizeable and active yards could be up to 400 fishers. Secondly, there were the women relatives of the fishers who traditionally looked after gutting and cleaning the catch on the beach and, when this was done, organised the sale of fresh fish in villages within a radius of half a day's walking. Thirdly, there were coolie labourers who were available to undertake tasks such as shifting the catch and the cured product or cleaning fish ready for curing and at times curing itself. Along with the officials and the ticket holders, the position of these groups was affected by the workings of the yards (Table 3.1).

The creation of the yards in Madras Presidency began in the mid-1870s and by the mid-1880s there were 63 yards in the Presidency. On the west coast, seven were in the South Canara district and twenty in Malabar district. On the eastern ('Coromandel') coast there were 36 yards by 1885, the majority of which were in the north-eastern coastal districts of Madras Presidency (which historically were known as 'the Northern Circars')—Ganjam, Vizagapatam, Godavari and Kistna and in the southern-most coastal districts of Tanjore, Ramnad and Tinnevelley.

The concentration of east coast districts in the northeast and southeast was brought about by the lack of yard development in the coastal areas of Chingleput

Table 3.2 Fish-Curing yards, Madras presidency 1885–1913: Number of ticket holders in FCYs. (Source: Govindan (comp.) 1916, pp. 7–9, 65–71)

Region	1885–1889	1890–1894	1895–1899	1900–1904	1905–1909	1910–1914
S. Canara						
Total	154	213	391	675	548	472
Average	22	24	33	45	37	28
Malabar						
Total	502	791	882	1,047	1,106	1,078
Average	25	33	33	33	35	34
E. Coast North						
Total	444	773	1,553	2,232	2,733	2,352
Average	17	24	32	41	42	37
E. Coast South						
Total	200	256	291	322	277	182
Average	20	26	22	21	16	11

and South Arcot, the districts closest to Madras because, given the urban presence in this central region, there was a strong demand for fresh fish and this discouraged curing (Govindan 1916). Over the three decades up to 1915 the total number of yards on both coasts increased to 129.

The strongest growth was in the period from 1890 to 1905 when the total went from 75 to 121. Although expansion took place to fill 'gaps' in the coastal coverage, the areas of strength in sub-regions remained clear. There was little growth in Madras Presidency after 1915 but over the 1920s and 1930s the Bombay Presidency developed 33 yards, all of them in the coastal areas of Ratnagiri and North Canara districts, the only areas which had shown promise in the early—discouraging—phase of Bombay's opposition to the FCY. By the time of India's independence the total number of FCY was 155 (Government of India 1945).[12]

The FCY 'experiment' aimed to establish separate yards in each chosen area but it is important to recognize that the yards were, in fact, very varied. One measure of the difference in size ('scale') is provided by the number of participant curers since the yards had to accommodate these curers and their working apparatus (see Table 3.2).

The first obvious difference between the yards was the number of ticket-holders present in different yards. Table 3.2 records the number of ticket holders registered over the 1885–1913 period and shows clearly the growth in the number of ticket holders in each region. What this does not show, however, is the difference in size between yards in these areas; indeed, the calculation of the 'average' number of ticket holders in each region at quinquennial stages in the 1885–1914 period rather glosses over the actual differences in the number of curers in each yard, which was

---

[12] The total was comprised of the following: Madras East Coast 49; Orissa 4; Madras West Coast 56; Cochin and Travancore 14; Bombay (Ratnagari and North Canara districts) 32.

Table 3.3 Fish-Curing yards, Madras presidency 1885–1914: Fish processed (maunds), 1910–1914. (Source: Govindan (comp.) 1916, pp. 7–9, 65–71)

Region	Fish total maunds	Fish average maunds	Range highest maunds	Range lowest maunds	No. less than 500 maunds	No. less than 1000 maunds	No report
(FCY)							
S. Canara (17)	151,039	8,884.6	19,164	2.408	Nil	Nil	Nil
Malabar (32)	403,525	12,610.2	75,210	596	Nil	2	Nil
East Coast 'North' (68)	97,429	1,546.5	12,302	38	7	15	5
East Coast 'South' (20)	41,526	2,422.7	7,231	717	Nil	3	3

Table 3.4 Fish-Curing yards, Madras presidency 1885–1914: Boats at FCY, c. 1915. Note: 'DoC' is an abbreviation for 'Dugout Canoe'; 'BuB' stands for 'Built-up Boat'; 'Cat' stands for 'Catamaran.' The sizes used are as follows: DoC: large up to 35 ft; medium up to 30 ft; small between 8 and 20 ft; BuB: large is 40 ft plus; small up to 35 ft; Cats: large up to 30 ft; small 15–25 ft. (Source: Govindan (comp.) 1916, pp. 7–9, 65–71)

Region	DoC	DoC	DoC	BuB	BuB	Cat	Cat	'fleet'
(FCY)	Large	Medium	Small	Large	Small	Large	Small	
S. Canara (17)	396	432	989	43	Nil	Nil	Nil	1,860
Malabar (32)	1,677	Nil	2,218	Nil	Nil	Nil	Nil	3,895
E. Coast 'north' (68)	Nil	Nil	Nil	Nil	768	Nil	2,351	3,119
E. Coast 'south' (20)	69	Nil	279	46	30	303	501	1,228
	2,142	432	3,486	89	798	303	2,852	10,102

the real measure of the activity in the yard and of the yard's importance in the curing industry of the region. Table 3.2 provides some measure of this difference, in terms of the number of curers in each yard (Table 3.3).

Table 3.4 gives details of the boats involved in fishing in the four regions. It shows the differences between the vessels used on the west and east coasts: the predominance of canoes on the west coast (with some additional canoes in the southeast coast) and the importance of catamarans on the east coast. It also makes clear that built-up boats were a small minority of working vessels (despite their importance when boosted by outsiders such as the Ratnagiri men).

Fishers and the Fish-Curing Yards

In areas where the yards were located, the yards became—for those with the varieties of fish used in the curing process—a key market for their catch. Furthermore, even though the fish-curing yards were set up as a means to assist fishers, they in fact had the effect of placing the curing operation at a more pivotal point within the chain of fish production-processing-marketing than before and they consequently gave those who became the curers a new and more powerful position in the industry than they had before. It was in the acquisition of this position that the fishers tended to miss out to local capitalists and merchants because early on the construction of the yards and the sheds and vats within the yards, and repairs and maintenance, was the responsibility of those who became 'registered fish-curers'—i.e., the people who could receive salt within the yard for curing purposes—and this told against fishers, who traditionally had little liquid capital and who had necessarily to invest what capital they had in their boats and gear. So, in this initial situation traders and merchants (who were referred to as 'local capitalists' in the reports) had an advantage, if they identified such openings into fish processing and the dried/cured-fish trade as an opportunity for investment. The evidence is clearly that some at least did from the earliest stages.

The initial reaction of fishers to the changes clearly signalled that they did not like them because they interrupted their traditional control of fishing. They tried, therefore, to nullify the changes by acting as though the changes did not need to be followed. In this, however, they underestimated the government by mistakenly thinking that the government would not persevere. This attitude seems to have been enhanced at least in some areas by 'mistakes' or 'misperceptions' on the part of fishers. In part they distrusted the idea of the yards: what did the *Sarkar*—the government—really have in mind? What restrictions were they really trying to impose? Even more directly, the fishers resented the fact that the Government, as a corollary to setting up the yards (which, it has to be understood, Government saw as a *concession* to the fishers) made the use of naturally-occurring 'salt-earth' outside the yards a criminal offence. The result was that fishers in a number of areas believed that if they refrained from using the yards, Government would 'back down' and scrap the yard and/or allow the use of salt-earth for curing. In both of these presumptions, the fishers were wrong and as a result some fishers were punished (including spending time in jail) for the use of salt-earth.

An example was given by the Madras Museum's ethnologist, Edgar Thurston, who investigated the fisheries on the west coast in the 1890s. He related an anecdote from Badagara in Malabar which illustrates this problem. A group of fishermen whom he met when he visited, he says, were 'nearly all' Mukkarans (i.e., those who fished in the sea, as opposed to Mukkayan who fished in the rivers). The curers were Mappilas ('Moplahs') who were traditionally traders on the coast and whom Thurston described as 'industrious, successful and prosperous'. Thurston wrote:

> A deputation of fishermen waited on me…. The main grievance … was that the Mukkarans are the hereditary fishermen, and formerly the Moplahs were only the purchasers of fish.

> [A few years ago] the Moplahs started as fishermen on their own account, with small boats
> and thattuvala (tapping nets) [which were considered unfair because the tapping of the sides
> of the boat to drive fish into the net scared off smaller fish] ... [Then] a veteran fisherman
> put the real grievance of his brethren in a nut-shell. In the old days, he stated, they used
> salt-earth for curing fishes. When the fish-curing yards were started, and Government salt
> was issued, the Mukkarans thought that they were going to be heavily taxed by the Sircar
> (Government). They did not understand exactly what was going to happen and were suspi-
> cious. The result was that they would have nothing to do with the curing yards. The use of
> salt-earth was stopped on the establishment of the yard and the issue of Government salt,
> and some of the fishermen were convicted for illegal use thereof. They thought that, if they
> held out, they would be allowed to use salt earth as formerly. Meanwhile, the Moplahs,
> being more wide-awake than the Mukkarans, took advantage of the opportunity (in 1884)
> and erected yards whereof they are still in possession. (Thurston 1900)

This account of change at Badagara highlights the nature of the shift in local power
and control within fishing communities and the local community in which they
were situated that the creation of the yards made possible. The Mappilas saw the
opportunity, which the yards present to them and, given their capital resources, they
were able to quickly move into the position of curers from which they could exert
considerable influence over the sale of an important part of the catch and its subsequent processing. Moreover, Thurston recounted how they moved quickly to effect
'vertical integration' by moving their fellow-Mappila fishers into fishing operations
for their yards. It seems clear that the yards created possibilities for those with
capital to enter the industry in ways which were formerly unattractive, or closed, to
them. The result was that the position and power of local fishing communities was
fundamentally altered.

This is further spelt out in the account of the Mappila success in the Fish-Curing Yard at Tellicherry (another Malabar yard), which is included in *Statistics and
Information*(Govindan 1916):

> Except a few, all curers are very poor entirely depending upon Mappilla merchants who
> finance them in return for the sale of the whole lot of their fish at a low price. Labourer-fish-
> ermen take advances of from Rs. 50 to Rs. 100 from the owners of boats and nets to work
> in their boats. No interest is charged on such advances but before repaying it he cannot go
> and work in another man's boat. No special wages for gutting or salting are fixed. Curing
> work is done by Mukkuva women curers themselves or in the case of Mappilla curers by
> hired labour. The fishermen are Hindus, Christians or Mappillas. An ordinary pair of fishing
> boats with a complete set of gear together with advances given to labourer-fishermen will
> cost Rs. 2,000 to Rs. 4,000. After investing so much money the owner of boats and nets
> may, owing to failure of fishing season or other causes be obliged to borrow a few hundreds
> from the fish merchants or other capitalists. In such cases all the fish caught by the borrower
> must be placed at the disposal of the lender, the latter has liberty to buy it for himself or sell
> it to others. The price in such cases is much less than that realised by other fishermen who
> have not borrowed money and who are free to sell their catches to anybody they like. Most
> of the boats of this place are thus controlled by a few capitalists.

And the report goes on to detail the full cost of these changes:

> The Mukkuvas of Tellicherry were at one time—-some forty years ago—the richest and
> most advanced among the fisher community on the Malabar Coast. They lived in well-
> built tiled houses, several of them being double-storied, owned landed and other immove-
> able property worth many thousands of rupees, and also carried on trade in dry fish with

Colombo and other places on the East Coast. Most of the males were literates and could read and write their vernacular, and a fair number of their young men also attended the English schools. At that time not only the fishing and curing industry, but also the landing and shipping business of the port were in their hands, for conducting which they had large cargo boats of their own each costing a couple of thousand rupees. But for several years past the shipping and landing business has gone out of their hands, and with the exception of two or three individuals none of this community is at present engaged in it except as lascars and coolies working in the cargo boats owned by Mappilla merchants.

This community used to have a strong and well organized caste Panchayet at Tellicherry but with the departure of the prosperity of the people this has also become weak and its voice is seldom heard and rarely respected.

What new factors did the fish-curing yards bring into the local situation that allowed the curers, who were also the traders, to take up the central position at the expense of the fishers? There is evidence that, in some cases, very able or 'enterprising' fishermen and curers would come to yards out of their own area if the fishing was attractive. One important indication of this was the entry of fishermen and curers from Ratnagiri district into yards on the Kanara and Malabar coasts. The yards left an opening for traders and fish merchants, particularly 'more enterprising' fishers from other districts, like Ratnagiri, to take up the curers' positions in Malabar or South Kanara. In this way they were able to establish a primary hold on the yards and vertically integrate the fish production process (catching-curing-marketing and distribution) which was to prove very valuable in the long run.

Thurston reported in 1900 that in both Calicut, in Malabar, and Malpe in South Canara, there were fisher-curers from Ratnagiri. In previous years, he reported further, there had been fishers from Goa as well. In the Malpe yard, in fact, there were 82 registered fish-curers from Ratnagiri compared to 130 local curers. Whether the Ratnagiri men and their *machwas* had traditionally come to South Kanara is not clear; but it seems at least possible that the yards opened new avenues for activity beyond some groups' normal fishing zones. This links with the argument that the situation in the fish-curing yards favoured curers rather than fishers. There was in fact recognition from the outset that merchants and financiers would be at the leading edge in the changes that would come about. Moreover it is possible to argue that Madras' faster and more positive response was brought about because it was prepared to see other fishers in yards and was primarily concerned to get better fish curing—rather than just protecting fishers.

A further difficulty confronted some groups of fishers. Hindu Mukkuvar groups of fishers traditionally relied on their curing to be done by their women. In the earlier situation, where this was done on the beach using salt-earth, the sexual division of labour (men fishing; women selling fresh fish and curing the surplus) worked well. However, in the yard situation, where male curers were present in the yard and where many of these curers had hired male labour ('coolies' in the literature) and had to be able to work at night, especially during the busiest part of the season, Hindu fisher groups ruled against their women working in the yards and so deprived themselves for some time of a place for their curers in the yards. As a result, when they needed to take fish for curing or dispose of their catch for curing, they had to deal increasingly with other curers who were in many cases Muslims or Christians or other outsiders.

The Tellicherry narrative cited above also highlights this positioning of Mukkavar women as part of the explanation of the weakening of the fisher community's control over curing (Govindan 1916):

> As for fishing and curing these were allied industries; the men caught the fish and their women either sold them as fresh or cured and kept them till they had a good demand. Curing of all the smaller kind of fish including mackerel and even medium sized cat-fish was done with salt-earth which they gathered free … When the collection of salt-earth was prohibited owing to the introduction of the salt tax and fish-curing yards were opened for enabling fish to be cured with duty-free salt, these people were reluctant to do so … It was at this time that the Mappillas, who were till then merely petty traders who purchased the cured fish from fisherwomen and sent it to the interior and distant markets for sale, stepped in and became ticket-holders in the fish-curing yard. With cheap salt at the disposal of the Mappillas the fisherwomen could not compete with them, and all the fish caught by the fishermen went to them at a very low price. After some years the fisherwomen also became ticket-holders but it was too late as the Mappillas had by that time practically monopolised the curing industry. These women were therefore obliged to serve as labourers under Mappilla curers and merchants. Hence a community who had lived in comfort on the income derived from the conjoint labour of their men and women, lost the major portion of the benefit derived from their industry, and being obliged to depend on the earnings of their male members alone, gradually lost their prosperity.

A Colonial Agenda for the 'Improvement' of Fishers

The reporter of the Telicherry narrative summed up the experience there as follows:

> That the fish-curing industry has brought in large fortunes to some people other than the fisherfolk is also a significant fact. In almost every fishing centre there are men of non-fisher castes who, beginning life as labourers or petty dealers, with hardly any capital have amassed considerable wealth in the short space of ten or fifteen years. It shows that the industry is a profitable one but the people who are now benefited most by it are not the fisherfolk, and so long as this disadvantage continues the fishermen themselves cannot develop their industry.

Moreover, he added the point that capitalists—in whatever guise—had won control of the situation to the disadvantage of the fishers:

> Many (and not only Hindu) fishers found that the situation in the yards also worked to subordinate them to the local capitalists/traders/merchants who were increasingly ensconced as curers in the yards. In the slack (i.e., 'non-fishing') season, many fishermen needed loans or advances to carry them through and to ensure their boats and equipment were ready for the season; and the most ready source of such loans or advances were the local 'capitalist'/curers. This may have always been the case but there was now a difference because the trader-merchant-capitalist-cum-curer now had an interest not only in the fishers but in the disposal of the catch of the fishers (who were his debtors)….

In fact, it had been clear well before 1916 that capitalists had secured considerable hold over the catch and limited the ability of fishers to control the supply chain.

Indeed, there might be a case for seeing *Statistics and Information* as a document designed to get action in the fisheries; that is, to have the *Bulletins* develop the materials which would clearly outline the plight of the fishers and the agenda needed to change their position. After all, Sir Frederick Nicholson's concern to have the situation recognized had become one of the oft-repeated cachets for the system:

> In Madras we have a vast existing industry worked for centuries in the most primitive fashion by a large population of ignorant but industrious men and we cannot ignore them and their interests, welfare and industrial conditions; we cannot jump at once from the catamaram to the steam trawler. (Madras Fisheries Bureau 1915)

In such a system, capital accumulation by even the strongest fishers was difficult. As a result there was a concern among some colonial officials that more concerted action should be taken within the framework provided by the FCY. The issue was taken up most cogently by Sir Frederick, who looked to officially-guided and supported 'community development' to build more effective fisheries and to 'improve' the marginalized men who they regarded as 'inadequate' fishers.

The cooperative programme in Mangalore, which was begun informally in 1910 by V. Govindan was held back until 1914 by local moneylenders who developed credit relations with individual fishers. Between 1917 and 1919, Nicholson submitted a comprehensive scheme for cooperatives designed to use them as channels for government loans and as a basis for training young men. In 1919 it became part of the programme at the Fishing Training Institute in Calicut.

The MFB was retitled the Madras Fisheries 'Department' (MFD) in the early 1920s, when it took over the responsibility of managing the FCYs from the Salt Department and was placed in a position to bring about the development of Madras fisheries. In doing this, the MFB followed the path laid down by Nicholson in 1907, which gave priority to research and the education and improvement of fisher communities. This approach linked such social campaigns as 'temperance', 'thrift' and 'cooperativeness' with education and with research on Madras fish species and on fishing methods, together with the development of new forms of preservation of fish and accounts of fisheries development in other parts of the world.

In a major paper, Hornell (1923) spoke of a number of ventures which pointed the way to deep-sea fishing:

> [i] 'Sea fisheries development by imported boats presented greater difficulties because of the conservatism of fisher-folk and the difficulty of getting teachers or supervisors'… [tried in a number of places] … At none of these was there decided success, but recently at both Calicut and Negaputtnam local men came forward with proposals to take over the boats and gear we have used for the demonstrating.
> [ii] 'Supervision of actual operations on the fishing grounds has been hitherto in the hands of Intelligent but otherwise uneducated men as no educated Indian fishermen at present are available, the material being faulty and lacking in initiative it is no wonder the results have been faulty in the main. Neither have these men been able to report intelligently upon the condition of the new waters they have worked in.'
> [iii] 'With the recent placing of an order for an experimental sea-going launch to be devoted to the purpose of introducing and testing new methods. This vitally important problem will be attacked in a fresh direction with greater resources than in the past, and with greater

concentration of effort. This fishing launch will be manned by an Indian crew who will be taught under the working direction of an experienced European master-fisherman able to report results and to vary the method according to the circumstances.'

The MFD also sought to expand the agenda further by attempts to develop cooperatives, education on practical lines in fishery schools and institutes and newer forms of preservation. In the mid- to late-1920s there was activity in these areas but no great change in how the fishers perceived their basic tasks.

The Madras Fisheries Committee of 1929, which comprised a group of commercial and professional figures, criticized the Bureau/Department on the grounds that it 'appears to have contributed nothing directly to touch the professional life of the sea-going fishermen'. To this the Government replied caustically that since fishing was so limited in Presidency waters there was little hope of development and concluded that it should not bother with recommendations for development of marine fisheries.

Such a meagre outcome can perhaps be explained by a feeling on the part of both officials and fishers that they had only limited support in the battles for 'improvement' because louder voices disparaged the fisher people and the campaigns of MFD as leading nowhere and as taking away time and assets which should be devoted to 'real' fisheries, which meant improvement by deep sea research, new mechanized boats and gear and the identification of new markets. It is possible on this basis to suggest that MFD personnel were cautious in approach because they recognized that they would have great difficulty in competing with the opinions of the wealthy commercial and political voices that, in the case of the Madras Fisheries Commission of 1929, refused to see any point in what MFD did for fishers. As a result, MFD tried to work with relatively simple changes and operations.

This reading of the nature of the impact of the yards is reinforced if we look at the way in which the regulations for issuing tickets to 'registered fish curers' in the *Madras Fisheries Manual* (1929) operated. Part II of the *Manual* concerned the Fish-Curing Yards and laid down in paragraph 4 are the regulations for issuing tickets of admission, which, from what we have seen over the first four to five decades of the FCY were a key part of the system. The regulations made it clear that those who would be most likely to become registered fish-curers were those with capital—and hence would be unlikely to be fishers. Para. 4(1) made the stipulation that curers had to agree to construct their own shed and concrete floors and tubs for curing, although it allowed for 'special cases':

> In special cases as when the applicant being poor but deserving a ticket of admission … the Inspector may, with the Special Sanction of the Divisional Office in each case, authorize the issue of a ticket without requiring the applicant to construct a permanent shed.

Despite its paternalistic concern for the odd 'special case', this was a clear indication that officials did not think that those without capital would be able to become curers. Moreover, the next paragraph, 4(2), very clearly pointed to what must have been the experience in the past:

> Tickets should also not be granted if the Inspector is satisfied that the issue of a ticket in a particular case is likely to prove detrimental to the interest of *actual fishermen* by affording

facilities for their exploitation by capitalists…Subject to these restrictions, [it concluded]…
tickets for admission should be freely granted.

The *Manual* also underlined the method by which curers gained an important part
of their control within the local situation. The regulations required the officials running the yard to keep notes on the fishermen and curers in the yard and their relations to 'capitalists and middlemen'. They had to record the number of boats using
the yard, big or small, owned or controlled (by means of advances) by each curer
and the number of men to a boat (Madras Fisheries Manual 1929).[13]

The notes were also to specify boats under caste or community names: 'Mappila', 'Mukkaran' and so on. These requirements clearly suggest that those who had
sufficient capital to take up positions as registered curers were able to put themselves in an important strategic position within the industry since, in addition to the
influence they had over the disposal of a key part of the catch, they had acquired a
crucial financial role as sellers of processed fish. Thus, they were in possession of
a cash flow ('liquid capital') which they could employ—in the form of advances
and credit—to increase their hold over the 'producers', i.e., the fishers, who were
already in debt to them.

The experience of the 'colonial agenda' was not entirely blank by any means;
but the hope that it would transform traditional 'artisanal' fishers into a new group
of 'modernised' fishers was not realised. In the end that transformation was to be
the result of the move towards capitalist interests who could afford to work through
the transition from the 1950s to 1980s ushered in by the growing importance of the
frozen and processed seafood trades to the high value markets of North America
and Europe, which called for mechanised boats, new high value gear and the ability to work in deep sea fishing areas using boats that could command the labour of
traditional fishers who found themselves increasingly subordinated to more capital-
intensive forms of production and extraction.

What was more telling was that the policy-makers for Indian fisheries in the new
era of independence continued to draw the conclusion that the fishers offered no
worthwhile base for the modernisation of fisheries. So spoke Prof K. T. Shah of the
National Planning Committee in 1946:

The traditional fishing sector is largely of a primitive character, carried on by ignorant,
unorganised and ill-equipped fishermen. Their techniques are rudimentary, their tackle
elementary, their capital equipment slight and inefficient. (Shah 1946)

And he was echoed by the *Times of India* yearbook for 1949:

The fish trades are universally relegated to low caste men who alike from their want of
education, the isolation caused by their work and caste and their extreme conservatism, are
among the most ignorant, suspicious and prejudiced of the population, extremely averse
to changing the methods of their forefathers and almost universally without the financial
resources necessary for the adoption of new methods, even when convinced of their value.
(Jehu 1949)

---

[13] For an overview of this process see: Kurien (1985); Klausen (1968); Pharo (1988); Sandven
(1959).

Conclusion

Although some fishers initially opposed and/or disregarded the FCYs, the yards became a central part of the structure of coastal fisheries in the Presidency and the regulations fashioned by the colonial government. Where, therefore, the yards brought fundamental changes to the working of the fisheries and the trade in cured fish these changes had major repercussions within the fishing communities. Effectively, local merchants and financiers—many of whom had no background in fisheries—found openings in the yards' structure and mode of working that enabled them to become, as the financiers and controllers of the fishing, curing and trading operations in the yards, the dominant element in an increasingly vertically-integrated local fishing industry in the areas served by the 'fish-curing yards'. As a result, the fishers found themselves in a position where they were likely to be tied to the curers, through combined debt and contractual obligations, in ways that circumscribed their role as potential entrepreneurs in an industry that, given population growth, the expansion of urban areas and the potential of export trade within the South Asian region, contained opportunities for expansion and profit in which the fishers could have shared but from which they were largely excluded.

References

Achaya KT (1994) The food industries of British India. Oxford University Press, New Delhi

Achaya KT (1998) A historical dictionary of Indian food. Oxford University Press, New Delhi

Barui B (1985) The salt industry of Bengal, 1757–1800: a study in the interaction of British monopoly control and indigenous enterprise. KP Bagchi, Calcutta

Day F (1865a) On the fishes of Cochin, on the Malabar coast of India. Proc Zool Soc Lond 33:286–318

Day F (1865b) The fishes of Malabar. Bernard Quaritch

Day F (1873a) Report on the fresh water fish and fisheries of India and Burma. Superintendent of Government Printing

Day F (1873b) Report on the sea fish and fisheries of India and Burma. Superintendent of Government Printing

Day F (1876a) The fishes of India: being a natural history of the fishes known to inhabit the seas and fresh waters of India, Burma, and Ceylon. Bernard Quaritch

Day F (1876b) The fish and fisheries of Bengal. WW Hunter's a statistical account of Bengal XX:1–20

Dodwell H, Wheeler SREM (1968) The Cambridge history of India. University Press

Government of India (1945) Report. Policy Committee no. 5 on Agriculture, Forestry and Fisheries, Fishery Sub-Committee (Chair: Sir A. Rahimtoola)

Govindan V (1916) Madras fisheries: fisheries statistics and information: west and east coasts, Madras presidency. Superintendent Government Press (Madras Fisheries Bureau Bulletin no. 9, Madras)

Hamilton F (1822) An account of the fishes found in the river Ganges and its branches. Printed for A. Constable and company, Edinburgh

Hornell J (1915) Marine fisheries. In: Playne S (ed) Southern India: its history, people, commerce and industrial resources. Foreign and Colonial Compiling and Publishing, London, pp 384–388

Hornell J (1923) Aims and achievements. Madras Fish Bull no.15, pp. 113–129

Jehu IS (ed) (1949) The Indian and Pakistan yearbook and who's who, 1949. Coleman & Co. Ltd, Bennett

Klausen AM (1968) Kerala fishermen and the Indo-Norwegian pilot project. Universitetsforlaget

Kurien J (1985) Technical assistance projects and socio-economic change: Norwegian intervention in Kerala's fisheries development. Econ Polit Wkly 13 (36): 70–88

Madras Fisheries Bureau (1915) Bulletin, papers from 1899 related chiefly to the development of the Madras fisheries bureau: extracts from an official note of 1899 on the desirability of developing the agriculture department. Superintendent, Government Press, Madras 1:2–7

Moxham R (2001) The great hedge of India. Constable Publishing, London

Oates E, Blanford W (1889) The fauna of British India, including Ceylon and Burma (Birds), vol I, 1st edn. Blanford WT (ed). Taylor and Francis, London

Oriental and India Office Collections (1874a) P/678, India agriculture, revenue and commerce. The British Library

Oriental and India Office Collections (1874b) P/678, India revenue, agriculture and commerce, July 1874, nos. 1–3. The British Library

Oriental and India Office Collections (1874c) P/349, Madras revenue consultations 22 July 1874, nos. 179–180. The British Library

Oriental and India Office Collections (1878) P/939, India agriculture, revenue and commerce. The British Library

Oriental and India Office Collections (1932) Madras Development Department GO1507: government response to Cttee P/11973: the Madras fisheries admin report for 1931–1932 GO 1664. The British Library

Pharo H (1988) The Indo-Norwegian project and the modernisation of Kerala fisheries, 1950–1970. In: Shepperdson M, Simmons C (eds) The Indian National Congress and the political economy of India, 1885–1985. Avebury, Aldershot

Reeves P (1995) Inland waters and freshwater fisheries: issues of control access and conservation in colonial India. In: Arnold D, Guha R (eds) Nature, culture and imperialism: essays on the environmental history of South Asia. Oxford, New Delhi

Royle JF (1842) On the production of isinglass along the coasts of India with a notice of its fisheries. W. H. Allen, London

Russell P, Bulmer W (1803) Descriptions and figures of two hundred fishes collected at Vizagapatam on the coast of Coromandel. Printed by W. Bulmer and Co. Shakspeare Press, Cleveland Row, for G. and W. Nicol, booksellers to His Majesty, Pall-Mall

Sandven P (1959) The Indo-Norwegian project in Kerala. Norwegian Foundation for Assistance to Underdeveloped Countries

Shah KT (1946) Address to newly-formed national planning committee, cited in: Kurien J (1985) Technical assistance projects and socio-economic change: Norwegian intervention in Kerala's fisheries development. Econ Polit Wkly 70–88

Strachey SJ (1911) India: its administration & progress edition revised by Sir T W Holderness, 4th ed. Macmillan

Strachey SJ, Strachey SR (1882) The finances and public works of India from 1869 to 1881. K. Paul, Trench & co

Sykes L (1839) On the fishes of the Dukhun. Trans Zool Soc Lond 2:349–378

Thurston E (1900) The sea fisheries of Canara and Malabar. Madras Government Museum, Bulletin 3

Watt SG (1908) The commercial products of India: being an abridgement of the dictionary of the economic products of India. J. Murray

Whitehead PJP, Talwar PK (1976) Francis Day (1829–1889) and his collections of Indian fishes. British Museum (Natural history)

Chapter 4
The History of Shark Fishing in Indonesia

Malcolm Tull

Abstract Indonesia's catches of elasmobranchs (sharks, skates and rays) grew rapidly from the 1970s, driven mainly by the demand for shark fins, and by the beginning of the twenty-first century Indonesia was the world's leading elasmobranch producer. The Indonesian fishery is effectively an open access one and overfishing has led to declining yields in Indonesian waters. Fishers have pushed the geographical catch frontier outwards and this has led to illegal fishing, especially in the Australian Fisheries Zone. Traditionally small scale fishers utilised most of the sharks for food and value-processes including the production of leather, but a large amount of shark is caught as by-catch in industrial fisheries for high value species such as tuna and this has increased the frequency of 'finning', a wasteful and cruel practice. The competition from industrial fishing has adversely impacted small scale fishers and their families; the main beneficiaries of the lucrative shark fin trade have been boat owners and traders rather than fishers and their families. A National Plan of Action is needed but complicated by fiscal constraints and the division of powers between the national, Kabupaten (district/regency) and provincial governments. Governance failures in fisheries are unfortunately a widespread problem in the Indo-Pacific Region.

Keywords Historical knowledge · Indonesia · Elasmobranch fisheries · Shark fins · Artisanal fisheries

Indonesia comprises more than 17,500 islands and has a coastline of about 81,000 km, making it the world's largest archipelagic nation. Due to its geography, Indonesia has long enjoyed access to a diverse range of marine environments and resources. Since 1980 Indonesia has had jurisdiction over an Exclusive Economic Zone (EEZ) of about 3.1 million km^2, the fifth largest in the world. Indonesia is a leading member of the Coral Triangle Initiative on Coral Reefs, Fisheries and Food Security (CTI-CFF) which was formed in formed in 2009 to help preserve the rich resources of the region. Its geographic size, combined with a population of about 243 million and GDP of US\$ 1.033 trillion (2010), makes it the largest country in Southeast Asia.

M. Tull (✉)
School of Management and Governance, Murdoch University, 90 South Street, 6150, Murdoch, WA, Australia
e-mail: M.Tull@murdoch.edu.au

J. Christensen, M. Tull (eds.), *Historical Perspectives of Fisheries Exploitation in the Indo-Pacific,* MARE Publication Series 12, DOI 10.1007/978-94-017-8727-7_4,
© Springer Science+Business Media Dordrecht 2014

From 1950 to 2009, total wild capture fisheries production grew at an average rate of 4.5% per annum or just over double the population growth rate. In 2008 Indonesia accounted for over 5% of the world's total wild capture fisheries production of 90.8 million t and was the third largest producer in the world after China and Peru. While total production is still growing there are increasing signs of depletion of fish stocks, especially in the Arafura Sea, Java Sea and Malacca Straits (Wagey et al. 2009; Williams 2007).

The long history of fishing activity in Indonesia has received surprisingly little attention from maritime historians. In many respects, the under representation of Southeast Asia is a reflection of the broader neglect of the region's maritime history. Although there are exceptions, especially for the Indian Ocean (McPherson 1993; Pearson 2003), historians have tended to focus on the history of the 'land' rather than the 'sea' (Emmerson 1980). John Butcher's *The Closing of the Frontier* (2004) provides a path breaking and valuable overview but does not claim to provide a comprehensive history of all the marine fisheries of Southeast Asia. Another reason for the neglect of fisheries history is that the historian undertaking research on Southeast Asian fisheries does not have an extensive secondary literature to draw on and the limitations of fishery statistics are especially acute in this region (Butcher 2004).

This chapter aims to make a contribution to correcting the gaps in the historical literature by outlining the history of elasmobranch (sharks, skates, rays and chimaeras) fishing in Indonesia since the middle of the twentieth century. Given the relative greater value of shark catches, mainly due to their fins, the major emphasis will be on the importance of shark fishing to small scale fishers and to the national and regional economies. This chapter also aims to help improve understanding of the current state of the fishery and could assist with the development of management policies directed towards the conservation of shark stocks. The chapter begins by providing an overview of Indonesia's fishing industry, and outlining the broad features of elasmobranch fisheries. It then traces the growth of Indonesian's elasmobranch fisheries and finally presents some conclusions.

Overview of Indonesian Fisheries

Indonesia's fish resources can be classified into three main ecosystems- the Sunda shelf, the Sahul Shelf and the Indian Ocean and other deep seas (Bailey et al. 1987). The shallow Sunda Shelf which includes the islands of Sumatra, Java and Kalimantan, were historically rich in marine life and account for about two thirds of the total catch. Offshore waters and the eastern part of Indonesia were less intensively exploited by both industrial and small scale fishers at least until the early 1980s. Since then large numbers of foreign trawlers and small scale fishers have—legally and illegally—exploited eastern waters, taking a heavy toll of fish stocks (Wagey et al. 2009; Fegan 2003; Heazle and Butcher 2007).

Fig. 4.1 Wild capture fisheries production, Indonesia: 1950–2009. (Source: FAO, Fishstat Plus- Capture production)

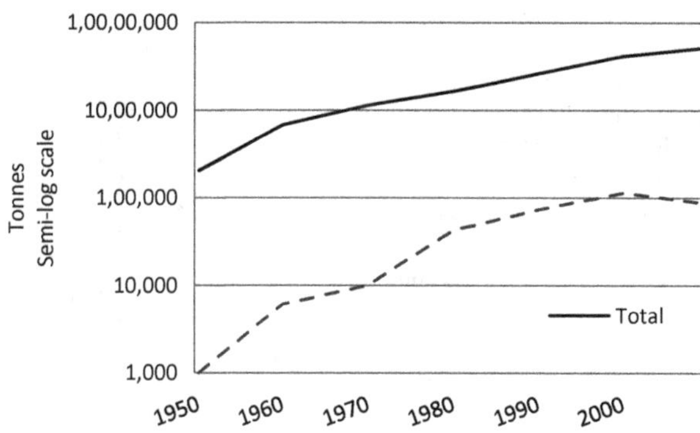

Data limitations limit our understanding of the long term historical impact of fishing activity in Indonesia. Before 1972 Indonesia did not have a national system for the collection of fishing records. Surviving records from the Dutch colonial era tend to be aligned with the specific regions that supported profitable export fisheries and do not provide national coverage. A search of Indonesian archives failed to uncover any records of fisheries during the Japanese Occupation (1942–1945). Each region or city appears to have slightly different record keeping practices. The contemporary records for Pekalongan, for example, include shark data under 'other' and do not differentiate between species. The data is more detailed in species and sub species in some other localities, but generally there is limited taxonomic detail (White et al. 2006). This is unfortunate as, for example, species of sharks vary in their susceptibility to overfishing. There is some concerning evidence, for example, that whale sharks are an opportunistic catch for small scale fishers (White and Cavanagh 2007). The national data, of course, suffers from the usual omissions and inaccuracies of fisheries data including the impact of illegal, unreported and unregulated (IUU) fishing (Bailey 1996; Bentley 1996a). A study of IUU fishing in Raja Ampat Regency, Eastern Indonesia, estimated that in 2006 the IUU catch exceeded the reported catch by 50 % (Varkey et al. 2010). A 2009 study of IUU fishing in the Arafura Sea estimated that the official statistics recorded only 0.9–19.4 % of the true catch (Wagey et al. 2009). It has been estimated that Indonesia's compliance with the data requirements of the FAO's Code of Conduct for Responsible Fisheries is only 26–32 % (Wagey et al. 2009). However, subject to the above caveats, the national data provide a useful source of information on trends in fishing activity.

The available data suggest that all fishing activity was relatively low in the years immediately after the Second World War (WWII) due to the temporary withdrawal of Japanese fishers and the disruption caused by the war of independence against the Netherlands which ended in 1949 (Pauly and Thia-Eng 1988; Butcher 2004). Figure 4.1 shows elasmobranch and total fish landings between 1950 and 2009. For over half a century, total catches grew at about 4.5 % per annum. As result of this strong growth, between 1950 and 2009 total catches grew from less than 0.2 million t to about 4.8 million t. Elasmobranch catches grew more rapidly at about

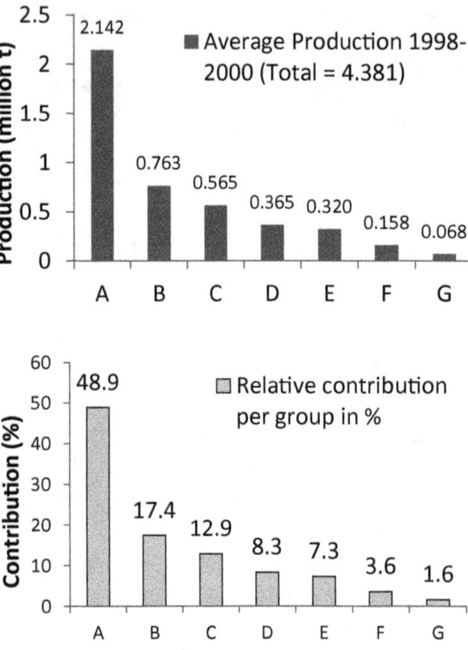

Fig. 4.2 Indonesia, average fish landings by species, 1998–2002. (*A*: Pelagic, marine fish; *B*: Demersal, marine fish; *C*: Marine fish, nei; *D*: Freshwater fish; *E*: Crustaceans; *F*: Molluscs; *G*: others). (Source: FAO, Fishstat Plus- Capture production)

6.9 % per annum, but after peaking at 118,000 t in 2003, plummeted to 89,000 t by 2009—an indication that overfishing was affecting catches.

Figure 4.2 illustrates the wide diversity of species captured by Indonesian fishers. Almost half of the catch is made up of pelagic species such as anchovy, scad, mackerel sardines, trevallies and tuna. Crustaceans (mainly prawns) accounted for only 7 % of total catches by weight but are more important in value terms. Elasmobranches have never accounted for a major part of the total catch and even at their peak in 2003 averaged only 3 % of the total catch.

Of the total catch landed, close to 50 % is sold and consumed as fresh product while the remaining proportion is processed or preserved through methods such as salting, drying, boiling, smoking and freezing (Suboko 2001). Aquaculture in coastal fish ponds (tambaks) has grown rapidly since the late 1970s and exceeded 1 million t by the early 2000s. The main cultivated species are Penaeid prawns but fish (e.g., milkfish) and seaweeds are also cultivated.

Given Indonesia's abundant marine resources, fishing has long been important as a source of food, employment and foreign exchange earnings. Fish consumption increased from about 6.5 kg per capita in the 1940s to 8.5–10 kg per capita in the 1970s, which was just below the world average (Polunin 1983). From the mid-1970s the Indonesian economy enjoyed two decades of steady growth; this led to the emergence of a sizeable middle class and altered the composition of the demand for food products (East Asia Analytical Unit 1994). By 2002 per capita fish consumption was 22.8 kg and above the world average but it was still less than in Singapore (80 kg), Malaysia (45 kg) and Thailand (35 kg). This was probably due to lower average incomes in Indonesia and the limited awareness of the health benefits of eating seafood (Kusumastanto 2004). It has been estimated that fish contribute about 60% of total protein intake and are the main source of protein for lower income groups (Tomascik et al. 1997). However, national averages conceal variations in per capita consumption within the country; for example, less fish is consumed in the interior of Java than on the coast. Indonesians have long been reliant on vegetable protein, using small amounts of fish and belacan or shrimp paste (also known as terasi) as a condiment.

As is common in developing economies, the Indonesian fishing industry is characterized by a dual structure. On the one hand, the small scale (often termed artisanal) fishery traditionally utilised small size fishing boats built from local materials, and operated predominantly in shallow, in-shore waters. However, some groups such as the Bajo Laut or 'sea nomads' have a long history of distant water fishing (Stacey 2007). Boats employed in small scale fishing are usually less than 20 t gross weight, made of wood, are rarely equipped with insulated fish holds or refrigeration, but increasingly utilise small engines. The small scale sector still accounts for about 90% of total production. Small scale artisanal fisheries offer major advantages for developing economies:

> The small scale fishery is labour and local-skill intensive; it is capital and fuel-saving (particularly with the option of multiple energy use). Its technology and mode of organisation and management are well mastered by local fishing communities and give rise to a decentralised settlement pattern. It does not promote large income disparities. (Kurien 1998)

On the other hand, the modern industrial sector of the industry is characterised by the use of well equipped boats, often owned by licensed foreigners or by joint ventures, to capture high value species such as prawns, tuna and other deeper off-shore fish resources, mainly for export markets. In practice, however, as Campbell and Wilson (1993) observed, the distinction between small scale and industrial fishing is less clear cut:

> The Indonesian artisinal fishing sector (often called the small-scale sector) embraces the people, technology, skills and knowledge of the indigenous Indonesian fishing industry. Within the artisinal sector the relation between owners of vessels or perahu and those who work on them is not just a matter of employer and employee but incorporates cultural norms influencing sharing of resources and access to employment. The term artisinal clearly excludes forms of industrial fishing that rely on Western metallurgy, wage labour and scientific navigation. At the other extreme, artisinal unequivocally includes small sail and paddle-powered fishing vessels constructed according to traditional Indonesian boat building techniques, and owned by families or villages. There is, however, a great deal of fishing endeavour that falls between these two extremes. (Campbell and Wilson 1993)

The differences between the small scale and industrial sectors are apparent in boats and equipment employed and also in financial and profit sharing arrangements. The beginnings of mechanisation can be traced to the 1920s when muro ami fishing developed (Polunin 1983), but by the mid-1960s powered boats still made up only 1.4% of the total fleet (Sulaeman 1969). On average, between 1986 and 2000, non-powered fishing boats and boats powered by out-board or in-board motors respectively accounted for 61 and 39% of the total fishing fleet. In general, the non-powered boats have the lowest fishing capacity, the out-board-engine boats a higher capacity, while the in-board-engine vessels, equipped with trawl and seine nets, have the highest capability. Although their numbers are falling, non-powered boats still constitute just under half of the Indonesian fishing fleet, consistent with the notion of a small scale industry (FAO 2012). The gear used depends on the scale of the operation: large scale fishers' tend to use purse seine, trawl and gillnets; small-scale fishers use gillnets, drift nets and longlines (Anak 1997).

Entry into the Indonesian fishing industry is relatively easy as only a minimum level of skills and capital is needed and historically fish was an easily accessible resource. From 1951 to 1967 the official data shows that the number of fishers increased from 315,000 to 836,000 or by 165%, the number of fishing boats from 80,400 to 245,200 or by 205% and the catch from 324,000 to 638,000 t or by 97%. Thus over the same period, catch per boat and catch per fisher declined from 4.0 to 2.6 t and from 1.0 to 0.8 t respectively (Sulaeman 1969). The expansion of output came from increased inputs of traditional equipment and skills rather than technological change, a phenomenon termed 'static expansion' (Butcher 2004).

Fishing is not only an important source of full time employment but it also provides seasonal work, absorbing 'slack time' in farm households. From 1986 to 2000, fulltime fishers averaged 50% of total employment but if those who spend a major part of their time fishing are included, the ratio is 86%. In 2005 over 6 million people were directly involved with fishing but they made up less than 3% of the total population (FAO 2012). In addition to providing direct employment as fishermen, the fisheries sector indirectly provides employment to over a million men and women, in fish processing, transportation, marketing, and support industries (such as boat building and fish gear manufacturing).

Indonesia's small scale fisheries are dominated by the Bajo, Bugis, Makassarese and Butonese peoples from south and southeast Sulawesi. While some groups such as the Bajo have engaged in distant voyaging for centuries (Stacey 2007), as their local fisheries have become overfished others have voyaged further afield. In the mid-1990s a Makassarese fisher admitted that he fished as far as Kalimantan, the Philippines, Irian Jaya and Australia (Bentley 1996a). According to the records of the Dutch East India Company, Indonesian fishers have, however, fished distantshore waters, including waters in what is now the Australian Fishing Zone (AFZ), since the early eighteenth century, when they began harvesting trepang for the Chinese market (Fox 2009). In 1974 a Memorandum of Understanding (MOU) between Australia and Indonesia allowed subsistence fishing by non motorised boats on a number of offshore reefs and islands in the Timor Sea (known as the MOU Box). Since the 1970s there have been regular sightings and detentions of Indonesian fishing craft off the North West coast of Western Australia that have breached the MOU. One of the major arguments then used by the Australian government to justify the exclusion of Indonesian fishers from the AFZ was that the Indonesians had changed from subsistence to commercial fishers and thus forfeited any usufructory rights of entry to the AFZ. However, Campbell and Wilson (1993) argued that contrary to popular belief in Australia, the Indonesian artisanal fishery operating in

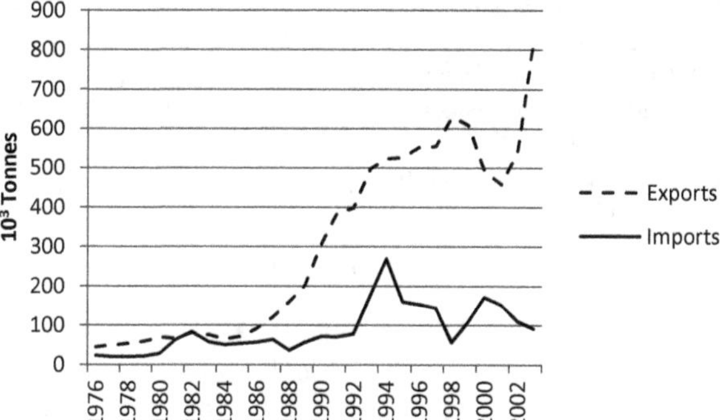

Fig. 4.3 Indonesia total fish exports and imports, 1976–2008. (Source: FAO, Fishstat Plus- Commodities production and trade)

what is now the AFZ was never a subsistence fishery but was always a commercial one and therefore government policy was based on a faulty premise.

The relative abundance of sharks, trochus, trepang, and other marine species in the AFZ has led to continued incursions and regular arrests of Indonesian fishing vessels by navy patrol boats (Fox 2009). Since the early 2000s there has been an upsurge in incursions by more sophisticated Indonesian fishing boats, mainly from Probolinggo, equipped with radar and ice facilities for storing catches. To reduce the risk of interception and loss of the poached fish, one of these 'ice-boats' would dash in and out of the AFZ with the fleet's catch. In addition, from about 2000 to 2004, when Australian naval patrols curtailed activity, small scale fishers based at Pepela on the island of Rote, have used specialist high speed shark-fishing vessels, known as *bodi*, for incursions into the AFZ (Fox et al. 2009). The increasing capital intensity of the operations has led to speculation that they are being funded by criminal groups attracted by the potential profits. Figure 4.3 shows total fish imports and exports from 1976 to 2008. Increased affluence has led to growing imports of high value sea food products, although changing economic conditions such as the Asian crisis of 1997 have caused some volatility in imports. Exports of fishery products, mainly tuna, prawns and shark fins, have increased substantially since the 1970s and by 2008 total exports were 825,000 t or about 35 times their level in 1976. The growing trade surplus in fish has made Indonesia the world's eleventh largest exporter of fish products (Williams 2007).

However, this export expansion has led to increased pressure on fish stocks and by the 1980s there was concern that a large number of fisheries were close to being fully exploited, although there is conflicting evidence on the state of the fisheries.

Figure 4.4 shows that since the 1970s fish production per fisher, a crude indicator of fishing effort has grown at about 1.6 % per annum but reached a plateau in the 1990s. This is similar to the situation in many other Asia-Pacific nations since the 1970s (Gelchu and Pauly 2007). A 1979 survey of the Kendal area found that fishers on non-powered boats worked an average of 9 h per day but caught nothing on 52–71 % of the days, leading to the conclusion that 'an important Indonesian life-support system is thus in a critical state' (Polunin 1983, p. 41). According to an estimate in the early 1990s, the maximum sustainable catch of marine fisheries in

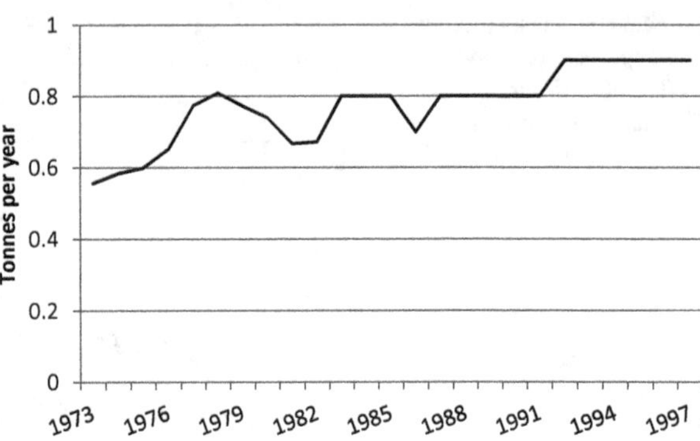

Fig. 4.4 Fishery production per fisher, Indonesia, 1973–1997. (Source: FAO Fisheries Circulars C929 various issues and FAO, Fishstat Plus. Production per fisher prior to 1983 calculated by author)

Indonesia was about 6.6 million t per year, about 57 % above the level in 2000 (Naamin 1995). But by 2003 the Ministry of Marine Affairs and Fisheries had concluded that most fish resources had been exploited 'beyond long-term sustainability' (Fox et al. 2009). Since the mid-2000s catches have reached a plateau at about 5 million t per annum. There is no doubt that pressure on stocks has been exacerbated by the degradation of the coastal environment caused by the pressures of population growth, agricultural, industrial activity and deforestation (Tomascik et al. 1997). As I will demonstrate, these broad trends are apparent in Indonesia's elasmobranch fishery.

Elasmobranch Fisheries

Elasmobranches provide a diverse range of products including meat (dried, salted or smoked), shark fins and skins for human consumption; skins for processing into leather; liver oil used in producing lubricants, cosmetics and vitamin A products; cartilage for medicinal purposes; and teeth and jaws for sale to tourists as curios (Rose 1996; Anak 1997). The wastes from processing sharks, skates and rays have been used to produce fishmeal for use in animal feed and fertilisers or oil for industrial purposes. Shark meat has long provided a source of protein for subsistence fishers but has a relatively low value compared to species such as tuna and has often been discarded by fishers. In Indonesia, for example, shark meat is not a popular item for human consumption (Bentley 1996b).

By contrast, shark fins are one of the world's most valuable fish products. In China shark fins have been regarded as a delicacy at least since the Sung dynasty (960-1279 AD). They form the main ingredient in shark fin soup; a prestige dish consumed by Chinese the world over. Consumption was discouraged in China after the communists took over in 1949 but by the 1980s a relaxed policy combined with increased affluence led to soaring demand. Hong Kong, as the gateway to China, is the centre of the world shark fin trade with most of the reminder shipped via Singapore (Clarke 2003).

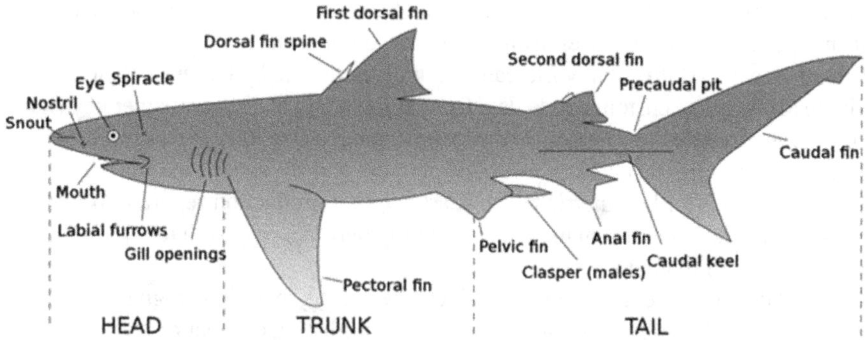

Fig. 4.5 Key features of a shark. (Source: http://commons.wikimedia.org/wiki/File:Parts_of_a_shark.svg)

A fin's value is determined by the number and quality of ceratotrichia or fin needles that can be derived from it. The most valuable fins are usually the caudal, dorsal and pectoral fins (see Fig. 4.5) but the species of shark, fin size and fin cut are also important determinants of value (Clarke 2003).

In 2005 a bowl of shark fin soup in Hong Kong cost up to US$ 75 and in London braised shark fin sold for up to US$ 150 per person, making them 'the white truffles and caviar of the Chinese culinary world, prized for their rarity, and for their alleged tonic properties' (Clarke 2003).

The increased demand for shark fin has obviously led to large increases in shark catches. Total world elasmobranch catches were below 300,000 t until the mid-1950s when they began to steadily increase passing 500,000 t in the early 1970s, 600,000 t in the late 1970s and 700,000 t in the early 1990s. A world record production of 900,000 t was recorded in 2003 but this was not sustainable and by 2009 production was down to 721,000 t. It is, however, widely recognised that the official statistics understate the real level of catches. FAO data is based on landings which include partially processed carcases, belly flaps, fillets, fins and livers; furthermore, reported catches do not include catches discarded at sea (Rose 1996).

A major reason for the underreporting is that sharks are mostly caught as by-catch in fisheries targeting higher valued species such as tuna, swordfish, shrimp, and squid. In the Canadian tuna and swordfish fishery, for example, the bycatch of blue sharks often exceeds the amount of tuna and swordfish caught (Kura et al. 2004). Due to the large bycatch incurred by driftnets– in which sharks figured prominently– their use by large scale operators in international waters was banned in December 1992 (Bonfil 1994). It is generally accepted that 'finning', that is, the cutting off of the fins and the dumping of the carcases, is widespread but the lack of accurate data makes it hard to accurately assess the magnitude of this practice (Bonfil 1994). Shark finning is very wasteful as it uses only 2–5 % of the shark and it also threatens the long term sustainability of the fishery (IUCN 2003). Kelleher (2005) has estimated at the global level, over 200,000 t of shark are discarded annually as a result of finning. In an attempt to reduce shark finning some countries, including Australia, Brazil, Canada, and the United States, have banned the landing of shark fins without carcasses. Bonfil (1994) estimates that the omission of bycatch discards mean that the total world catch is underestimated by between 30 and 50%. Clarke (2003) used customs trade data from Hong Kong and several other key shark fin trading countries to calculate estimates of global shark trade. She estimated that from 17 to 89 million sharks were traded per year and the shark biomass was in the range of 0.5–4.5 million metric t. The shark capture production reported to FAO in 2000 was about 0.35 million metric t or roughly between one tenth to three quarters of the estimated biomass (Clarke 2003).

Regardless of the exact level of shark catches, it is generally recognised that due to their delayed sexual maturation and low fecundity rates, sharks are especially vulnerable to overfishing. By 2002 fishing pressure had led to 22 shark species being included on the International Union for the Conservation of Nature and Natural Resources (IUCN's) Red List of Threatened Species and the basking shark and whale shark were listed under Appendix II of the Convention on International Trade in Endangered Species of Wild Fauna and Flora (CITES) (Kura et al. 2004). There is growing concern about the sustainability of shark fishing and a desire to develop management strategies to help ensure the continued viability of the fishery. The FAO and CITES have developed an International Plan of Action (IPOA) designed 'to ensure the conservation and management of sharks and their long-term sustainable use' (FAO 2012). Having briefly sketched the main features of elasmobranch fisheries, I now turn and consider the Indonesian situation.

Indonesia's Elasmobranch Fishery

Elasmobranches have been caught in Indonesian waters for centuries. Specialist shark fishers from southeast Sulawesi caught them using bamboo poles with rattles made from coconut shells as shark attractors (Osseweijer 2001). In 1800 a Dutch civil servant reported that the Bay in Batavia was 'crawling with sharks, which the Javanese feared but caught in large numbers' (Osseweijer 2001, p. 144). The Dutch colonial authorities supported shark fishing as a new business opportunity:

> When the public reads about sharks, the reader ordinarily thinks of a frightening fish, which, preceded by a pilot fish, makes the environment of ships at sea unsafe and provides material for piles of fascinating adventures in popular boys' books. Rarely does one suppose…that the shark is a fish, which of its kind can be as useful to humans as a cow. (Osseweijer 2001)

In the late 1920s the threat of foreign competition spurred the authorities to action. In 1928 the Batavia News reported that an Australian company, using three motor boats with about 30 men, in 8 months caught about 3,000 sharks with an average weight of 200 lbs. The skins were used for leather, the fins exported to China, oil from the livers used for medical proposes, and the meat for food. The authorities were quick to support Dutch companies wanting to enter the industry and exploit this abundant marine resource.

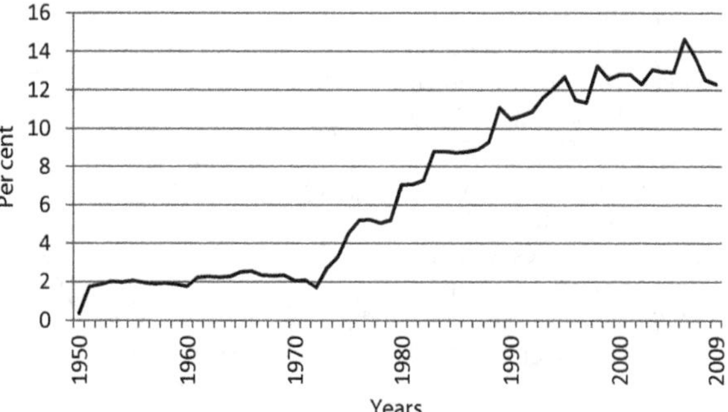

Fig. 4.6 Indonesia's share of world elasmobranch production. (Source: Calculated from data in FAO, Fishstat Plus- Commodities production and trade)

Despite these pioneering efforts, shark fishing has remained a minor part of the total catch (Bentley 1996a). For example, from 1950 to the early 2000s elasmobranches averaged only about 2 % of the total catch. Despite their high value, shark fins at their peak in 1992 accounted for about 2 % of the total value of fish exports and from 1976 to 2006 averaged less than 1 % of the total. However, Fig. 4.1 shows that for over half a century catches grew at 7.7 % per annum; as a result of this impressive growth, elasmobranch catches increased from 1,000 in 1950 to a peak of 117,600 t in 2003. They then declined to only 98,300 t in 2006.

As Fig. 4.6 illustrates, Indonesia's share of world elasmobranch production started to takeoff in the early 1970s. By the beginning of the twenty-first century Indonesia was the world's leading elasmobranch producer accounting for about 13 % of the world total. The fishery is dominated by sharks which averaged about two thirds of catches. While some shark fins are consumed locally, most are destined for export. Appendix 1 shows the volume and value of shark fin exports from 1976 to 2008. Until the opening up of the Chinese market in the 1980s most exports went to Singapore. The increased Chinese demand led to large increases in the price of shark fins: the average unit price of fins imported into Hong Kong increased from US$ 11 in 1980 to US$ 40 in 1992 (Butcher 2004). Export values increased rapidly peaking at over US$ 17 m million in 1992 but tonnages continued to rise and reached a record 1,550 t in 2005.

Figure 4.7 suggests that the real unit value of shark fins exports (2,000 prices) showed an almost symmetrical pattern of growth and decline around the peak value of US$ 190 in 1992. This is puzzling in view of the global trend of fin prices. Apart from the possibility of data errors and omissions, it is likely that overfishing exacerbated by an increase in illegal, unreported and unregulated (IUU) fishing, has led to catches becoming increasingly dominated by lower value species of shark. By 1992 the giant guitarfish *Rhynchobatus djiddensis*– the main source of fins for the Hong Kong market– were becoming increasingly scarce in Indonesian waters (Suzuki 1997). In 2006 shark fishers from the small port of Dobo on Aru Island reported that they could no longer catch the higher valued species (Fox et al. 2009). In the early 2000s even illegal fishers in the AFZ reported that valuable hammerhead sharks and shovelnose rays were becoming harder to catch (Fox 2009). This

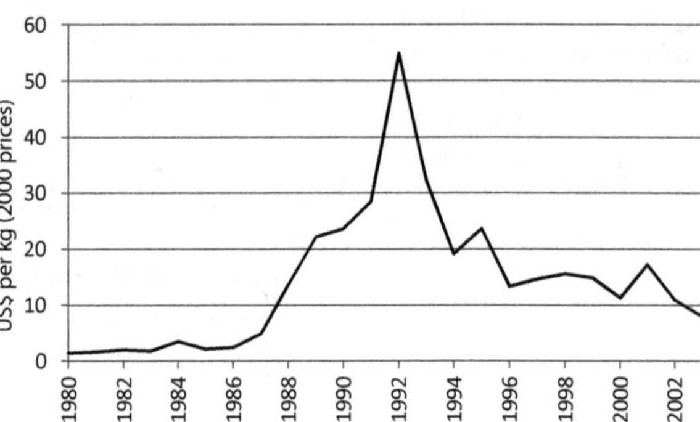

Fig. 4.7 The estimated value of Indonesian shark fin exports, 1980–2006. (Source: Calculated from data in FAO, Fishstat Plus- Commodities production and trade. Deflated using the GDP deflator for Indonesia from the International Monetary Fund, World Economic Outlook Database, October 2008)

trend helps explain why fin size average and consequently unit prices have fallen (Bentley 1996b).

Foreign boats, especially from Taiwan, have made large catches in the Arafura Sea fishing both off the Indonesian and northern Australian coasts. Shark are caught both as bycatch when trawling for demersal species and as target species. Following the introduction of the AFZ in 1979 and restrictions on the size of gillnets, the Taiwanese appear to have fished Indonesian waters more intensively. Small scale fishers, however, complained that industrial trawlers were destroying their livelihoods and by 1981 the government had banned trawling from the waters surrounding Java, Bali and Sumatra and by 1983 throughout Indonesia's EEZ. However, exemptions for shrimp dragnets in parts of eastern Indonesia and limited enforcement activity meant that the restrictions had little effect (Bentley 1996b; Butcher 2004). In 1984, in another attempt to curtail fishing effort, a licensing system was introduced for foreign vessels wanting to operate in Indonesia's EEZ. By 1994, 55 Taiwanese boats were licensed to fish in the zone, but an unknown number ran the risk of being arrested and fished illegally (Butcher 2004). As these catches were never landed in Indonesia it is clear the official data seriously underestimate the real level of exploitation of sharks.

The geographical focus of shark fishing has shifted over time. During the 1970s most shark landings were in the central and western provinces, in the Java Sea, Strait of Malacca and the Indian Ocean. By the 1990s many of these regions had experienced falling catches and landings were increasing in the eastern provinces including Maluku, North Sulawesi, West Nusa Tenggara and Irian Jaya. Currently, the main centres for sharks and ray catches are ports such as Palabuhanratu (West Java), Cilacap (central Java), Kedonganan (Bali) and Tanjung Luar (eastern Lombok) (Dharmadi et al. 2004). Sharks and rays are a major target species for small scale fishers based at locations in eastern Indonesia such as Tanjung Luar but are also frequently captured as bycatch by tuna fishers. The shifting geographical catch frontier is evidence that overfishing has occurred; falling catches and sizes of sharks have forced fishers to travel further afield to fill their holds with sharks, including into the AFZ (Bentley 1996b; Moss and Van Der Wal 1998).

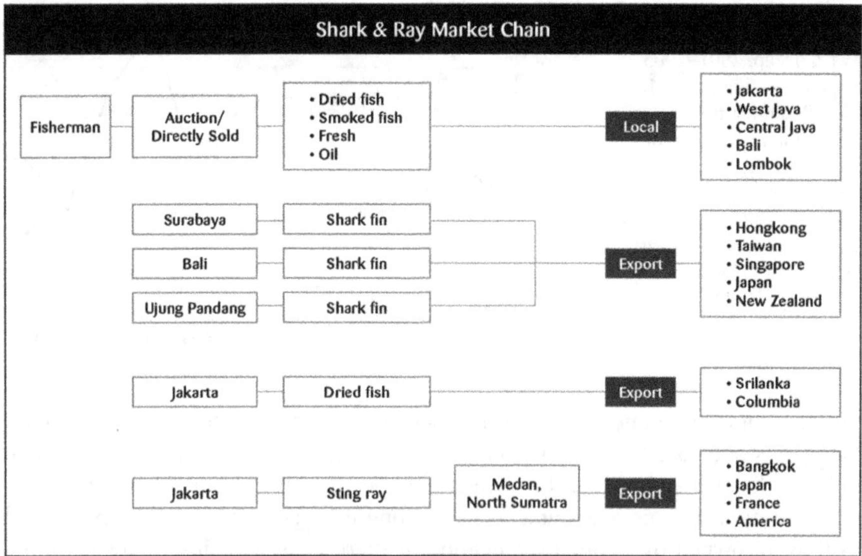

Fig. 4.8 The marketing chain for elasmobranches in Indonesia. (Source: Dharmadi et al. 2004)

A variety of gear types are employed including drift gill nets, longlines (surface longlines and bottom longlines) and trammel nets (Dharmadi et al. 2004). Surface longlines are used mainly by fishers from Cilacap and Tanjung Luar to catch pelagic sharks. Each boat spends between 7 and 15 days at sea and carries two longlines. Details of the gear vary from site to site but surface longlines usually have a main line of 8 mm diameter synthetic rope at least 3,000 m in length with 300–500 branch lines of 4 mm diameter rope from which from which hooks are hung. The hooks are baited with small fish previously caught using gill nets. Boats operating from Tanjung Luar can spend 6–7 days of a 15 day trip catching bait. Bottom longlines which target bottom dwelling species are similar in design.

The proceeds of the catch are distributed on a share basis after the deduction of operating expenses. Increasingly, fishing activity is dominated by 'fishing bosses' known as punggawas who provide credit to fishers in return for purchasing their catch at a low price. In the mid-1990s a punggawa in Ujung Pandang had a fleet of 43 boats which employed members from 200 families; he took two thirds of the catch leaving the balance for the fishers (Bentley 1996b). Most of the catch is processed locally, with the flesh being salted and dried and the skins used to produce leather for fashion accessories such as belts and wallets. Shark fins are not normally processed before export but are exported dried or frozen as importers prefer to do the processing themselves. The marketing chain is shown in Fig. 4.8.

Suzuki's (1997) case study of longline shark fishing in Karangsong village in Indramaya, about 200 km east of Jakarta, casts valuable light on the impact of shark fishing. Shark catches at Karangsong increased in the 1970s following the

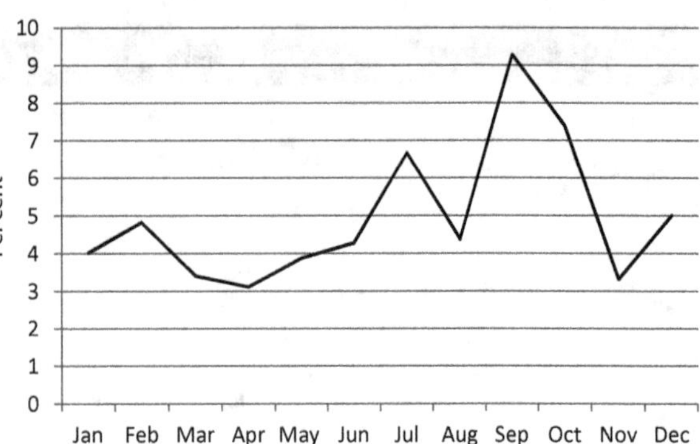

Fig. 4.9 Average proportion of catch from sharks and rays, Cilacap, 2001–2003. (Source: S. Vieira and M. Tull, 'Potential impacts of management measures on artisanal fishers in Indonesian shark and ray fisheries: a case study of Cilacap', Asia Research Centre Working Paper No. 127. Murdoch University, 2005)

introduction of nylon nets and drift gill net fishing. Shark flesh was salted aboard the vessels and sent for sale in interior. In the 1980s rising demand for fins, both within Indonesia and overseas, led many boat owners to borrow from punggawas to add motors to their boats and focus on longline shark fishing. However, most boats were owned by punggawas dealing in shark meat and fins– 7 owned more than ten boats in the Karangsong fleet. The boats were usually had a crew of 4 men including the captain, who fished all year round shifting grounds depending on the monsoon season. A study of 22 voyages in 1989 found that the average catch was 690 kg of salted shark and ray meat and 16.6 kg of shark fin, worth respectively Rp 290,000 rupiah and Rp 470,000. After operating costs were deducted the balance was divided up as follows: owner 50 %, captain 20 and 10 % for each crew member. The catch was very variable and depended heavily on the skill of the captain: in eight of the 22 voyages the value of the catch was less than the operating costs. Most fishers borrowed money to cover family expenses while at sea so an unsuccessful trip meant that the loans were added to cost of the next voyage. Punggawas paid an average of 50–65 % of the selling price for the catch. Unfortunately, large increases in operating costs between from 1987 to 1991 cancelled out increasing catch values so returns to crews plateaued or even fell. Stacey's (2007) detailed study of the economics of Bajo shark fishing also shows that while shark fishing generated valuable returns, the main beneficiaries were boat owners and traders rather than fishers and their families.

As we have seen in this chapter, increased fishing pressure is threatening the sustainability of the shark and ray fishery in Indonesia. Currently, there are few effective management controls in the Indonesian Fishing Zone so the fishery can be regarded as approximating an open access one. The low incomes of small scale fishing communities means that any restrictions on catch can have potentially serious impacts on the incomes and well-being of fishers, their families and local communities.

A recent case study of the artisanal fishery at Cilacap, an important fishing port located on central Java's southern coast, casts light on the potential impact of catch restrictions (Vieira and Tull 2008). Figure 4.9 shows the percentage contribution of elasmobranches to the total catch value at Cilacap from 2001 to 2003. There is

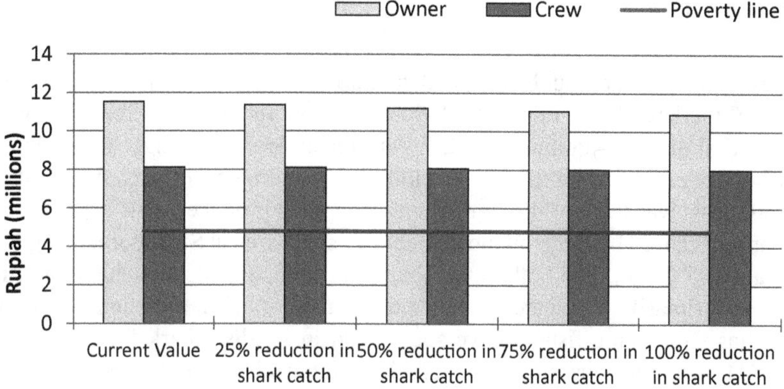

Fig. 4.10 Impact of reductions in shark catch on the household income of boat owner and crew at cilacap. (Notes and source: Cash returns to owner are used in calculation of owner's household income. 2001 poverty line is from BPS-Statistics Indonesia (2002, 2003). See Vieira and Tull 2008)

some evidence of seasonality with higher shark and ray catches in June/July and September/October; effort drops from November onwards due to the onset of the wet season. The key point, however, is that revenue from sharks and rays averaged only 5 % of total catch value.

In order to assess the potential impact of management controls on the artisanal fishers at Cilacap, Fig. 4.10 shows the estimated effects of reducing catch by 25, 50, 75 and 100 % (Vieira and Tull 2008). According to the estimates, if a 100 % reduction in shark catch is imposed the household income levels of both boat owners (based on cash returns) and crew members will be reduced by only 7 and 3 % respectively. Income levels appear to be quite high in both the boat owner's and the crew member's case when compared with the 2001 household poverty line for Indonesia. This suggests that reductions in shark catch would impact on boat revenue and profitability only marginally. But there may be households that that are more reliant on shark and ray catch, and thus could potentially be seriously affected even by small reductions in catches. Furthermore, if non-monetary capital costs are taken into account, with no change in shark catch the boat owner's household income is only just above the poverty line at Rp 4.9 million; with a 100 % reduction in shark catch it falls to about 25 % below the poverty line. These findings are based on Cilacap only and may not apply elsewhere especially at locations like Tanjung Luar, Lombok, where sharks and rays are major target species.

Catch restrictions will also reduce downstream processing of shark products, an important source of employment, especially for females. The forward linkage multiplier for agriculture, livestock, forestry and fisheries, suggests that a decrease of 10 % in output from fisheries will decrease total output by about 9 %. Clearly, while the precise impact is uncertain, any restrictions on fisheries output will create ripple effects throughout the entire economy.

Conclusions

Indonesia's fisheries output has expanded enormously since the 1950s, although at least until the late 1960s, this was 'static expansion' driven by increased inputs of traditional skills and equipment rather than technological change. Elasmobranch catches grew rapidly from the 1970s and by the beginning of the twenty-first century Indonesia was the world's leading elasmobranch producer. This expansion had been largely caused by the growing demand for shark fins in Singapore, Hong Kong and mainland China. For both small scale and industrial fishers sharks and rays have been primarily a by catch rather than a target species, although as the price of fins has risen some fishers have specialised in sharks. Shark fin exports have provided a valuable source of income and foreign exchange earnings but the main beneficiaries of the shark trade have, however, been boat owners and traders rather than fishers and their families.

There is growing worldwide concern about the state of elasmobranch stocks. Since 2003 recorded elasmobranch landings in Indonesia have fallen; exports peaked in the early 1990s and by 2008 were about US$ 7.0 m or about 40 % of their peak level- clearly, overfishing is taking its toll. It is generally accepted that due to the practice of 'finning' and other omissions, official data understate the real level of catches and trade by up to 50 %.

Fishers have encountered declining yields in the central and western areas of the EEZ and have been forced to fish the eastern waters more intensively and, increasingly, venture into Australia's territorial waters. The outwards shift of the geographical catch frontier is approximate evidence that fishing pressure has seriously depleted shark stocks in Indonesia. In the last decade high prices for shark fins, trepang and other marine species and the lure of high returns has led to a surge in illegal fishing in the AFZ.

Since 1979 Indonesia has subscribed to CITES but despite this and other measures, including the establishment of marine protected areas, 'Indonesia is experiencing rampant exploitation of many living marine resources' (Moss and Van Der Wal 1998). There are few management controls in Indonesia's elasmobranch fishery; in practice, the fishery is an open access one. In order to avoid a 'tragedy of the commons' effective management measures need to be introduced. Ideally, Indonesia should adopt a National Plan of Action to ensure that fishing effort is kept at a sustainable level but this is complicated by funding constraints and since 2002 the division of powers between the national, Kabupaten (district/ regency) and provincial governments. Unfortunately, inadequate governance arrangements are a widespread problem in the Indian Ocean Region (Rumley 2009). Another problem is that the increase in large scale industrial fishing has increased the potential for conflict with small scale fishers. There is a risk that catch restrictions could have serious negative direct and indirect impacts on the livelihoods of small scale fishers and their families so it is suggested that a cautious approach be adopted, that is, any gear or catch limitations should start at low levels and be gradually phased in. Curbing overfishing will not, of course, restore fisheries affected by environmental factors such as global warming or excess nutrient inputs.

This chapter has shown that, like many developing countries, Indonesia has faced a complex task in trading-off the potentially conflicting aims of fishery development which include providing food, employment and export earnings. Governance failures mean that the future of most fish stocks and the livelihoods of the millions of Indonesians who depend on them, remain uncertain. In the long run, the continued growth of the economy, based on its natural resources and labour supply, offers the best hope of improving the living standards of all Indonesians and providing the means to save its emptying oceans.

Appendix 1

Shark Fin Exports from Indonesia, 1976–2008

Year	Tonnes	US$'000	US$ per kg
1976	277	177	0.6
1977	87	63	0.7
1978	134	155	1.2
1979	186	202	1.1
1980	179	259	1.4
1981	225	363	1.6
1982	249	497	2
1983	334	600	1.8
1984	232	797	3.4
1985	329	677	2.1
1986	444	1,048	2.4
1987	573	2,762	4.8
1988	473	6,422	13.6
1989	475	10,473	22
1990	422	9,949	23.6
1991	376	10,680	28.4
1992	316	17,338	54.9
1993	367	11,900	32.4
1994	498	9,491	19.1
1995	447	10,523	23.5
1996	894	11,841	13.2
1997	676	9,867	14.6
1998	231	3,601	15.6
1999	615	9,148	14.9
2000	1,166	13,095	11.2
2001	479	8,220	17.2
2002	771	8,414	10.9
2003	1,288	10,204	7.9
2004	943	10,936	11.6

Year	Tonnes	US$'000	US$ per kg
2005	1,554	8,065	5.2
2006	1,073	9,174	8.5
2007	801	7,303	9.1
2008	1,320	7,047	5.3

Source: FAO, Fishstat Plus- Commodities production and trade

References

Anak NA (1997) An overview of sharks in world and regional trade. In: Fowler SL, Reed TM, Dipper FA (eds) Elasmobranch biodiversity, conservation and management: proceedings of the international seminar and workshop, Sabah, Malaysia, July 1997. Occasional paper of the IUCN Species Survival Commission

Bailey C (1996) Chapter 2. In: The world trade in sharks: a compendium of TRAFFIC's regional studies (vol1). TRAFFIC International, Cambridge

Bailey C, Dwiponggo A, Marahudin F (1987) Indonesian marine capture fisheries. International Center for Living Aquatic Resources Management, Manilla

Bentley N (1996a) Indonesia. In: Chen HK (ed) An overview of shark trade in selected countries of Southeast Asia. TRAFFIC Southeast Asia, Petaling Jaya

Bentley N (1996b) An overview of the exploitation, trade and management of chondrychthians in Indonesia. In: Keong CH (ed) Shark fisheries and the trade in sharks and shark products of Southeast Asia. TRAFFIC Southeast Asia, Petaling Jaya

Bonfil R (1994) Overview of the world elasmobranch fisheries. Food and Agriculture Organization of the United Nations, Fisheries Technical Paper No. 341, Rome

Butcher JG (2004) The closing of the frontier: a history of the marine fisheries of Southeast Asia, c.1850–2000. ISEAS Publications, Singapore

Campbell BC, Wilson BVE (1993) The politics of exclusion: Indonesian fishing in the Australian fishing zone. Indian Ocean Centre for Peace Studies, Perth

Clarke SC (2003) Quantification of the trade in shark fins. PhD Thesis, University of London

Dharmadi F, White W, Wdodo J (2004) Some aspects of shark and ray artisanal fisheries in south of Indonesia. ASEAN Fisheries Forum, Penang

East Asia Analytical Unit (1994) Subsistence to supermarkets: food and agriculture transformations in South-East Asia. Department of Foreign Affairs and Trade, Canberra

Kurien J (1998) Small-scale fisheries in the context of globalisation: Ingredients for a secure future, citing Report of the International conference of Fishworkers and their Supporters (1984), Rome. In: Eide, A and Vassal, T (eds) (1998) Proceedings of the 9th International Conference of the International Institute of Fisheries Economics and Trade, vol 1, University of Tromso, Norway

Emmerson DK (1980) The case for a maritime perspective on Southeast Asia. J Southeast Asian Stud 11:139–145

FAO (2010) International plan of action for the conservation and management of sharks. Food and Agriculture Organization of the United Nations, Rome

FAO (2012) Fishery and aquaculture profiles: Indonesia. Food and Agriculture Organization of the United Nations, Rome

Fegan B (2003) Plundering the sea. Inside Indonesia 73:21–23

Fox JJ (2009) Legal and illegal Indonesian fishing in Australian waters. In: Cribb R, Ford M (eds) Indonesia beyond the water's edge: managing an archipelagic state. ISEAS, Singapore

Fox JJ, Adhuri D, Therik T, Carnegie M (2009) Searching for a livelihood: the dilemma of small boat fishermen in Indonesia. In: Resosudarmo BP, Jotzo F (eds) Working with nature against poverty: development, resources and the environment in Eastern Indonesia. ISEAS, Singapore

Gelchu A, Pauly D (2007) Growth and distribution of port-based global fishing effort within countries' EEZs from 1970–1995. University of British Colombia

Heazle M, Butcher JG (2007) Fisheries depletion and the state in Indonesia: towards a regional regulatory regime. Mar Policy 31:276–286

IUCN (2003) Shark finning. IUCN Marine Programme, Malaga

Kelleher K (2005) Discards in the world's marine fisheries: an update. Food and Agriculture Organization of the United Nations, Rome

Kura Y, Revenga C, Hoshino E, Mock G (2004) Fishing for answers: making sense of the global fish crisis. World Resources Institute, Washington

Kusumastanto T (2004) The challenge of improving fish consumption in a developing country: the case study of Indonesia. Twelfth Biennale Conference of the International Institute of Fisheries Economics and Trade (IIFET), International Institute of Fisheries Economics and Trade, Tokyo

McPherson K (1993) The Indian ocean. Oxford University Press, London

Moss SM, Van Der Wal M (1998) Rape and run in Maluku: exploitation of living marine resources in Eastern Indonesia. Cakalele 9(2):85–97

Naamin N (1995) An overview of fisheries and fish processing in Indonesia. In: Champ BR, Highley E (eds) Fish drying in Indonesia. Proceedings of an international workshop, Jakarta

Osseweijer M (2001) Taken at the flood. Marine resource use and management in the Aru Islands Maluku, Eastern Indonesia. PhD Thesis, University of Leiden

Pauly D, Thia-Eng C (1988) The overfishing of marine resources: socioeconomic background in Southeast Asia. Ambio 17:200–206

Pearson M (2003) The Indian ocean. Routledge, London

Polunin NVC (1983) The marine resources of Indonesia. Oceanogr Mar Biol Annu Rev 21:475

Rose DA (1996) An overview of world trade in sharks and other cartilaginous fishes. TRAFFIC International, Cambridge

Rumley D, Chaturvedi S, Sakhuja V (2009) Fisheries exploitation in the Indian ocean: threats and opportunities. Institute of South East Asian Studies, Singapore

Stacey N (2007) Boats to burn: Bajo fishing activity in the Australian fishing zone. ANU E Press, Canberra. Http://epress.anu.edu.au/boats_citation.html

Suboko B (2001) Country profile: Indonesia- fisheries sector survives economic crisis. Infofish Int 2:14–20

Sulaeman K (1969) The economic development of Indonesia's sea fishing industry. Bull Indonesian Econ Stud 5:49–72

Suzuki T (1997) Development in shark fisheries and shark fin export in Indonesia: case study of Karangsong village, Indramayu, West Java. In: Fowler SL, Reed TM, Dipper FA (eds) Elasmobranch biodiversity, conservation and management: proceedings of the international seminar and workshop, Sabah, Malaysia, July 1997. Occasional paper of the IUCN Species Survival Commission

Tomascik T, Mah AJ, Nontji A, Moosa, MK (1997) The ecology of the Indonesian seas. Part II. The ecology of Indonesia series, vol VIII. Periplus editions, Hong Kong

Varkey DA, Ainsworth CH, Pitcher TJ, Goram Y, Sumaila R (2010) Illegal, unreported and unregulated fisheries catch in Raja Ampat Regency. Eastern Indonesia. Mar Policy 34:228–236

Vieira S, Tull M (2008) Restricting fishing: a socio-economic impact assessment of artisanal shark and ray fishing in Cilacap. Bull Indonesian Econ Stud 44:263–288

Wagey GA et al (2009) A study of illegal, unreported and unregulated (IUU) fishing in the arafura sea, Indonesia. Research Centre for Capture Fisheries, Ministry of Marine Affairs and Fisheries, Jakarta

White WT et al (2006) Economically important sharks & rays of Indonesia = Hiu dan pari yang bernilai ekonomis penting di Indonesia. Australian Centre for International Agricultural Research, Canberra

White WT, Cavanagh RD (2007) Whale shark landings in Indonesian artisanal shark and ray fisheries. Fish Res 84:128–131

Williams MJ (2007) Enmeshed. Australia and Southeast Asia's fisheries. Lowy Institute for International Policy, Double Bay

Chapter 5
A History of Whaling in the Philippines:
A Glimpse of the Past and Current Distribution of Whales

Jo Marie V. Acebes

Abstract Little is known about the history of whaling in the Philippines. This chapter aims to document the nature and extent of whaling in the Philippines from the nineteenth century to recent times, identify past whaling grounds and compare these with current distributions of the species of whales hunted. The comparison illustrates changes in the historical abundance and distribution of these species, enhancing understanding of what brought about these changes and providing a basis for further investigations into the ecological and social impacts of changing abundance and distribution. Field studies were conducted at known former whaling locations, and historical records, published works, popular accounts, town records and news articles were also examined. Given the limited available data examined, the practice of whaling in the country is only traced back to the nineteenth century. Decrease in the frequency of sightings and encounters are attributed to a number of factors which include past and recent commercial and small-scale whaling activities by foreign and local whalers and other disturbances such as increased boat traffic and habitat degradation due to development of coastal areas. Further research is needed to make a more comprehensive illustration of these changes and understand the factors that caused it.

Keywords Humpback whale · Sperm whale · Philippines · American whaling · Whaling history

The history of whaling in the Philippines is poorly understood. Accounts of the 'whale jumpers of Pamilacan' have attracted popular interest (Severin 1999; *The Scotsman* 6 November 1999), but little is known about the origins and extent of indigenous whaling practices, or of the situation today. Foreign whalers frequented the Philippines in the nineteenth and nineteenth centuries, but the extent and impact of their activities remains unclear.

J. M. V. Acebes (✉)
Asia Research Centre, Murdoch University, 90 South Street,
6150, Murdoch, WA, Australia
e-mail: jomacebes@yahoo.com

J. Christensen, M. Tull (eds.), *Historical Perspectives of Fisheries Exploitation in the Indo-Pacific*, MARE Publication Series 12, DOI 10.1007/978-94-017-8727-7_5,

In modern times, the country has had only a brief career in commercial whaling. The Philippines attended the annual meetings of the International Whaling Commission (IWC) from 1981 to 1986 (Barut 1994), first as a 'non-whaling nation' then to a 'coastal whaling nation' between 1982 and 1987 (Ohmagari 2005). This engagement in coastal whaling was a contentious one as it involved Japanese nationals apparently owning the company and operating all activities, and violated IWC rules and CITES regulations (Jiji Press Ticker Service 1984a, 1984b; Japan Economic Journal 1984; Davies 1986, Day 1993; Barut 1994). These whaling operations were supposed to have ended after the IWC issued the moratorium in 1986 but it is believed in some quarters to have continued for a few months thereafter (Davies 1986). According to the Department of Agriculture—Bureau of Fisheries and Aquatic Resources (DA-BFAR), the company withdrew from the fishery because of declining catches, after which the government ceased issuing commercial whaling licenses.

Whaling again attracted scientific and media attention when reports of killing and the by-catch of cetaceans in fishing gear in Palawan, Central Visayas and Northern Mindanao caught the attention of the DA-BFAR during the early 1990s. Investigations were conducted which soon resulted in the issuing of Fisheries Administrative Order (FAO) 185 (DA-BFAR 1992), banning the taking or catching, selling, purchasing and possessing, transporting and exporting of dolphins in 1992 (Barut 1994; Buckley 1994; Dolar et al. 1994). This did not necessarily stop the killings and by-catches but instead drove the activities and the market underground, making data collection more difficult (Dolar et al. 1994; Barut 1999). Although this did not affect the Bryde's whale hunting in Pamilacan, given the ban only included dolphins, it was not long after, in 1997, that FAO 185-1 was issued, amending the first law to include 'all cetaceans', including whales and porpoises. This, as far as it has been documented, marked the end of whaling in the Philippines.

Yet rumours still circulate that some opportunistic whaling still exists, if not in Bohol, then in other areas of the Philippines (Barut 1998, 1999; Severin 1999; Japan Times 2004). In an archipelago with more than 7,000 islands, about 1.3 million fishers and their families depend on fisheries for their livelihood (Green et al. 2003). With declining fisheries in the region (Cruz-Trinidad et al. 2002; Green et al. 2003) it is not surprising to find fishers increasing their fishing effort, shifting their target species or becoming less selective in fishing, changing their fishing gear, searching for new fishing grounds, or engaging in the opportunistic capture of other, larger marine animals. Opportunistic whaling is, of course, difficult to prove with certainty due to the fact that the ban is well-known and illegal whalers are presumably quite aware of the consequences of being caught (Fig. 5.1).

The objectives of this chapter are to document the nature and extent of whaling in the Philippines from the nineteenth century to the present, to investigate the origins and development of whaling in the Philippines, and to identify areas of past whale occurrences and compare these with currently-known areas of distribution of the species of whales previously exploited. The chapter focusses particularly on the islands of Bohol (including Pamilacan Island), Camiguin and Salay in Misamis Oriental, where local whaling have previously been known to occur

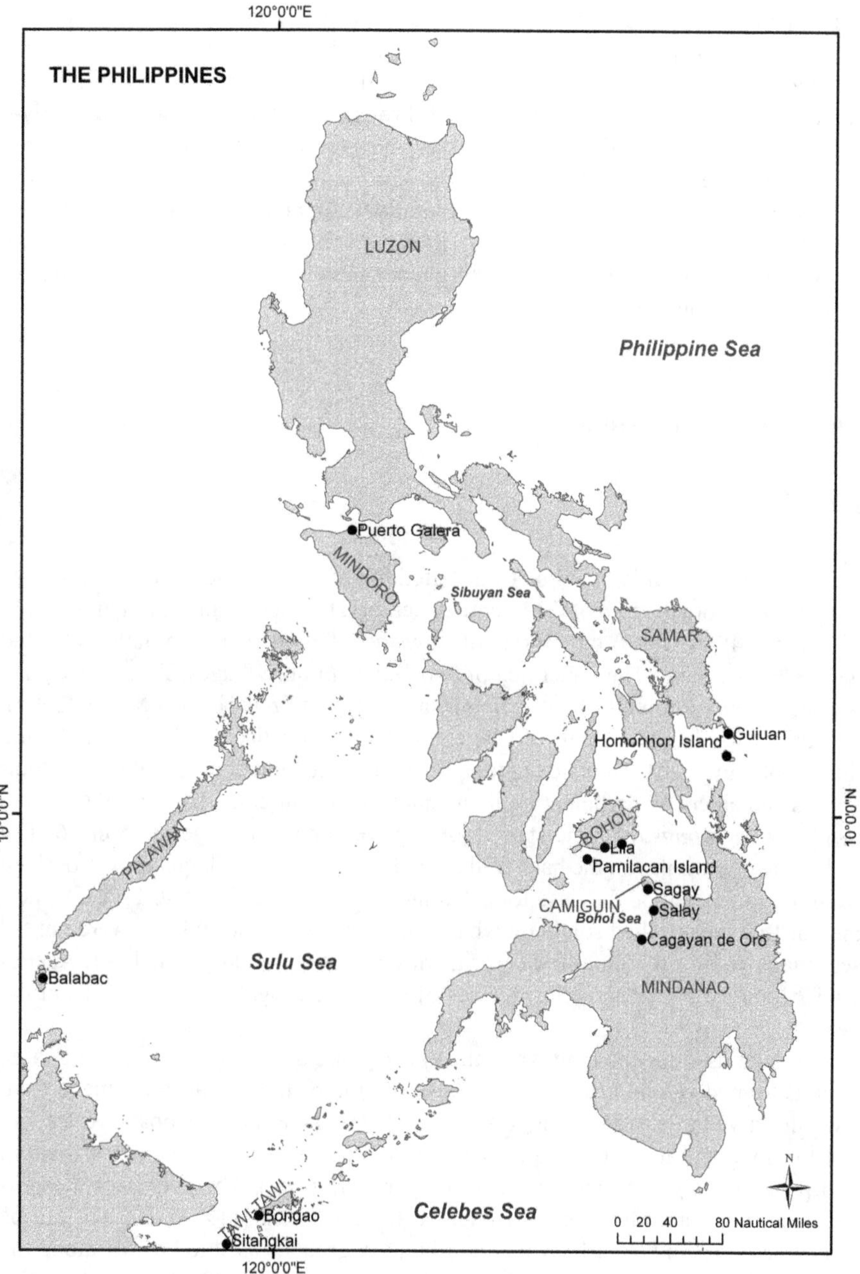

Fig. 5.1 Map of the Philippines, showing places named in the chapter

(Dolar et al. 1994), and on the Sulu-Sulawesi seas which were frequented by nine-teenth century American whalers. Evidence has been obtained through a series of interviews with local fishers and residents of coastal communities, through archival research in municipal records, and through literature searches and archival research in a number of foreign libraries and museums, and has been reported upon elsewhere (Acebes 2009). Enhanced knowledge of the size and distribution of populations in the past, of the origins and development of indigenous whaling, and of the effectiveness of recent regulations banning whaling in the archipelago, can serve as a better guideline in making management decisions and assessing the impact of existing policies.

Local coastal whaling

Lila, Bohol

Residents of the municipality of Lila started hunting for whales, locally known as *bongkaras* before the turn of the twentieth century. Former whalers born in the early 1900s (i.e. 1916, 1926) tell stories of how their fathers and grandfathers hunted whales before them. The fishermen of two *barangays* or villages, Taug and Tiguis, are specifically known to hunt *sanga* (manta ray) and *bongkaras* (Acebes 2009). All former whalers and locals of Lila claim that the practice of whaling (including the technology used) originated in their town. They learnt to hunt whales from their fathers and elders by participating in the hunt as apprentices, whose usual role is to paddle or *taga-bugsay*. Some started joining the hunts at the age of 15 or 16. The technique of jumping on the back of the whale and thrusting a large hook is derived from the same method they use for catching manta rays (Dolar et al. 1994). Fishermen of Lila never relied solely on whales for their livelihood. Whales were hunted seasonally during the months of April, May and June (Dolar et al. 1994; Acebes 2009), but manta rays and other species of fish were caught in other months of the year.

The boat used for the hunt was called *pilang* and the technique of using a large hook is termed *pamimilak*. Prior to the use of motors, these boats were made from a whole carved tree, with outriggers on both sides and a sail. The boats were about 9–10 m long and were manned by six men: three rowers or oarsmen (*taga-remo*), two paddlers (*taga-bugsay*) and one jumper (*manuung*). The details of the procedure of the hunt have been documented by Dolar et al. (1994) and will not be discussed here.[1] In the early years, their hunting grounds used to be only along the shores of Lila, between 5 to 15 km from shore (Acebes 2009). Blows of whales swimming close to shore signal the start of the season. Whalers put out to sea early

---

[1] Note however, that Dolar et al. (1994) focused on the more recent method of using motorized boats hence, the difference in crew members and roles.

in the morning and return as soon as a whale is caught. The method of distributing the catch among the crew varied slightly among some of the fishermen. The jumper receives two shares, the owner of the boat also receives two shares and the rest receives one share each (Acebes 2009). Alternatively, the oarsman, usually the older and stronger member of the crew, also receives two shares while the rest will divide equally among themselves whatever remained of the whale (Acebes 2009). Not only crew members partook of the catch, but whoever helped in the catch or the cutting up of the whale also received a share. Members of the crew did as they wish with their shares. They either 'bartered' or exchanged it for local produce, sold it to the market or to their neighbours, or took it home to eat (Acebes 2009). More often though, during the early years, there would be so much meat for the crew and their families to eat that it will simply be given away to everyone in the *barangay* or village. Essentially the entire whale is utilized, except for the bones.

With the introduction of motorized boats, fishermen became more efficient in getting a catch. They could go faster, could catch up with the whales and go further out to sea. Once they have caught one, they could tow it back to shore to cut-up and possibly go out again to catch another.

The town of Lila is dubbed as the 'Town of Adventurers' and is known for its 'peddlers' (Acebes 2009). They venture off to nearby Cagayan, Camiguin, and Leyte, 'peddling' or selling tobacco and other products (Acebes 2009). They would spend weeks in Samar and Leyte selling tobacco and when they returned they would resume whaling (Acebes 2009). This lifestyle could have made it easy for them to transfer the practice or technology to other islands.

The *bongkaras*, or the whale, got its name from a political candidate. According to the story, a politician with a surname Bongcaras was campaigning for office when a whale beached itself along the shores of Lila. Taking advantage of the crowd gathered to see the spectacle Mr. Bongcaras stood on top of the dead whale and delivered his speech. Ever since, that type of whale has been called *bongkaras*.

When the whales no longer frequented their shores, and motorized boats became available, fishers ventured further away from Lila towards Pamilacan, where apparently the whales were more plentiful (Acebes 2009). Although they stopped whaling in 1986, they continued to be involved in the buying and selling of the meat from the Pamilacan fishery.

Pamilacan Island, Bohol

Pamilacan Island is a *barangay* of the municipality of Baclayon. Some say the island derived its name from the practice by islanders of catching manta rays using a hook locally called *pilak* (Dolar et al. 1994, Tan 1995) and referred to as *pamilak ug sanga* (Acebes 2009; Moral 2003). It may have also been because Lila fishers who ventured in their boats away from their town's shores opened the surrounding waters of the island as a new hunting ground of manta rays and whales and started landing their catch on the island. The whaling practice in Pamilacan was

Fig. 5.2 A toggle-harpoon used to catch Bryde's whale in Sagay, Camiguin

clearly derived from the whale fishery in nearby Lila. The oldest respondent from the island, born in 1924, said that residents of the island didn't use to catch *bugangsiso* (the name they call the whale) or *bongkaras* but instead people from Lila did (Acebes 2009). He recalls that it was in 1939 that fishers from Lila started catching whales around their island. They would land the animal there, cut it up and sell the meat in Lila. Sometimes when islanders helped, meat is given to them for free.

However, some also believe that the fishers of Pamilacan learned to catch whales on their own (Acebes 2009). The boats and techniques of hunting used in Pamilacan are essentially the same as that used by Lila fishers. The meat from the catch is also either consumed locally or sold to the people of Lila. Furthermore, Pamilacan fishers did not rely on whales as their source of livelihood. Their main target species are manta rays and whale sharks (*balilan*), caught using the hook, while other fish species are caught using nets and hook and line (Acebes 2009). Whales were caught opportunistically and seasonally.

The practice came under media attention in 1990 and with pressures from national officials, the BFAR came out with FAO 185 then 185-1 bringing an end to the whaling practice in 1997. After the declaration of the ban on whaling, NGOs came in to assist the communities with alternative livelihoods. WWF-Philippines introduced dolphin-whale watching to the community in 1997. However, not all former whalers were willing, at least initially, to take part in this, and apparently continued to hunt whales in protest.

Sagay, Camiguin

Whalers in Sagay used a locally manufactured iron toggle harpoon called *isi* similar to that used by American whalers in the middle of the nineteenth century (Verrill 1923; Spence 1980; Dolar et al. 1994), except it was shorter and the toggle blade was slightly different in shape (Fig. 5.2). This harpoon is attached to the end of a wooden pole about 3 m long and secured with a rope. The other end of the rope is tied to the boat. The harpoons were only made by one blacksmith from the nearby town of Mambajao (Acebes 2009). Dolar et al. (1994) stated that this harpoon is most likely an adaptation from the American whalers and gives some indication that this fishery may have been influenced by this source.

However, locals believe the practice originated in Sagay. They learned to hunt from their fathers and grandfathers (Acebes 2009). The date of practice could be put to at least the early 1900s. One whaler indicated that his grandfather, who was also a whaler, came from Bohol and moved to Sagay to settle. This points to another possible origin of the practice.

The boat used was called a *pamilacan*, and unlike the boats used in Bohol it did not have a sail. A sail would have made it dangerous to use when going after a whale against the wind. There were also five crewmen (*taga-bugsay*) and one jumper (*taga-bangkaw*). Just like in Pamilacan and Lila, the jumper leaps off the boat to thrust the harpoon (instead of a hook) into the whale. Their original hunting grounds were along the shores of Sagay, extending to the towns of Guinsiliban and Catarman. Using a motorboat, they were later able to reach across Balingoan in Misamis Oriental. The season for hunting was mainly during March, April and May, although for some it extended from as early as January until June.

The fishers divided the catch in the following way: half was given to the man who harpooned the whale (*nakabangkaw*), and the other half was divided among all the other people who helped or participated in the catch (Acebes 2009).[2] This includes a share for the owners of the implements used in catching the whale, namely: harpoon, boat, *pisi* (rope), and gangso (hook). Another way of sharing the catch was by dividing the whale equally among the people who took part in the hunt (including owners of whaling implements) but the *taga-bangkaw* received two shares. As in Lila and Pamilacan, participants were free to do whatever they wished with their shares. Just like in Bohol, all parts of the whale are eaten, including the baleen. Whaling here ended in 1997 (Acebes 2009).

Salay, Misamis Oriental

The origins and extent of whaling in Salay is unclear. The practice or technology may have likely originated from Bohol (Acebes 2009). Whalers whose origins are from Camiguin where they used to hunt in the 1920s, speak of their fathers (also whalers) spending years living in Bohol in the mid-1800s. A harpoon used by a whaler was reportedly made and imported from the USA probably in the late nineteenth century. More recent harpoons were crafted in Bryg. Lakas, Mambajao, Camiguin in 1948. A couple of whalers were even believed to have been taken to Hawaii by Americans in 1941 to catch whales. Apparently, the technique used in killing whales in Salay was faster compared to the technique Americans used at that time. Although an actual harpoon was not available for inspection, the respondent

---

[2] This description corresponds with Dolar et al. (1994).

Fig. 5.3 A toggle-harpoon
used to catch manta rays in
Jagna, Bohol

was able to make a clear illustration of what it looked like (Acebes 2009). It is almost the exact likeness of the toggle iron harpoon used by whalers of Sagay, and thus, bears close resemblance to the harpoon used by American whalers.

The municipal profile shows no indication of a whaling tradition or any references to whales as marine resources, although fishing is one of the main sources of livelihood for local peoples. Whaling apparently ended here also in 1997.

Whale consumption in other areas

The municipality of Jagna in Bohol is well-known for its tradition of manta ray fishing, the same fishery from which the practice of whaling in Lila is suspected to derive from (Dolar et al. 1994). Barangay Bunga Mar is the centre of the manta ray fishery in Jagna. The people here have been hunting manta rays, also known as *sanga*, for over a century.

The harpoon or *isi* used by the fishers to catch *sanga* is identical to that used by the whalers in Sagay, Camiguin and Salay, Misamis Oriental (Fig. 5.3). The hook or *gangso* used by the Jagna-anons to secure the manta ray after it is struck with the harpoon is similar to the hook used by the whalers of Lila and Pamilacan, except that it is more slender. Fishers claim that the harpoon and hooks are all locally designed and made by blacksmiths in the nearby village of Cantagay. Interestingly, the boat used to hunt manta rays is locally called *pamilacan*, just like in Sagay.

The similarity in techniques, hunting instruments, and boats used, as well as the corroborating stories of Boholanos migrating or travelling to Camiguin and Misamis Oriental, may lead one to conclude that the fishery in Lila, Pamilacan, Sagay and Salay may all have been derived from the manta ray fishery in Jagna.

In Jagna whales were never hunted directly with the *isi*. However, once there was a whale seen circling near shore and was struck, dragged to Lila and butchered there. This happened sometime between 1970 and 1972, long after the people of Lila started hunting whales. Fishers have also caught two sperm whales with a net

30 km offshore, in 1979 or 1980, which were butchered and eaten by the locals. Eating whale meat appears to be quite common among locals although the meat would come mainly from stranded animals.

There are no indications of whaling in Puerto Galera, Mindoro however, stranded whales are known to be butchered and eaten. Whales are locally called *bugangsiso*, just like in Pamilacan, and are so named because the whale 'blows out' (*buga* in Tagalog) small fishes (*isdang pino*) from 'the hole on the top of its head' (Acebes 2009). In Balabac, Palawan whales are locally called *kahumbo*, or for migrants from Tawi-Tawi, *gadja mina*. There was no indication of history of local whaling and direct catches of dolphins. Locals do not eat them because they have traditional and religious beliefs that discourage this type of fishery. In Bongao and Sitangkai, in Tawi-Tawi the Badjao and Samal people do not hunt whales either. Their traditional beliefs inscribe fear unto the people. However, they do hunt and eat dolphins. Similar to Balabac, the Tausugs of Tawi-Tawi (migrants from Jolo) do not eat dolphins as it is forbidden.

Kühlmann (2000) wrote about the municipality of Guiuan in eastern Samar and refers to the historical utilization of whales as food and a source of income by the coastal population of Samar. The municipality's website refers to a historical link to the Marianas and Guam which were American whaling grounds; 'Guiuan was convenient emergency stop for the galleon and from the late nineteenth century was a take off point for the Marianas' (Panublion Heritage Sites 2005). American whalers frequented the Marianas in the 1850s and 1860s to hunt for humpbacks (Jenkins 1921).

Whales were never hunted along the coasts of this municipality however, some people have eaten whale and dolphin meat in the past (Acebes 2009). One respondent recalls of a stranding of a whale in Homonhon Island after a big storm in the 1950s, wherein 'residents of all eight barangays on the island took part in it…cutting up the whale' (Acebes 2009).

Offshore whaling in the Philippines

Foreign whaling in the pre-colonial and colonial eras

Documents pertaining to pre-colonial and colonial whaling by the Spanish, Americans, and other foreigners within Philippine waters are scarce, if not elusive. The Spanish (Basques) were some of the earliest whalers in the world's oceans (Birnie 1985) and one would presume that they would have brought this practice with them when they visited and colonized the Philippines. Information on this has not however been found. Similarly, although Spanish missionaries have written volumes about Filipinos during their stay in the Philippines, Jesuit letters that were examined did not indicate any reference to whaling or activities relating to cetaceans, whether by the Spanish or the native peoples.

The Chinese possibly had some involvement in whaling in and around the Philippine as they too practiced whaling 'even though the documentation in Western literature is scarce' (Ellis 1991). The Chinese hunted whales near and around Hainan Island in the South China Sea, and the harpoon used had only one barb:

> '…and about fifteen inches from the point of the iron it is made with a socket; above which, an eye is wrought, with a cord attached to slack along the wooden shaft so that when the fish is struck, the iron and the line tightens, the shaft draws out, and leaves less chance of the iron cutting out of loosing its hold on the skin of the fish.' This is similar to the harpoon head used in Lomblen, Indonesia, which resembles the harpoon head excavated in Cagayan de Oro, Mindanao, Philippines. (Richard Ellis, quoted in Heritage Conservation Advocates 2005)

The earliest contact with the Chinese probably dates back to 982 AD (Patanne 1996). Documents of the Sung trade indicate that among hundreds of items of import, one of the most important were aromatics and drugs such as 'ambergis' (Patanne 1996). Ambergis comes from sperm whales. Although this may suggest the presence of sperm whales, it is possible to find ambergris without having a whaling industry (Ellis 1991).

American and British whaling

In Townsend (1935), reference to the extent and area of whaling in the Philippines can be found:

> Sperm whales were taken in great numbers during all seasons of the year in whaling grounds known as…'Sulu sea', 'Celebes Sea'…

Based on records of logbooks of American whalers examined by Townsend, the sperm whaling grounds were much larger than what was strictly known as the 'Sulu' and 'Celebes Seas'. Sperm whales were also taken from the South China Sea, from west of Manila Bay and Mindoro all the way down to south western Palawan, within the Sibuyan Sea, Bohol Sea and to the east along the Philippine Sea.

Fifteen American whaling logbooks from the period of 1838–1839, 1853–1855, 1863–1864 and 1868–1869 were examined during research for this chapter. Several of the logbooks recorded catches of sperm whales within Philippine waters, mainly in the Sulu and Celebes seas. A document referring to American whalers and traders in the Philippines from 1817 to 1899 indicated the presence of American whalers in the country however, no detail was given on the exact nature and location of whaling activities aside from serving as a drop-off for the sick and rowdy whalers (Wuerch 1987; Mak 2005).

The significance of the Philippines in the history of American whaling is further substantiated by Clark (1887) where he notes that:

> Sperm-whaling was formerly carried on with good success…also in the Sooloo or Mindora Seas, and around the East India Islands, where ships continued to cruise until within about three years.

The 'Sooloo Sea' is again identified by Scammon (1969) as one of the sperm whale grounds in the Pacific. The Philippines, specifically Zamboanga, appears to have been one of the preferred stops of American whalers for replenishing supplies (Legarda 2002). Based on these different accounts, American whaling in the Philippines had probably started by 1825 and ended by 1880 when the Sulu grounds were finally abandoned.

British whaling in the Philippines most likely started in 1820 and ended in 1840, if not earlier. Similar to the Americans, the British frequented the whaling grounds known as the 'Sooloo Sea' (Rhys Richards, personal communication with the author). The extent of this whaling could not be determined, as no British whaling logbooks were examined during research for this chapter.

Modern commercial whaling

The so-called whaling station on the rocky coast of Barangay Inapulangan, Homonhon Island, a few kilometres off the coast of Guiuan, was clearly too small to serve as a commercial whaling landing site. A summarized account of this whaling period was written by the Greenpeace Environmental Trust (Davies 1986). It describes how the First International Sea Harvest Corporation (FISH) was set up in Manila in May 1982 and operated a whale-catcher/factory ship called the MV *Faith 1*. The *Faith 1* hunted in the waters around Palau Islands, New Guinea and East and west of the Caroline Islands during the winter and in the Marianas Islands Chain and 200-300 nautical miles south of Japan during the summer (Davies 1986). The vessel took Bryde's whales (Davies 1986, Barut 1994) and also reportedly humpback whales (Ellis 1991). Although the ship was based in Cebu and the meat was exported to Japan from Manila, there was no clear indication of hunting within Philippine territorial waters. It is not known if (and where) records have been kept since no whaling license and landing records were found at the DA-BFAR and National Fisheries Research and Development Institute offices (Acebes 2009). It can therefore be concluded that records dating back more than 10–20 years must have been disposed of.

Past and Current distribution of whales

Written documentation of the occurrence or distribution of whales before the 1980s is scant. Published reports include the following: Slijper et al's (1964) distribution of sperm whales, humpback whales and rorquals in the Pacific Ocean; Herre's (1925) brief description of the stranding of a 32-ft baleen whale in Bacoor, Cavite Province which he identified as *Balaenoptera rostrata;* and Townsend's (1935) charts showing the distribution of certain whales based on logbooks of

American whaling ships between 1761 to 1920. Data on the current distribution of these whales (indicated by black dots in Figs. 5.3, 5.4, and 5.5), although still incomplete, has been increasing for the past 10 years (Bautista 2002). The majority of this data on current distribution of whales was obtained from sightings on actual vessel surveys conducted by WWF-Philippines and stranding reports compiled by WWF-Philippines.

Sperm Whale distribution

Sperm Whales *(Physeter macrocephalus)* have been hunted in the Pacific (and around the world) for centuries, including in the Philippines, and identification of historical whaling grounds can give a good indication of their past distribution. Unfortunately, sperm whaling records in the Philippines are elusive. So far, it is only from Townsend's (1935) charts that one can obtain a vivid imagery of the extent of the distribution (and exploitation) of this species. Although it is only the Sulu Sea that is mentioned as one of the sperm whaling grounds, it can be clearly seen from the logbooks of American whaling ships that sperm whales were found and hunted in almost every sea in and around the archipelago throughout most of the year (Fig 5.5).[3]

Slijper et al's (1964) illustrative account showed locations of sightings of sperm whales, and, although the Philippines was not mentioned in the text, depicted sightings on the eastern coast of Luzon, southwest of Palawan, Sulu sea, and all along the south and east coast of Mindanao (Fig 5.5). Sperm whales were found year-round in the Indonesian archipelago, mostly calves (Slijper et al. 1964).

Currently, sperm whales appear to be found in almost all major seas within the Philippine archipelago, however most sightings are of solitary animals. The occurrence of this species is usually recorded in the country through stranding reports. So far only a handful of accounts document sightings of a group of more than two animals, along the eastern coast of Luzon (Van Lavieren 2001, Wang 2004), and in the South China Sea close to the Philippines (Miyashita et al. 1996). Observations of sperm whales in 1999 during a cruise from Mauritius to the Philippines suggest that the Balabac Strait might represent a migration route for sperm whales between Sulu Sea and the South China Sea (De Boer 2000).

Humpback Whale distribution

Similarly, Slijper et al. (1964) illustrated locations of humpback whales (*Megaptera novaeangliae*) and other rorquals around the Philippines. Their maps showed humpback sightings southwest of Palawan, Sulu Sea and along the south, west and east

---

[3] These figures draw on (Townsend 1935). Charts A and B. Digitized maps courtesy of Beth Josephson

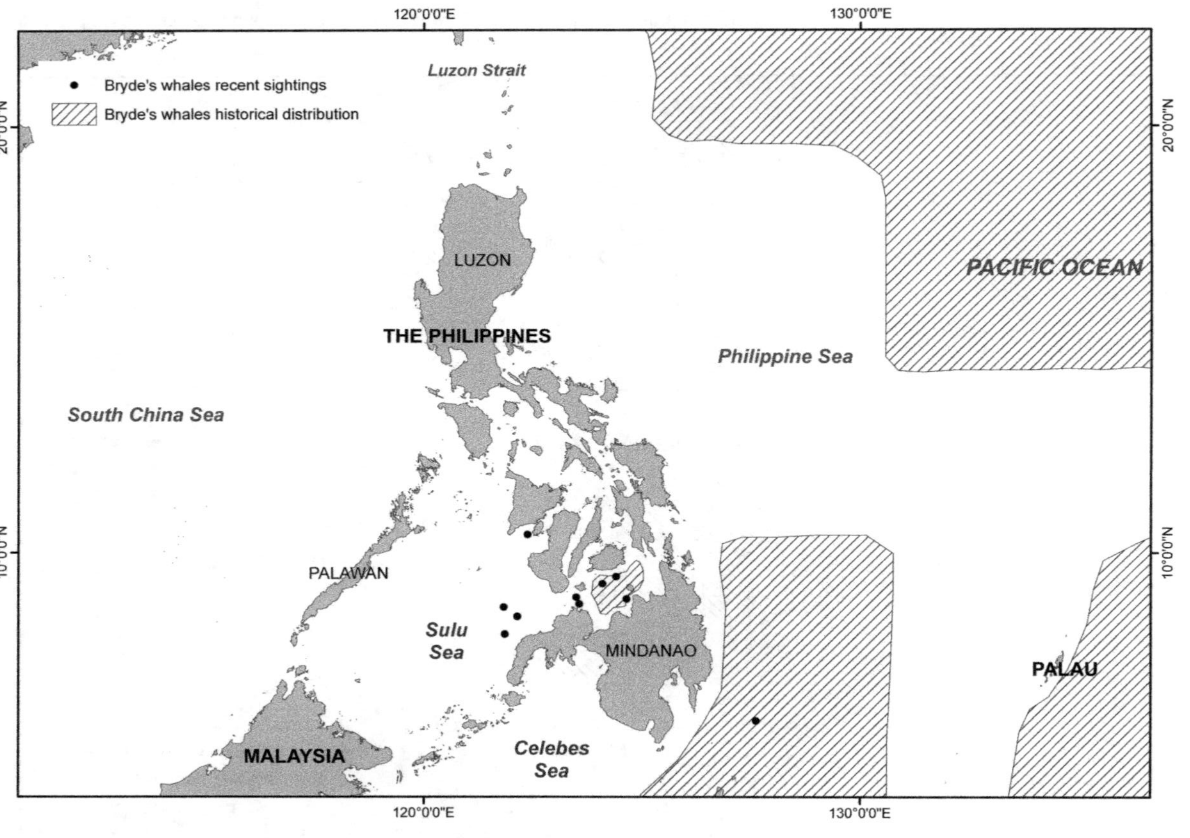

Fig. 5.4 Past and current distribution of Bryde's whales in and around the Philippines

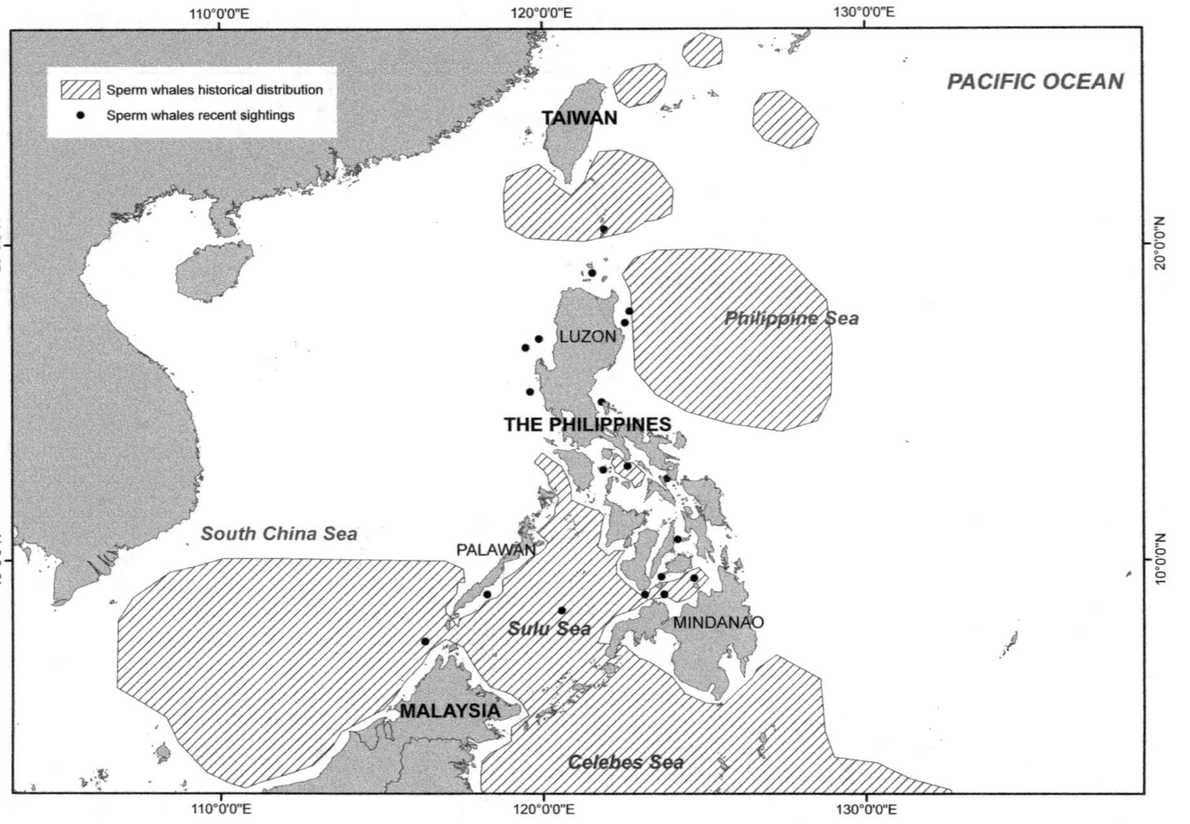

Fig. 5.5 Past and current distribution of Sperm whales in and around the Philippines

coast of Mindanao (Fig 5.6). Taiwan is known as one of the historical wintering grounds of humpback whales in the North Pacific (Nishiwaki 1959, 1960, 1961; Tomlin 1967). Based on whaling records (Yu 2002), the best-documented breeding grounds were off the southern end of Taiwan, just 325 km north of the Babuyan Islands (Darling and Mori 1993). This is further supported by records of an American whaling ship *Corinthian* that indicate humpbacks seen along the eastern coast of Taiwan and East China Sea (*Corinthian* logbook 1854–1857). Also noted was a sighting of a humpback whale as they approached Hong Kong from Guam, after passing the Batanes Islands.

Humpback whales used to be only known publicly in the Philippines through a stranding reported in a local newspaper (Dolar 1994). Although there have been a few other unconfirmed reports of sightings of the species in northern Luzon, it wasn't until 1999 that a breeding ground in the Babuyan islands, northern Luzon, was verified (Yaptinchay 1999; Acebes 2001). Humpbacks are now confirmed to occur around the Babuyan Islands going down along the eastern coast of Cagayan and Isabela, northern Luzon (Acebes and Lesaca 2003; Acebes et al. 2007) (Fig 5.6). Yet there are still several unconfirmed sighting reports in northern Palawan, Albay and northern Mindanao.

Bryde's Whale (Balaenoptera edeni) and other rorqual distribution

Past distribution of Bryde's Whales can be based on the hunting grounds of fishers of Bohol, Camiguin and Salay, Misamis Oriental. These fishers found the species to be abundant in the Bohol or Mindanao Seas (Fig. 5.7). Slijper et al. (1964) also indicates sightings off the eastern coast of Mindanao.[4]

The data on the current distribution of rorquals such as the Bryde's whales is minimal. In 1994 and 1995, there were only three encounters of Bryde's whales in the eastern Sulu Sea (Dolar 1999; Dolar et al. 2006) and another in 1994, in the Pacific coast, off eastern Mindanao (Miyashita et al. 1996). This area is considered part of the distribution of Bryde's whales in the Pacific (Kishiro 1996). Recent surveys in the Bohol Sea did not indicate sightings of the species (Sabater 2005). However, Lila and Camiguin locals have recently seen the species not far from their shores (Fuentes 2004; Acebes 2009). More recent sightings are off Puerto Princesa Bay and Balabac Island in Palawan (Dolar 2006; Lory Tan, personal communication with the author).

Rorquals referred to by Slijper et al. (1964) included Sei, Bryde's, fin and blue whales. They noted the occurrence of rorquals in the eastern coast of northern Luzon, southern Palawan and southern Mindanao (Fig. 5.7). An American whaling ship recorded a sighting of a 'finback' in the coast of Luzon, near the Lingayen Gulf (Avola logbook 1875–1899). 'Finbacks' and 'sulphur bottoms' were also found as whalers cruised the 'Sooloo isles' (*Stafford* logbook 1867–1870).

---

[4] As indicated in the map of Slijper et al. (1964)

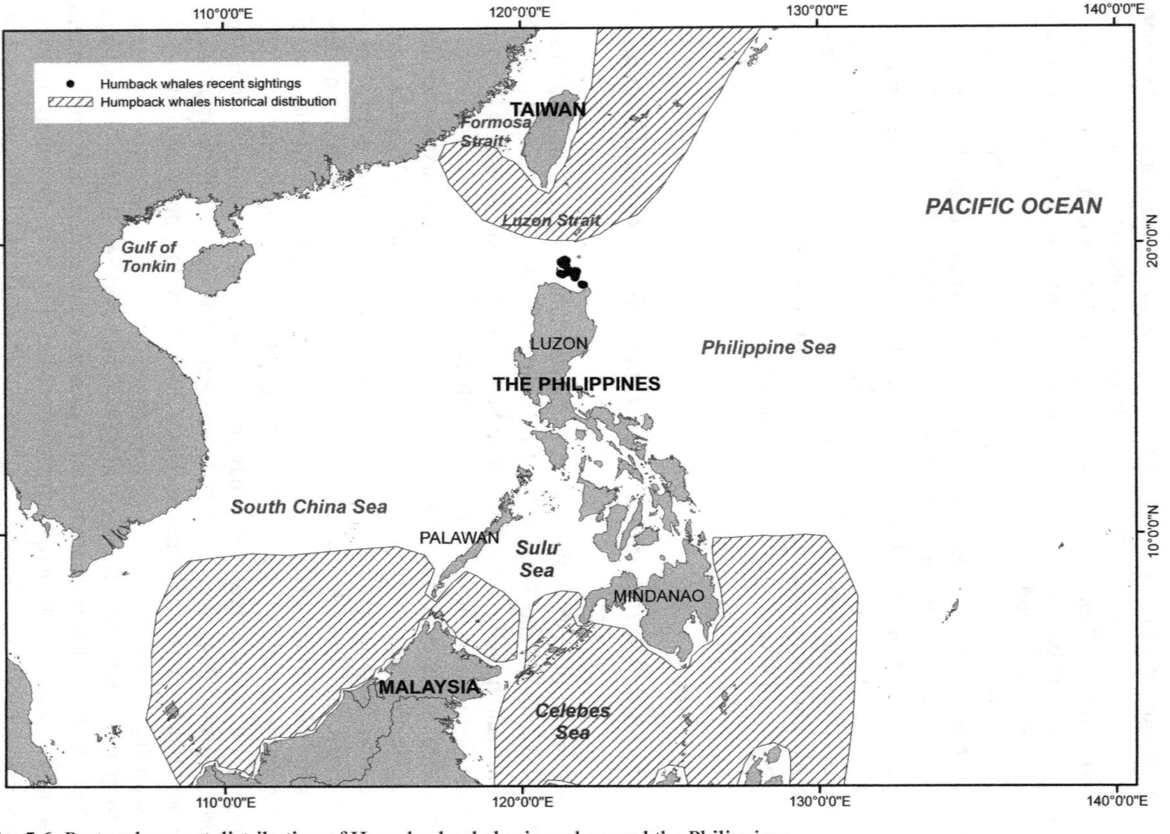

Fig. 5.6 Past and current distribution of Humpback whales in and around the Philippines

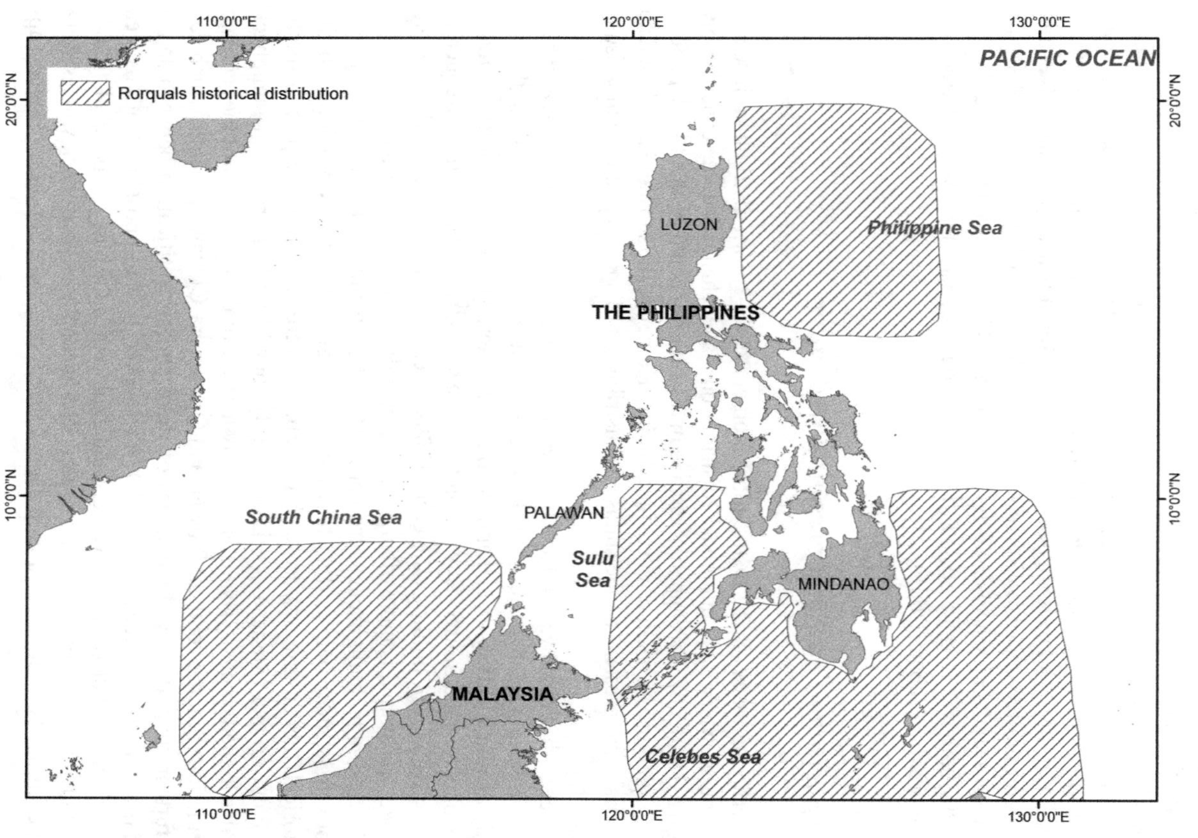

Fig. 5.7 Past distribution of Rorquals in and around the Philippines

Recent surveys have not encountered other species of baleen whales except for humpbacks and Bryde's whales mentioned in the preceding paragraphs and fin whales (*Balaenoptera physalus*) in the Sulu (Dolar et al. 2006) and South China Sea (De Boer 2000). The Balabac Strait has been suggested as a migration route not only for sperm whales but for fin whales between the Sulu Sea and the South China Sea (De Boer 2000). However, in 2005 a baleen whale was photographed near Palaui Island (Sta. Ana, Cagayan, Northern Luzon) (Lory Tan, personal communication with the author). Identification of the species has not been confirmed, but it is definitely not a humpback whale.[5] In the Bohol Sea, a sports and adventure television production crew filmed their encounter of a baleen whale mother and calf pair near Pamilacan Island in 2004 (Edna Sabater, personal communication with the author),[6] later identified through photographs as a blue whale *Balaenoptera musculus* (Sabater 2005).

Comparison of past and present distribution

In overlaying maps of past distribution upon currently-known distribution it seems that sperm whale and humpback distribution have changed quite considerably (Figs. 5.5 and 5.6). This apparent change in distribution could be due to a decline in the population as a result of intensive whaling. However, the substantial lack of data (both past and present) on the status and distribution of species makes comparisons difficult.

The southern Palawan and southern Mindanao regions, identified as areas where sperm whales were sighted, are lacking recent cetacean survey data. Given the currently available survey data, it appears that the eastern and western coasts of Luzon, the Bohol Sea and Sulu Sea are the areas where sperm whales can still be frequently sighted. However, unlike in the early 1800s when 'schools' of sperm whales were sighted, the animal is more often encountered solitary or in small groups.

Interestingly, humpbacks were not historically documented to occur along the Babuyan Islands chain and eastern coast of Luzon, despite its proximity to the historical Taiwan grounds. Instead, humpbacks were sighted in the past in southern Palawan and Mindanao. This apparently 'new' breeding ground of humpback whales in the Babuyan Islands is arguably a result of the movement of the remnant population from the southern Taiwan and Ogasawara-Okinawa grounds which can be further supported by the matches of fluke photo-identification between Japan and the Philippines (Acebes et al. 2007). However, it is also possible that this area simply remained unknown to whalers and explorers in the past. Slijper et al. (1964) concluded that the northern and southern Pacific stocks of Humpback whales were wintering in the Indonesian archipelago. Current data does not, or no longer, sup-

---

[5] The author has seen the photographs and is absolutely certain that it is not a Humpback whale because it lacks the characteristic hump on the back, shape of the dorsal fin, and bumps on the head region.

[6] The author has also seen the video footage of the whales.

ports this conclusion. Humpback whales have not been sighted recently in the Celebes Sea (Sulawesi sea), along the coast of Malaysia, Borneo and northern Sulawesi (Saifullah A Jaaman, personal communication with the author).

The decrease in the frequency of sightings or encounters of Bryde's whales in the Bohol Sea (Acebes 2009) may lead some to conclude that this decline was a consequence of over a century of local hunting (Fig. 5.7). It could be argued that other factors such as disturbance due to increased boat traffic and habitat degradation due to coastal area development could have caused it (Gordon and Moscrop 1996). Another aspect for consideration is the consequence of the recent commercial whaling along the eastern coast of the Philippines. It is thought that Bryde's whale populations migrate from Japan moving down along the eastern coast of the Philippines (Barut 1994), so it is possible that this population is the source of the animals that used to frequent the Bohol Sea. It can be speculated that a large-scale and efficient hunting of the species along this route may have caused the decline in numbers in the area.

Conclusion

The documented history of whaling in the Philippines can, at the time of writing, be traced back to the era of American whaling (1800s). Given the abundant local terms relating to whales and whaling and the recollections of some older residents in the area, it seems likely that American whaling occurred alongside local whaling in Bohol. The exact beginnings of this locally initiated fishery cannot be determined given the present data. The development of the whaling practice in Bohol seems to have come about as a derivation from manta ray fishing, most likely from the town of Jagna. Based on available evidence and oral history, the whaling in Lila, Bohol began in the late 1800s and ended in 1986 while the whaling practice in nearby Pamilacan Island was derived from the Lila fishery, probably from about 1939 (or earlier but not earlier than 1900) and ended in 1997. The whaling in Sagay, Camiguin was most likely derived from the Bohol fishery in the early 1900s and ended in 1997. The whaling in Salay, Misamis Oriental may have also been derived from Bohol but its beginnings are still unclear. The fishery also ended at about 1997. The whaling practices in Camiguin and Misamis Oriental may have originated from the manta ray fishery in Jagna. It is also possible that the Jagna fishery developed all on its own.

How the change in hunting instruments (from toggle-harpoon to hook) came about cannot be determined for certain, but it can be speculated that Lila fishers tried using the hooks (used by Jagna-anons to secure the manta rays) and found it a more effective method. This whale fishery, including the practise of catching the whale by jumping on its back with a hook, was transferred to nearby Pamilacan Island. Through the years of frequent seafaring trade and 'adventures' of Boholanos, the practice was most likely further transferred to Camiguin and probably to Misamis Oriental. The use of the toggle-harpoon can be speculated to have been copied from either the Boholanos who migrated to the islands, or through contact with American whalers engaged in sperm whaling in the region.

The extent of foreign whaling in the Philippines during the nineteenth century is still unclear given the present data. Links to whale fisheries in neighbouring countries such as China, Taiwan, Japan, and Indonesia are yet to be established. However, it is probable that the Filipinos learned or adapted this type of fishery during earlier contacts with other South and Southeast Asian peoples.

Other areas in the Philippines still need to be investigated for any history (ancient or recent) of whale hunting or cetacean utilization in general. The 'whale harpoon head' found in the archaeological site in Cagayan de Oro city could be an indication of an ancient practice of hunting marine animals including cetaceans. Archaeological sites such as the Duyong Cave have great potential for providing historical information on use and presence of marine mammals. The Bolobok cave in Tawi-Tawi should also be investigated for cetacean and other marine mammal remains.

Comparison of available data on past and current distribution of sperm whales, humpbacks and Bryde's whales shows an apparent decrease in abundance and area of distribution. These changes can be attributed to a decline in populations due to intensive whaling. However, the lack of historical and current data on species abundance and distribution made comparisons difficult. In order to illustrate a more comprehensive historical distribution map of exploited whale species in the Philippines all possible records should be investigated. Key information may be found in pre-colonial and Spanish colonial trade records as well as pre-World War Two records. Records of all American and British whaling ships that frequented the Pacific and Indo-China should be examined. For more recent data on whale distribution, vessel survey data should be updated and should cover previously less studied areas such as the south and western coast of Palawan, southern and eastern Mindanao, eastern coast of central Luzon, and north western Luzon.

Accessing historical data in the Philippines can be very difficult. Data has often either been lost (i.e. discarded, burnt, or current location unknown) or was not created in the first place. If it does exist, it is kept outside the country in foreign archives and is may not be readily accessible. The national ban on the catching or killing of cetaceans also influenced the accessibility of information, particularly from interviewing fishers and local officials. Most people are aware of the ban and would more than likely be cautious in giving information that will show that they (and their town's people) are conducting illegal activities or that they are not aware of the law.

It is evident that people in different regions in the Philippines have been eating whales and dolphins for centuries. Different tribes with varying beliefs and practices, utilize these marine mammals in different ways. The origin of each practice is difficult to determine given the present available data. Each group of people cannot be taken in isolation from one another, as they are inextricably linked socially, culturally and geographically. To this day, whales and dolphins in the Philippines are caught directly, incidentally and opportunistically. Although national legislations prohibiting these acts are present and well-known, they are obviously not adhered to by the locals and not enforced by the responsible government agencies.

Given the inadequacy of pertinent data, namely baseline and current data on whale distribution and stock abundance, it is not recommended for the Philippine government to lobby for the resumption of commercial whaling. A thorough examination of the socio-economic impacts of the ban on the hunting and killing of cetaceans on the communities of Pamilacan, Lila, Sagay, and Salay should be conducted. Although a lifting of the ban is not imminent, the apparent decrease in numbers of Bryde's whales within the waters of these areas alone indicates that necessary precautions need to be taken before reconsideration of the current policy.

References

Acebes JMV (2001) Photographic identification of humpback whale Megaptera novaeangliae in the Babuyan islands, Northern Luzon, Philippines. Poster presented at the 14th Biennial Conference on the Biology of Marine Mammals, Vancouver, Canada, 28 Nov–3 Dec 2001

Acebes JMV (2009) Historical whaling in the Philippines: origins of 'indigenous subsistence whaling', mapping whaling grounds and comparisons with current known distribution: A HMAP Asia project paper. Working paper no. 161. Murdoch University, Perth, Western Australia

Acebes JMV, Darling JD, Yamaguchi M (2007) Status and distribution of humpback whales *Megaptera novaeangliae* in northern Luzon, Philippines. J Cetacean Res Manag 9:37–43

Acebes JMV, Lesaca LAR (2003) 'Research and conservation of humpback whales and other cetacean species in the Babuyan Islands, Northern Luzon'. In: Ploeg Jvd, Masipiquena A, Bernardo EC (eds) The Sierra Madre Mountain Range: Global relevance, local realities papers presented at the 4th regional conference on environment and development. Golden Press, Tuguegarao City

Avola logbook (1875–1899) Notes taken from the logbook of Bark Avola/Ship Emma C. Jones, 1875–1899. Log 25, New Bedford Whaling Museum Research Library

Barut NC (1994) Policy and Management of Dophins and Whales in Philippines. In: Marine Science Institute UoP, Diliman, Quezon City (ed) Philippines marine mammals proceedings of a symposium-workshop on marine mammal conservation. Bookmark, Manilia

Barut NC (1998) Philippines. World Council of Whalers 1998 General Assembly report, Victoria, Canada. Quality Color Press

Barut NC (1999) Whale and dolphin hunting in the Philippines. World Council of Whalers 1999 General Assembly report, Rekyavick, Iceland. Quality Color Press

Bautista ALS (2002) Cetaceans in the Philippines: an update for 2002 A working paper of the second workshop on the biology and conservation of small Cetaceans and Dugongs of Southeast Asia. CMS/SEAMAMII/Doc. 25. Dumaguete, Philippines

Birnie P (1985) International regulation of whaling: from conservation of whaling to conservation of whales and regulation of whale-watching. Oceana Publications, New York

Buckley D (1994) Whales 'extinct in 100 years'. Courier Mail, 18 May 1994. http://www.bodley.ox.ac.uk/oxlip. Accessed Dec 22 2004

Clark AH (1887) The whale fishery. History and present condition of the fishery. In: The fisheries and fishery industries of the United States. Section V: History and Methods of the fisheries. Government Printing Offices, Washington

Corinthian logbook (1854–1857) Notes taken from the Ship Corinthian logbook, 1854–1857. Log 61, New Bedford Whaling Museum Research Library

Cruz-Trinidad A, White AT, Gleason M, Pura L (2002) Philippine fisheries in Crisis: a prescription for recovery. OverSeas: the online magazine for sustainable seas 5(10)

DA-BFAR (1992) Fisheries administrative order no. 185 series of 1992(FAO 185), Canberra

Darling JD, Mori K (1993) Recent observqtions of humpback whales (*Megaptera novaeangliae*) in Japanese waters, Ogasawara and Oakinawa. Can J Zool 71:325–333

Davies GH (1986) Japanese whaling in the Philippines. Calvert's Press, London

Day D (1993) A giant leap for survival. Daily Mail (London), 15 May 1993. http://www.bodley.ox.ac.uk/oxlip. Accessed 22 Dec 2004

De Boer MN (2000) A note on cetacean observations in the Indian Ocean Sanctuary and the South China Sea, Mauritius to the Philippines, April 1999. J Cetacean Res Manag 2:197–199

Dolar MLL (1994) Cetaceans of Philippines. In: Marine Science Institute UoP, Diliman, Quezon City (ed) Philippine marine mammals, Procedings of a symposium-workshop on marine mammal conservation. Bookmark, Manilia

Dolar MLL (1999) Abundance, distribution and feeding ecology of small Cetaceans in the Eastern Sulu Sea and Tanon Strait, Philippines. University of California

Dolar MLL (2006) Report on the cetacean survey off Balabac Island, Palawan in 2006. Unpublished report

Dolar MLL, Leatherwood SJ, Wood CJ, Alava MNR, Hill CL, Aragones LV (1994) Directed fiesheries for Cetaceans in the Philippines. Rep Int Whal Comm 44:439–449

Dolar MLL, Perrin WFP, Taylor BL, Kooyman GL (2006) Abundance and distributional ecology in the central Philippines. J Cetacean Res Manag 8:93–111

Ellis R (1991) Men and whales. Knopf, New York

Fuentes CA (2004) Whales, dolphins 'visiting' Bohol early, says tour exec. Philippine Daily Inquirer, 5 April 2004

Gordon J, Moscrop A (1996) Underwater noise pollution and its significance for whales and dolphins. In: The conservation of whales and dolphins: science and and practice. Wiley, Chichester

Green SJ, White AT, Flores JO, III MFC, Sia AE (2003) Philippine fisheries in crisis: a framework for management. Department of Environment and Natural Resources, Cebu City, Philippines

Heritage Conservation Advocates (2005) Heritage conservation advocates website. http://heritage.elizaga.net/huluga/artifacts/harpoon.html. Accessed 24 June 2005

Herre AW (1925) A Philippine rorqual. Science 61:541

Japan Economic Journal (1984) Gov't decides to restrict whale imports from the Philippines, 13 March 1984. http://www.bodley.ox.ac.uk/oxlip. Accessed 11 June 2005

Japan Times (2004) The sustainable whaling option, 29 July 2004. http://www.bodley.ox.ac.uk/oxlip. Accessed 11 June 2005

Jenkins JT (ed) (1921) A history of the whale fisheries from the Basque fisheries of the tenth century to the hunting of the finner whale at the present date. Port Washington, NY

Jiji Press Ticker Service (1984a) Japan impounds 47-ton whale meat shipped from R.P, 18 January 1984. http://www.bodley.ox.ac.uk/oxlip. Accessed 11 June 2005

Jiji Press Ticker Service (1984b) Govt to shut out Filipino whale meat from Japan, 2 March 1984. http://www.bodley.ox.ac.uk/oxlip. Accessed 11 June 2005

Kishiro T (1996) Movements of marked Bryde's whales in the Western North Pacific. Rep Int Whal Comm 46:421–428

Kühlmann K-J (2000) Between the slaughterhouse and freedom: a matter of man's choice or a whale's right? OverSeas: Online magazine for sustainable seas 3(8)

Legarda BJ (ed) (2002) After the galleons: foreign trade, economic change and entrepreneurshipin the nineteenth century. Centre for Southeast Asian Studies, Quezon City

Mak A (2005) Email message to Lynette Furusahi, forwarded to the author, 15 August 2005

Miyashita T, Kishiro T, Higashi N, Sato F, Mori K, Kato H (1996) Winter distribution of Cetaceans in the Western North Pacific Inferred form sighting cruises 1993–1995. Rep Int Whal Comm 46:437–441

Moral CV (2003) Whale of a difference. Philippine Daily Inquirer, 11 February 2003. http://www.bodley.ox.ac.uk/oxlip. Accessed 22 Dec 2004

Nishiwaki M (1959) Humpback whales in Ryukyuan waters. Sci Rep Whal Res Inst 14:49–87

Nishiwaki M (1960) Ryukyuan humpback whaling in 1960. Sci Rep Whales Res Inst 15:1–16

Nishiwaki M (1961) Ryukyuan whaling in 1961. Sci Rep Whales Res Inst 16:19–28

Ohmagari K (2005) Whaling conflicts: the international debate. In: Indigenious use and management of marine resources. National Museum of Ethnology, Osaka

Panublion Heritage Sites (2005) Guiuan, Panublion. http://www.admu.edu.ph/offices/mirlab/panublion/r8_guiuan.html. Accessed 19 June 2005

Patanne EP (ed) (1996) The Philippines in the 6th and 16th centuries. LSA Press, San Juan

Sabater ER (2005) Cetaceans of the Bohol marine triangle area, Bohol, Philippines: assessment and monitoring. Poster presented at the 16th Biennial Conference on the biology of marine mammals, San Diego, California, 12–16 Dec 2005

Scammon CM (1969) The marine mammals of the North-western coast of the North America and the american whale fishery. Riverside, California

Severin T (1999) In search of Moby Dick: quest for the white whale. Little, Brown and Company, London

Slijper EJ, Van Utrecht WL, Naaktgeboren C (1964) Remarks on the distribution and migration of whales, based on observations from Netherlands ships. Bijbragen Tot De Djerkunde XXXIV 34:3–93

Spence B (1980) Harpooned: the story of whaling. Conway Maritime Press, London

Stafford logbook (1867–1870) Notes taken from the logbook of bark stafford 1867–1870. Log 614, New Bedford Whaling Museum Research Library

Tan JML (1995) A field guide to whales and dolphins of the Philippines. Bookmark, Makati

Tomlin AG (1967) Mammals of the USSR and adjacent countries, vol 9. Cetacea. Israel Program for Scientific Translations, Jerusalem

Townsend CH (1935) The distribution of certain whales as shown by logbook records of American whaleships. Zoologica (NY) 19:1–50

Van Lavieren H (2001) Marine mammals and endangered species survey report: Northern Sierra Madre Natural Park. Plan Philippines NSMNP-CP, Cabagan

Verrill AH (1923) The real story of the whaler: whaling, past and present. Appleton & Company, New York

Wang JY (2004) Research and conservation of Indo-Pacific bottlenose dolphin Tursiops aduncus in the coastal waters of Northern Sierra Madre Natural Park Final report to WWF-Philippines. NSMNP-CDP, Quezon City

Wuerch WL (1987) American whalers and traders in the Philippine Islands, 1817–1899. A guide to references to America vessels contained in the National Archives microfilm publication. Micronesian Area Research Center, University of Guam

Yaptinchay AA (1999) New humpback whale wintering ground in the Philippines. Poster presented at the 13th Biennial Conference on the Biology of Marine Mammals, Maui, Hawaii, 28 Nov–3 Dec 1999

Yu H-Y (2002) Songs of a humpback whale (*Megaptera novaeangliae*) in Taiwan and evolution of communication in Cetaceans. Masters in Marine Biology thesis, National Sun-Yat Sen University

Chapter 6
Brackish Water Shrimp Farming and the Growth of Aquatic Monocultures in Coastal Bangladesh

Bob Pokrant

Abstract One of the most significant changes in marine and coastal environments since the mid–twentieth century has been the growth of coastal shrimp aquaculture in many tropical and sub–tropical regions of the world. This chapter, which draws on the author's own archival and field research and the published works of other students of the global shrimp market, examines the growth of brackish water shrimp production from the 1970s to the present in Bangladesh's coastal belt and its social and ecological impacts. It shows that for most of this period shrimp production was encouraged by the Bangladesh Government to expand in a fragmented and uncoordinated way with varying environmental, economic and social consequences. These included higher levels of soil salinity, increased risk of flooding, loss of agricultural land, a decline in biodiversity, contraction of various traditional occupational activities, growth in new non-agricultural work, a shift to diversified employment strategies among households, higher incomes for shrimp farmers and land renters and economic and social dislocation for others. Government, business and international aid agencies supported the expansion of mono–cultural forms of shrimp production integrated into global trading networks at the expense of local resource extraction activities such as artisanal fishing and forestry.

Keywords Aquaculture · Bangladesh history · Shrimp farming · Shrimp production · Ecological impacts

One of the most significant changes in marine and coastal environments since the mid–twentieth century has been the growth of coastal shrimp aquaculture in many tropical and sub–tropical regions of the world. This chapter, which draws on the author's own archival and field research and the published works of other students of the global shrimp market, examines the growth of brackish water shrimp production from the 1970s to the present in Bangladesh's coastal belt and its social and ecological impacts. It shows that for most of this period shrimp production was encouraged by the Bangladesh Government to expand in a fragmented and uncoordinated way

B. Pokrant (✉)
Department of Social Sciences & International Studies, School of Media, Culture & Creative Arts, Curtin University, GPO Box U1987, 6845, Perth, WA, Australia
e-mail: B.Pokrant@curtin.edu.au

J. Christensen, M. Tull (eds.), *Historical Perspectives of Fisheries Exploitation in the Indo-Pacific,* MARE Publication Series 12, DOI 10.1007/978-94-017-8727-7_6,
© Springer Science+Business Media Dordrecht 2014

with varying environmental, economic and social consequences. These included higher levels of soil salinity, increased risk of flooding, loss of agricultural land, a decline in biodiversity, contraction of various traditional occupational activities, growth in new non-agricultural work, a shift to diversified employment strategies among households, higher incomes for shrimp farmers and land renters and economic and social dislocation and displacement for others. Government, business and international aid agencies supported the expansion of mono–cultural forms of shrimp production integrated into global trading networks at the expense of local resource extraction activities such as artisanal fishing and forestry.[1]

Shrimp Culture in Bangladesh History

For many centuries the people of Bangladesh (formerly part of Bengal in pre-partition India and of East Pakistan from 1947 to 1971) have engaged in the open water capture of inland and marine finfish and the cultivation in perennial and seasonal tanks, bhunds (special tanks designed to mimic riverine conditions) and ponds of fin fish and of various species of freshwater and brackish water shrimp (Das 1931, 1932; Hora 1948; Bagchi and Jha 2011). Since ancient times, fish has been a central component of the Bengali diet for all classes and castes of people, although there were some restrictions on types of fish and crustacea eaten by Brahmanical castes and Muslims (Ray 1994). Over 200 different species of fish were caught, traded and consumed. Open water fishing was dominated by caste Hindu fishers with some Muslim participation (Pokrant et al.1997, 2001). Traditionally, shrimp grew alongside finfish in a polycultural system integrated into the seasonal social and ecological rhythms of village life and characterized by family and community ownership of ponds and tanks, integration of production with food cropping and livestock activities, risk-averse strategies of spreading food risks across different food sources, provision of off-season work for farmers, and production largely for local consumption. There were also extensive methods of more commercialised closed culture brackish and salt water fish production (*bhasabadha* and *bheri* in Bengali), which used few additional inputs other than the fry or fingerlings obtained from natural sources. Shrimp were also caught wild in rivers by fishers and farmers and in the Bay of Bengal by Hindu professional fishers.

Like open water capture fishing, closed culture fishing involved both natural and artificial stocking. Unlike open water capture fisheries, which until recent times were dominated by Hindu fishers (Pokrant et al. 1997), there was a limited development of professional closed culture fishers, with most of the labour being supplied by local villagers fishing their own ponds, working for wealthier landholders or drawing on common pool resources. Such culturing involved a wide variety of fish and, to a lesser extent, crustacea in both freshwater and brackish water

---

[1] Shrimp monoculture in this chapter refers to the practice of cultivating a single or limited number of shrimp species in a systematic way over several seasons.

environments. While there was trade in fish, it was largely for domestic consumption at the local and regional levels.

Many villages contained private and community water tanks some of which were used to raise fish (Gupta 1984). Formerly, tanks had been a main form of irrigation but by the end of the eighteenth century many had silted up and been abandoned. Under the British (1793–1947), while there were some attempts to promote pond culture and breed carp in captivity, the colonial state paid relatively little attention to culture fisheries as they were considered to be under the control of the *zamindars* (mainly Hindu landlords), their subordinates or in the hands of village communities. The main production was of finfish rather than shrimp, which were produced as a by-product of pond and tank culture or were taken wild from rivers, estuaries and the sea for domestic human consumption, manure, and as a regional trade item (Pokrant et al. 1997; Reeves 1995).

At the beginning of the twentieth century, fish culturing was done by the wealthy in ponds largely for their own domestic use rather than any commercial purpose. There was some leasing of tanks and many poorer people worked as fry collectors for tank owners and lessees. In addition, many of the rural poor had access to various common pool resources which included community ponds and tanks as well as seasonal *beels,* ditches, canals and pits (Gupta 1908; Webster 1911; Jack 1916; O'Malley 1925). There was a strong trade in fish fry and spawn for tanks. During the twentieth century, the most commercialised culture fisheries were the *bheris* in the Salt Lake area of what is now Kolkata in India's West Bengal (formerly known as Calcutta). Bheris were sewage-fed saltwater ponds that were controlled by the municipality and were let to lessees who sub-contracted to others. The bheris were flooded at the height of the rainy season and small fish and fry were brought in through sluices. These ponds were also nurseries for fish and prawns. Leases were annual so by February all fish had been caught, limiting any longer term expansion of the industry.

From the early twentieth century to the 1950s attempts were made to breed carp in ponds and enclosed spaces. By 1940 The Government of Bengal had made recommendations to improve closed culture fisheries, including shrimp, and there was a regional trade in open capture shrimp (*sutki chingdi/chingri*) from coastal estuaries with Burma and East Bengal. Shrimp were boiled and crushed to make into manure for sale to foreign and local companies. By 1945 fry and spawn were reared by local government authorities for sale (Rahman 1945). In 1949 a number of fish nursery units were established and fry and fingerlings were sold to private fish farmers. In addition, 25 demonstration fish farms were either under operation or being established by 1950. However, none of these were specifically devoted to shrimp.

During Pakistani times (1947–1971), there was little processing of either cultured or wild-caught fish apart from salting, drying and the production of shrimp paste, although there was extensive trade in fish fry between Chittagong and Kolkata. For example, the 1951 FAO-funded report (Kesteven and Ling 1951) on the fisheries of East Pakistan, prepared for the Pakistan Government by G L Kesteven, FAO Regional Fisheries Officer for Asia and the Pacific, and Assistant Regional Fisheries Officer S W Ling, refers to 'brackish water trapping pond operations' (bheries) in the Khulna area in which mullet, *bhetki* (ocean perch), *chanda* (pomfret) and shrimp were the main catch. Some 9,000 fishermen using seine and cast nets operated on the ponds alone or as part of paddy cultivation. Kesteven and Ling discuss freshwater tank culture, noting that tanks were supplied with wild spawn and fry, which supplemented the capture of fingerlings from inland water depressions (beels). Over 20,000 people carried some 15,000 containers of spawn and fry by train from Rajshahi and Chittagong in East Bengal/Pakistan to other parts of the province for local stocking. The report points out that tank cultivation for commercial reasons were limited and '…in the majority cultivation is for subsistence and pleasure; consequently the operations are not carried out with any efficiency and in fact many of the tanks are neglected and even derelict' (Kesteven and Ling 1951).

Yet apart from a small regional trade in fish and shrimp, most fish and crustaceans were consumed domestically. These early forms of culture fisheries are best described as 'proto-aquacultural' or a form of stock enhancement with limited intervention in faunal life cycles (Beveridge and Little 2002). As Bagchi and Jha (2011) put it in their survey of fisheries and pisciculture in India's history:

> Prior to India's independence in 1947, fish culture primarily consisted of purchasing some spawn from the market, putting them in the pond, and reaping a harvest at the end of the year.

The Pakistan Government paid limited attention to shrimp production during the 1950s but the beginnings of a more specialised export-oriented shrimp industry can be traced to that period when two factories were established in 1954 to export frozen and canned shrimp to the USA and Western Europe and in 1959 the first shrimp and fish processing and freezing plant was established in Chittagong to export headless, shell-on freshwater shrimp to Europe and the USA. In 1960/1961, fish canning plants were set up in Khulna District in East Pakistan (Pakistan 1961) and from that time Khulna and its environs became the main centre of wild shrimp processing. By 1970, there were five fish canning plants in Chittagong engaged in the freezing of prawn and frog legs (Rizvi 1970). By the mid-1960s, wild shrimp (of which there are 22 species) were being sold for the domestic market both fresh and in preserved and cured form by means of sun-drying, boiling and sun-drying and smoking (Ahmed 1967). Sources of shrimp for local consumption continued to be river estuaries, canals, *beels*, rivers, tanks, paddy fields and ponds. Both freshwater and brackish water shrimp species were consumed locally but the latter came mainly from open water river, estuarine and marine fisheries rather than shrimp farms.

The Bangladesh Shrimp-Export Sector and the International Seafood Industry Since 1971

The modern Bangladesh shrimp export sector is a product of the post-1950 growth of an international seafood industry. Traditionally, the industry was dominated by open capture fisheries but as wild fish stocks declined or levelled off, aquaculture has increasingly filled the global gap in the supply of fish, crustacean and molluscs.

Aquaculture accounted for almost 50 % of world fish food supply in 2008 (Bostock et al. 2010; FAO 2010) of which freshwater fish was the major contributor. As a result of the growth of the international industry, many marine artisanal fishers and coastal agricultural communities with traditional livelihoods rooted in local systems of fishing and crop cultivation have been incorporated into global networks of commodity flows which increasingly dictate standard and type of product, price, and other conditions of production, marketing and sale (Humphrey 2006). Seafood production and consumption has become increasingly freed from seasonal fluctuations and distance constraints as people living in widely separated localities have been linked electronically, organisationally and psychologically through international networks of commodity exchange, extensive air and sea transportation services and the gaze of private and public governance agencies. The agencies setting product and process standards are the governments of major importing countries, global governance agencies such as the WTO and the FAO, and a growing number of private third party certification agencies such as GLOBALG.A.P, the Global Aquaculture Alliance's Best Aquaculture Practices (BAP), and the Aquaculture Stewardship Council (ASC). These regulatory agencies put pressure on the seafood industry to respond quickly to biosecurity concerns of consumers in the richer countries and to middle class consumers in Asia and Latin America. These changes apply to both open capture and closed culture fisheries but reach their most developed form in the latter, a shift that represents a closing of the fishing frontier equivalent to the domestication of wild plants and animals and the emergence of terrestrial agriculture 11,000 years ago (Duarte et al. 2007).

Wild-caught and cultured shrimp is one of the highest valued commodities in the international seafood industry and while capture fisheries continue to supply the bulk of shrimp products traded internationally, a growing proportion of the volume and value supply comes from cold and warm water aquaculture with Penaeid shrimp, particularly Pacific White Shrimp (*Penaeus vannemai*) and Black Tiger Prawn (*Penaeus monodon*, dominating tropical brackish water aquaculture. Today, Asia is the main region of warm water shrimp production with the industry providing foreign exchange and direct and indirect employment to the main producer countries such as China, India, Bangladesh, Vietnam and Thailand. Bangladesh ranked eighth among world shrimp producing nations in 2008 (FAO 2010) and by 2010 export production reached 109,000 t compared with 19,224 t in 1993 (BBS 2011). Shrimp exporting countries have followed several accumulation strategies in an effort to meet market demands and increase financial returns. These include increased intensification of cultivation, expansion of the land and aquatic areas under cultivation, improved transport and storage, shortened supply chains, species switching in favour of fast-growing varieties, shift to value-added products, use of chemicals and antibiotics, government support (such as tax breaks and technical assistance) and an extension of the shrimp growing season. The most developed form of shrimp production today is that of the super-intensive raceway and 'bio floc' shrimp rearing systems found in the US, although they remain limited in extent and experimental. In most shrimp-exporting countries in the tropics, the dominant forms of shrimp cultivation continue to range from modified extensive to intensive.

The extension of export-oriented production into rural areas has resulted in many changes to agrarian societies and ecologies in the tropical world. There has been the spread of commodified relations of production and exchange, land and water use changes and attendant conflicts, the growth of new sources of local wealth with a concomitant expansion of demand for products and services, shifting political alliances as shrimp-based economic capital is translated into political capital, the physical alteration of land and waterscapes, reduced agro- and bio-diversity, and a shift away from traditional forms of village production and cultivation such as rice farming (Fig. 6.1).

Shrimp Farming and the Bangladesh Shrimp Sector

The Bangladesh shrimp farming sector lies at the lower end of a domestic and international commercial and governance network extending from the collection of broodstock from the Bay of Bengal to the restaurants and supermarkets of the developed world (75 % of export value comes from the EU) and increasingly to market segments in the rapidly growing developing economies such as China and India. Farms are relatively undeveloped by international standards with most extensive or improved extensive in function with low stocking densities, limited or no artificial feed use, and poor water quality management. The two main types of shrimp cultivated for export are brackish water shrimp (*Bagda chingri* or Black Tiger Prawn) and freshwater prawn (*Golda chingri* or the Giant Freshwater Prawn). Approximately 95 % of these two culture species are exported with brackish water shrimp accounting for the greater part of exports but exports of freshwater prawn have been growing steadily in recent years. Also, processors and exporters have pressed for greater production of Pacific White Shrimp as they grow quicker and considered more competitive internationally. There are over 120,000 shrimp farms covering over 217,000 ha concentrated mainly in the Southwest and Southeast of the country (GoB 2010; DoF 2010. See Table 6.1). Some 80 % (170,000 ha) of shrimp farm land is under brackish water shrimp and consists of some 37,397 shrimp farms run by 40,000 farmers with an average farm size of 4.5 ha, of which over 50 % are less than one hectare. In the southwest, the most common shrimp regime consists of shrimp-rice rotation compared with the southeast where shrimp-salt rotation and shrimp only production are most common. Employment is largely seasonal drawing on both local and migrant and predominantly Bangladeshi labour. Shrimp farmers buy shrimp fry from wild and hatchery fry traders who in turn rely on several hundred thousand wild shrimp fry collectors operating along the coast, estuaries and rivers and 48 shrimp hatcheries mainly located in Cox's Bazar. The farms sell to thousands of traders who supply over 10,000 shrimp depot owners. The depots sell on to independent traders and commission agents who supply the 148 processing plants (75 of which are EU approved) located mainly in Khulna and Chittagong. There are several feed companies but most farmers provide their own feed (Uddin 2008). The Bangladesh Government has constructed 21 modern shrimp landing and

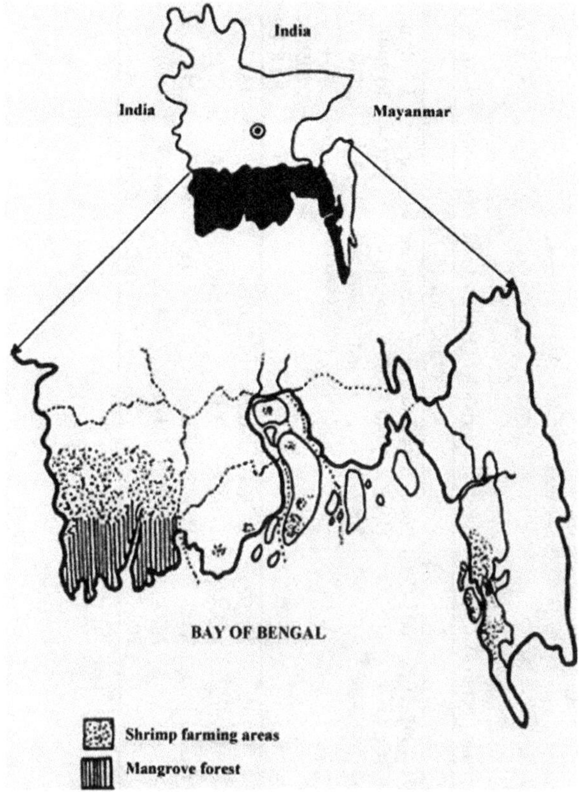

Fig. 6.1 Shrimp Farming Areas in Bangladesh. Source: Md Shahidul Islam (2003) Perspectives of the coastal and marine fisheries of the Bay of Bengal, Bangladesh. Ocean & Coastal Management, 46 (8): 763-796. http://dx.doi.org/10.1016/S0964-5691(03)00064-4.

service centers at a cost of 100 million taka (US$ 1.2 m.) to improve post-harvest quality and safety of shrimp raw materials. Production levels are low at less than 200 to 300 kg/ha compared with countries such as Thailand where yields can reach up to 5,000 kg/ha.

From its inception in the 1970s, shrimp farming expanded in a fragmented and uncoordinated way with varying environmental, economic and social consequences (Rahman et al. 2006). This expansion took place during a period of domestic economic and political turmoil and instability. In 1970 the country experienced one of the most devastating cyclones in its history, which resulted in major destruction to the country's limited physical, economic and social infrastructure and the death of up to 500,000 people, and in 1971 the country became independent from Pakistan after a bloody liberation struggle. Between 1971 and 1975 Bangladesh took an authoritarian and quasi-socialist direction under liberation hero, Sheik Mujib Rahman, which in combination with near famine conditions in 1974 resulted in a decline in national living standards. During the Mujib era, the government sought

Table 6.1 Year-wise shrimp/prawn farm area statistics of Bangladesh (area in hectares). (Source: Statistical Yearbook of Bangladesh, 2008–2009. Vol 26, no. 1)

Division	1984–1985	1985–1996	1987–1988	1988–1989	1993–1994 Up to 1996–1997	1997–1998 Up to 2002–2003	2003–2004	2004–2005 Upto 2008–2009	Rate of increase (%)			
									Compare to 2002– 2003 and 1996–1997	Compare to 2003– 2004 and 1996–1997	Compare to 2002– 2003 and 2003–2004	Compare to 2002– 2003 and 2004– 2005 to 2008–2009
Chittagong	23,437	24,781	24,781	25,514	29,792	29,792	29,792	34,704	0.00	0.00	0.00	16.49
Khulna	39,976	62,448	69,053	80,418	104,624	107,962	163,848	171,506	3.19	56.61	51.76	58.86
Barisal	833	71	176	348	3,341	3,357	9,189	11,425	0.48	175.04	173.73	240.33
Dhaka	–	–	–	–	239	242	242	242	1.26	1.26	0.00	0.00
Total	64,246	87,300	94,010	108,280	137,996	141,353	203,071	217,877	2.43	47.16	43.66	54.14

to distribute public lands to the landless and marginal farmers with limited success (Adnan 1993). In 1975 the government was deposed by the army and the new military government led by Ziaur Rahman adopted a more liberal and export-oriented economic policy continued by his successors to this day (van Schendel 2009). From the late 1970s there was a growth of private investment in brackish water shrimp farming and in the Bangladesh Second Five Year Plan (1980–1985), the Government supported many initiatives to improve cultured shrimp production technologies through agencies such as the Food and Agricultural Organisation/Swedish International Development Agency (FAO/SIDA) Bay of Bengal Programme, the First Aquaculture Development Project (Asian Development Bank/ADB), the Shrimp Culture Project (International Development Agency), the second Aquaculture Development Project (ADB), the Third Fisheries Project (World Bank) and from 1999, the Fourth Fisheries Project (World Bank). The main emphasis in these projects was technical development of nurseries for post-larvae, screening of pond sluices, selective stocking with brackish-water shrimp, water quality, and maintenance of appropriate water levels in shrimp farms (Karim and Aftabuzzaman 1995).

Military rule persisted until 1991 when the country returned to civilian rule and democratic politics. Despite the shift to democratic politics, attempts at democratic consolidation have generally failed. What we see instead is a circulation of the political elites each of which when in power took on the characteristics of patrimonial rule in which access to positions of authority and influence was shaped by political loyalty rather than political, technical or administrative competence (Islam 2008). The close relationship between government, business and the military during both periods was reinforced by World Bank Structural Adjustment finance amounting to US$ 1.76 billion aimed at creating an export-oriented and market driven economy. This close relationship extended to the shrimp sector, including shrimp farming.

During the early development of the shrimp export sector when international prices for shrimp were high, some members of the business community used their economic power and political connections to pressure rice farmers, particularly in the country's southwest, to lease out their lands for shrimp farming and to convert public (*khas*) lands to shrimp farming. The World Bank and Asian Development Bank promoted shrimp farming as an important new source of foreign exchange earnings and Bangladesh experienced a rapid growth in shrimp farms, shrimp depots, processing plants and a labour force engaged in wild shrimp fry collection. The shrimp export sector was declared a thrust industry in 1991 and shrimp farmers and hatchery operators were given, among other things, a tax holiday and reduced rates of bank interest on loans. A year later, the Bangladesh Government introduced the Shrimp Mohal Management Policy (Habib n.d.), which supported the turning over of suitable public land to shrimp farming, which went against existing legislation protecting land against salinity (Afroz and Alam 2012). The sector grew rapidly to become Bangladesh's second largest official earner of foreign exchange. However, it accounts for only 5 % of export earnings compared with garment manufacture, which accounts for over 80 % of total gross export earnings. Almost 90 % of the US$ 527 million foreign exchange earnings from the country's seafood sector come from farmed shrimp exports (2006/2007). The Bangladesh government, business and international aid agencies continue to support the expansion of export-oriented shrimp production and several million people rely directly or indirectly upon it for work and income.

Debates About Shrimp Farming in Bangladesh

The changes brought about by shrimp farming have been the subject of increasing scholarly and policy debate and analysis. The debate and analysis have generated a considerable body of literature of variable theoretical sophistication, methodological clarity and empirical rigour. It consists of a mix of official, academic and grey literature such as official Bangladesh government and national and international NGO reports, academic synchronic studies usually of single village or shrimp farming sites, consultancy reports for government or private business and newspaper reports on particular events and developments in shrimp farming areas. Many published sources include shrimp farming as one component of a wider study of the industry as a whole often with a focus on the technical aspects of shrimp farming rather than wider social and environmental concerns. Few studies have taken a longitudinal, cross-village and cross-regional, controlled comparative approach to gauge the particular impact over time of specific drivers on the social and ecological fabric of rural communities (Pokrant 2006). Most studies have been done in the southwest of the country where some 80 % of all shrimp farms are found and the bulk of these deals with brackish water shrimp farming, although increasing scholarly attention has been given to freshwater prawn farming (Ahmed et al. 2010; Ahmed and Garrett 2011).

Much of the literature on shrimp farming is descriptive and often technical, focusing on farm preparation, yields per hectare and costs of production. However, there is a growing body of academic and policy literature, which focuses on the social, economic and ecological aspects of shrimp farming. In this literature, two main approaches can be identified (Béné 2005; Pokrant 2006).

The first is the Radical or Political Ecology approach taken by some sectors of the national and international NGO movement and several academic commentators. Political Ecology focuses on the unequal distribution of political and economic power within and across countries, which is considered to determine access to land, technical inputs, and capacity to meet international product and process standards set by the major shrimp importing countries. This approach regards what is often referred to as industrial shrimp farming as unsustainable and proposes national and local changes in policy direction favouring small farmers and more ecologically sound farming practices, a reform of international trading regimes and a change in rich country consumption habits. Some proponents of this approach have a radical political agenda that seeks to replace or constrain neo-liberal capitalism with alternative forms of political and economic organisation or to drastically restrict the power of international corporations and global governance agencies and to shift the centre of political gravity to local communities or to networks of local communities within bio-regional contexts. This approach is found among a number of small advocacy-based national NGOs such as the Bangladesh Environmental Lawyers' Association, *Nijera Kori, Unnayan Bikalper Nitinirdharoni Gobeshona* (UBINIG 'The Policy Research for Development Alternatives'), and *Uttaran* supported by overseas NGOs such as the Environmental Justice Foundation (2003, 2004) and the Swedish Society for Nature Conservancy (2011).

The second approach is the Mainstream or Ecological Modernisation approach, which starts from the premise that environmental sustainability is attainable through the application of scientific, technological and organisational processes within existing political and economic frameworks. It is this view of sustainable development that is supported by the Bangladesh Government, national and international business, some service-oriented national and international NGOs such as the Bangladesh Shrimp and Fish Foundation and Caritas, and several key global governance agencies such as the World Bank and the Asian Development Bank. Supporters of this approach consider that the negative social and ecological effects of shrimp farming are the result of poor management and bad planning, which can be rectified by improved farmer education, better farm management and more efficient supply chains at the level of the shrimp farm. The greater financial resources, political and intellectual capital available to those who work within the Ecological Modernisation paradigm means that much of the literature on shrimp farming falls within this category. Drawing on sources from both approaches, including the author's own field work over 15 years, the following general observations can be made about the social, economic and ecological changes brought about by shrimp farming.

Social-Ecological Systems and the Impact of Shrimp Farming

The distinction made between socio-economic and agro-ecological changes should be regarded as an analytical or heuristic device rather than independently existing or isolated phenomena. Humans are part of nature and human life is entangled with a world of things, including natural things (Hodder 2012). As such, rural communities in general and shrimp farming in particular are embedded within ecological systems and the interaction between them is one of dependence and mutual constitution (Berkes et al. 2003). Modern shrimp farming was introduced into a society, which historically was (and remains) highly inegalitarian in terms of land ownership, distribution of political power, and gender and ethno-religious relations (Bose 1993). Before modern shrimp farming, Bangladesh had already undergone substantial environmental change over many centuries (Eaton 1990; Iqbal 2010; Richards and Flint 1990). Huge tracts of Bengal were cleared of forest to make way for rice farming and other uses so that by the 1970s forest land, other kinds of land cover and water bodies had been altered to accommodate new agricultural and urban populations (Mukerjee 1938). By the early 2000s, forest cover had been reduced to less than 10 % of the country's land area (Choudhury and Hossain 2011). The clearing of land also led to the decline and extinction of many terrestrial and aquatic floral and faunal species. This has given the remaining forest cover and water bodies great political, economic and ecological significance in contemporary debates regarding the impacts of shrimp farming.

Since the 1970s, shrimp farming has acted singly and in combination with other historical and contemporary drivers of change to affect rural people's livelihoods. These drivers include national government policies aimed at promoting rice production through conversion of land and water bodies, consumer country regulation and control of shrimp imports, the building of embankments and polders and the introduction of High Yield Varieties (HYV) of rice, local, national and international water control and diversions ranging from the Farakka Barrage in India to filling in of water depressions (*beels*) and ponds across the country; climatic and human-induced changes to water and soil quality, increased coastal populations, land fragmentation, and the growing competition for inland and marine fish and forest products.

Shrimp farming is one part of a global network of seafood processing activities and its introduction to rural communities has made those communities increasingly subject to the actions of rich country governments, international trading and processing companies, and global governance agencies. Shrimp farmers are dependent on global demand for shrimp and shrimp products, which in good years has meant high incomes for some but in poor years resulted in heightened indebtedness and even loss of land and livelihoods. The power of global and regional regulatory agencies is particularly apparent as they can (and do) shut down the industry overnight if contaminated shrimp are found in export consignments. For example, since the early 1990s, the EU has warned the Bangladesh Government that the shrimp sector needed to improved its health and safety regime. This came to a head in 1997 when an EU Inspection team inspected conditions in several Bangladesh processing plants, which led to a 1998 ban on exports to the EU resulting in substantial financial losses throughout the sector (Cato and Lima dos Santos 1998; Alam and Pokrant 2009; Alam 2010). At this time, the main cost of restructuring was borne by processing plants but subsequent inspection visits by the EU Commission resulted in growing surveillance of operations at shrimp farm and hatchery levels.

In some areas and over time smallholders have benefited economically from expanded shrimp production but often at the expense of consumption crops such as rice and a growing inequality in landholdings, forcing marginal landowners out of production (Ali 2006; Islam 2009; Karim 2000). Many farmers switched to shrimp as returns were higher compared with either rice farming or salt farming or rice farming alone, with shrimp-salt rotation being the most profitable (Islam et al. 2003). A key problem for the sector has been disease, which increased local farmers' perceptions of risk, causing many to seek other kinds of work as a hedge against disease epidemics and other risks such as uncertain foreign markets, irregular supply of shrimp fry and dependence on informal credit arrangements in long domestic supply chains (Ahsan 2011). These domestic supply chains consist of many intermediaries such as fry traders, hatcheries, shrimp traders, money lenders, and shrimp depot owners who supply essential inputs to farmers and sell their products. Such dependencies shape farmer perceptions of the viability of shrimp farming.

Government investment and assistance to the shrimp sector has been greatest in the processing and hatchery sectors, which are dominated by wealthy Bangladeshis who exert most economic influence on the supply chain. For 20 years processing plant owners and some hatchery operators have pressed for farming land to be turned over to them to allow them to engage in more controlled contract farming and to shift towards more intensive forms of production. At present, processing plants have considerable over-capacity as a result of poorly planned expansion. Governments have resisted meeting their demands as they fear rural unrest from smallholders losing markets and resistance from those poor and landless parts of rural society dependent on public lands and common pool resources. While some shrimp farms are unable to meet processor demand, which gives them some advantage in bargaining over price, this is limited by long supply chains with many intermediaries and a lack of powerful national shrimp associations to act as bargaining agents. Some smallholders have sought to enhance their economic security through cooperation in water sharing.

Several studies point to a decline in land area devoted to rice farming, traditional forms of livelihood and employment opportunities as a result of shrimp farming. There has been conversion of common pool resources to private use, a reduction of sharecropping opportunities in rice farming, reduced access to grazing land, and lower labour requirements of shrimp farms compared with those of paddy production (Rahman et al. 1997; Rahman et al. 2006). However, the shrimp sector as a whole has generated new jobs and income opportunities in shrimp processing, trading and distribution as well as the multiplier effects of increased incomes on local communities.

In areas of mixed rice and shrimp farming found mainly in the southwest and south central coastal zones, opportunities for sharecropping have declined as rice farmers switched to the more lucrative shrimp farming, which requires lower labour inputs per land area and where workers are paid daily or short-term contractual wage rates rather than a share of shrimp harvests (Datta 2006; Tutu 2006). Maniruzzaman (1998) notes in his study of three villages in one union in Khulna District the decline of pre-shrimp forms of labour. These included *bebaira* (engagement of guest labour) when friends and relatives of the farmer were given food in return for work; *badla* or labour exchange under which arrangement workers worked on each other's land and no cash wages (*kamla*) or food payments were given; sharecroppers who worked for a landlord as cultivators but were expected to provide labour services to the household such as catching fish, house repair etc. This *begar khata* or work without payment was a means by which landlords could retain labour for the next season's sharecropping. Similar changes have occurred in the country's southeast but there mixed salt and shrimp or shrimp only production is more common and workers alternate between shrimp farm and salt pan work. Recruitment of labour took place through local market places, direct recruitment and the use of labour contractors. Shrimp farmers also enter into harvesting arrangements with local fishers and others to allow harvesting of fin fish in shrimp ghers (Pokrant and Reeves 2003).

Shrimp farming has contributed to growing pressures on marginal farmers and the landless to migrate from rural areas to the cities or overseas. There is also evidence of in-migration in shrimp farming communities as people seek to take advantage of perceived work and investment opportunities (Maniruzzaman 1998). There have been multiplier effects of rising incomes from shrimp farming. For example, Ito (2002) reports from the freshwater prawn farming areas of Southwest Bangladesh that landless workers have secured jobs as farm guards, shrimp harvesters, shrimp farm building workers, mud snail shell breakers and traders, van drivers and transporters of prawns, fry, ice and shrimp feed. He argues that the availability of such work has strengthened the bargaining power of workers employed on traditional annual labour contracts by richer farmers. Pokrant and Reeves (2003) report from brackish water shrimp farming areas of Southeast Bangladesh that many landless labourers and marginal farmers worked as shrimp farm labour, shrimp fry collectors, fishers, salt workers, petty traders, short distance transporters, rice farm labourers and sharecroppers, rickshaw pullers, snail de-shellers, wood collectors and other small-scale artisanal and petty commodity activities. Their survey of 958 shrimp farms in Chakoria Sub-district in Southeast Bangladesh revealed that total employment on these farms was approximately 5,394 with over 80% employing six or fewer workers, some of whom were able to negotiate long-term contracts of a year or more. The sub-district is noted for its salt production and some workers alternated between shrimp and salt work, the latter being more remunerative.

Women have been affected in different ways by the growth of shrimp farming. The majority of women who physically work in shrimp farming and related activities such as fry collecting come from lower socio-economic groups. For these women, their work and income opportunities are gendered such that they are confined to particular types of shrimp farm and related tasks. On shrimp farms, adult women and female children are restricted to pond preparation and repair, some harvesting and snail collecting, although this varies across the country. Poorer women from functionally landless households are more likely than other women to be found collecting shrimp fry from the ocean, rivers and estuaries.

Physical work on shrimp farms and in fry collecting is considered by the more well-off members of both Hindu and Muslim local communities to be of low status as it is poorly paid, often carried out in unhygienic and dangerous conditions, socially demeaning and morally suspect for women who are forced to work in public spaces (Delap and Lugg 1999; Gain 2005; Pokrant and Reeves 2003). The wealthier and politically influential village elites and middle class consider fry collecting to be a threat to the moral order as women move about freely uncontrolled by men (Jalais 2010). However, female fry collectors often reject such negative views and assert their right and need to work, pointing out that it provides a degree of autonomy and a capacity to work with other family members, something unavailable in the now banned shrimp de-heading sheds where they were often monitored by employers (Delap and Lugg 1999). Women have also benefited financially from work in shrimp processing plants located in the large urban centres, although the benefits remain gendered with male processing workers earning more, on average, than female ones. Some landless people have taken up crab collection as an alternative livelihood strategy. These are usually among the poorest and possess limited bargaining power in the supply chain dominated by intermediaries and buyers in regional and national urban centres (Zafar and Ahsan 2006). Those with greater assets are able to invest in crab fattening and enjoy higher returns.

There is evidence that shrimp (and prawn) fry collection for shrimp farms has threatened coastal ecosystems with declines in black tiger prawn fry, freshwater prawn fry and other aquatic species (Ahmed and Troell 2010; Hoq 2007; Islam and Wahab 2005) as it takes large quantities of bycatch which affects aquatic species diversity. The Government ban in 2000 on fry collection was motivated, in part, by environmental considerations but was also a response to pressures from shrimp hatchery owners who saw wild fry collectors as competitors (this is discussed further below).

One of the most dramatic impacts of shrimp farming on local social and ecological environments is the destruction and clearing of the Chakoria Sundarban mangrove forest in Southeast Bangladesh during the 1980s and 1990s (Pokrant 2009). Up until the early 1980's, the forest was in public ownership under the control of the forestry department, which had recommended a resting of the land for 10 years to recover from what it saw as poor land use in the past. It was turned over to private investors, many outsiders to the area, who rapidly replaced the forest with shrimp farms and who then sub-leased the farms illegally to local lessees. Other public lands intended for use by the landless as common pool resources were appropriated by politicians, their supporters and urban business interests who sought to profit from a growing global demand for tropical shrimp.

Social and environmental impacts have been felt across much of the coastal belt as shrimp farms have encroached upon private rice farming land and public lands, often used as common pool resources by local farmers, fishers and the landless. The modified extensive nature of shrimp farming has meant that increased production has been brought about by conversion of rice fields and other lands to shrimp farms and the decline of several traditional non-shrimp livelihood practices and associated land and water uses. These include cattle grazing lands, fishing sites, and vegetable growing areas, and the decline or disappearance of local timber and plant varieties (Giasuddin et al. 2003).

A significant by-product of the shift to shrimp farming has been the massive growth in the number of wild shrimp fry collectors drawn largely from the poorer, landless sections of the rural population. Whether landless prior to shrimp farming or made so by the removal of paddy land from production in favour of shrimp cultivation, at their peak in the 1990s, fry collectors numbered over 400,000 (Azad et al. 2007; USAID 2006), spread across the coastal region of the country. Working in family teams often financed by small-scale fry traders, fry collectors came into conflict with fishers, some environmental and social NGOs and shrimp hatchery owners for a variety of reasons. For fishers, the indiscriminate collection of shrimp fry reduced aquatic biodiversity, including fish stocks. For some environmental NGOs, fry collectors undermined biodiversity and the capacity of local ecosystems to sustain themselves. For social NGOs, fry collecting meant the exploitation of minors required to work long hours waist deep in water. For the hatchery owners, fry collectors were a source of competition as they sought to establish themselves as the main source of shrimp fry for farmers. They lobbied government to restrict collectors' activities on the grounds of their threat to biodiversity and some local livelihoods and their use of child labour. The result was a ban on fry collecting in 2000 (SRO No. 289/Act/2000) but the ban has been poorly implemented and fry collection continues. Government commitments to providing alternative employment have not been realised. The decision to ban fry collecting met with opposition from fry collectors themselves, fry traders who stood to lose business and shrimp farmers who preferred wild fry to that of hatchery fry. It was also criticised by some NGOs and donor agencies on the grounds that it made fry collectors scapegoats for a much wider problem of environmental destruction caused by such factors as deforestation, overfishing, and inefficient shrimp trawling in the Bay of Bengal. Also, it was argued that any ban should be directed at traders and shrimp farmers themselves who dominate the local supply chains (Azad et al. 2007; Frankenberger 2002).

Agro-Ecological Impacts

There has been an increase in levels of soil and water salinity through changes in groundwater quality and deliberate flooding of rice fields, which has affected rice farming in contiguous areas. Shrimp investors and farmers have legally and illegally appropriated common pool resources, which have displaced fishers, landless labourers and marginal farmers. The conversion of land to a single use has resulted in threats to biodiversity and the spread of shrimp diseases (particularly White Spot SyndromeVirus) into shrimp farming areas (Shahid and Islam 2003). Shrimp disease was introduced early into shrimp farming as a result of the import of diseased shrimp fry from Thailand. It is now endemic across practically all brackish water shrimp farming areas.

The spread of shrimp disease illustrates that shrimp farmers themselves are highly vulnerable to human-induced hazards and also natural hazards. The latter is vividly demonstrated by the impacts of two cyclones on Southwest Bangladesh in 2007 and 2009. In the 2007 cyclone (Sidr) it was reported that over 90 % of shrimp farms were destroyed in some sub-districts of Bagerhat District in the country's southwest (The Fish Site 2007). In the 2009 cyclone (Aila) it was estimated that over 40 % of shrimp farms were affected or destroyed and the livelihoods of thousands compromised and undermined (Kumar et al. 2010). The capacity of shrimp farmers to respond to this cyclone depended on their income and assets, familial and other social networks, and links to government officials. Some small shrimp farmers heavily dependent on small loans obtained at high interest rates were least resilient and for many their futures became uncertain (Kartiki 2011). As a result of the cyclone, there is some evidence of both a backlash against shrimp farming and a more positive view that shrimp farming could provide an alternative livelihood on land that had become so saline it could not support rice farming (Daily Star 1 May 2011a; 26 May 26 2011b). The cyclone also forced some families to take up shrimp fry collection.

Shrimp farming has reduced soil quality, increased social and water salinity levels, increased acidity of soils, reduced the area under rice cultivation, reduced rice yields and reduced fodder supply for livestock (Ali 2006; FAO 2009; Haque Muniral et al. 2010). Swapan and Gavin (2011) report changing land use, village organisation, increased susceptibility to cyclone and other hazards, shifts to salt tolerant rice strains and reduced fish availability. Food habits have changed as local people are less able to grow or afford fruits, vegetables, duck and beef as a result in part of increased soil salinity caused by shrimp farming (Rahman et al. 2011).

Ali's work in Southwest Bangladesh sums up well some of the consequences:

> Shrimp farming has affected the village rice ecosystem in several ways. It has brought major changes in soil properties and caused soil degradation that affects rice yields. Transformation of rice fields into shrimp ponds has reduced the total area under rice and fodder production and has created food shortage for both human and livestock population. Toxic chemicals and effluents in shrimp ponds have disrupted the habitat for fresh water fish and aquatic species inherent in rice ecosystem (Ali 2006)

Local Resistance to Shrimp Farming

The changes described above have met with various kinds of resistance largely directed at the appropriation of public lands, decline in sharecropping opportunities, and the coercive treatment of rice farmers unwilling to rent out land for shrimp. The early years of shrimp farming from the 1980s to the 1990s were marked by a spurt of investment by urban and rurally-based business people in response to rising world prices for shrimp and government incentives. They leased out land from local rice farmers or were able to acquire public lands often at the expense of local landless people and marginal farmers (Guimaraes 1989; Adnan 1993). During this period, referred to as an 'era of resistance' by Islam (2009), there was considerable controversy and public debate over the spread of shrimp farming and the newspapers of the time are full of reports about land appropriation, strong-arming of rice farmers to lease out land for shrimp, flooding of paddy fields with salt water, and looting of shrimp farms. There are numerous examples of local resistance to shrimp farming in Khulna and Satkhira districts. One of the most well-known was the Horinkhola movement in Paikgacha in Khulna District in 1990 when a local business man sought to establish a shrimp farm on land owned by absentee landlords. Fearing dispossession and destruction of their livelihoods, the landless organised a protest led by a widow named Korunamoyee Sarder who was a member of Nijera Kori Mohila Bhumiheen Samity (Nijera Kori Women's Landless Society) and who was killed by supporters of the farm owner. After this killing, a big protest meeting was held in Horinkhola with the participation of Nijera Kori leader, Khushi Kabir, and members of the Communist Party and the main national opposition party, the

Awami League.[2] A memorial was built to her and a memorial service is observed every year.

Other movements include the Jaliakhali Movement in Dacope sub-district of Khulna, which was formed in 1987 when a group of shrimp farmers tried to start a shrimp farm on 70 ha of land. Local people under the leadership of the Communist Party started a movement against them, which resulted in the farmers leaving the farm. The farmers attempted to stop the local protest by using a hired gang called 'Hunda-Gunda Bahini' to get control of the farm but were unsuccessful. The Bhaina Beel movement was formed in 1988 in Dumuria sub-district of Khulna to oppose the establishment of a shrimp farm on local land. The Tala movement operated during the period from 1996 to 1998 in Tala sub-district of Satkhira when a group of local influential people established shrimp farms on several thousand acres of land in seven publicly owned canals (*khas khals*).

The newspapers of the time are full of stories about land-grabbing and conflicts over land use, particularly in the south west of the country where most shrimp farms are located close to rice farming lands. For example, the National Daily, Dainik Ajker Kagoj, reported in March 1994 that: 'Law and order situation [has] deteriorated due to shrimp farming in Southern Khulna'. The Dainik Ittefaq reported in May 1994 a 'Clash between two groups for capturing shrimp project land in Chakaria [southeast Bangladesh]: one murdered and 6 injured'. Manik Saha, a fearless local journalist, wrote a report entitled: 'struggle against gher [shrimp farm or enclosure] owners and terrorism by the gher owners in South Khulna & socio-economic and environmental impacts of shrimp culture in Khulna region' in which he described the many conflicts over control of land and water in the region. His report, based on observations from 1990 to 1995, describes lootings, killings and rape, which he summarised in Table 6.2. He also reported on the killing of a local landless leader against land appropriation by wealthy local business people in November 1990 (Saha 2000; n.d.a; n.d.b) (Table 6.2).

Saha was murdered on a public street in January 2004 although his killers have yet to be brought to justice and it is unclear if his death was retaliation against his reports of the activities of 'shrimp mafia' or part of a wider attack on journalists for reporting various cases of extortion and other rackets in rural Bangladesh.

Similar English-language and Bengali-language newspaper reports from the Cox's Bazar region appeared in the National and local press from the early 1990s to the mid-2000s. For example, the *Daily Star* of August 9, 2002 ran a headline: 'Shrimp lords destroy coastal mangrove: Local BNP [Bangladesh National Party] MP's men fell trees, exposes Sonadia island to tidal wave'. This long article outlines some of the changes brought about by land grabbing in the area, summarised in the opening paragraph:

> The ecologically critical Sonadia Island under Moheshkhali Police Station fast loses thousands of acres of state-owned natural mangroves. Hundreds of workers engaged by the local BNP MP Alamgir Mohammad Mahfuzullah Farid are chopping down the mangroves

---

[2] See http://www.nijerakori.org/documents/The_harin_khola_movement.pdf.

Table 6.2 Saha report on violence and displacement in Southwest Bangladesh, 1990–1995

Total news reports published on attacks/clashes	50
Total murdered (in 30 incidents)	40
Total injured	525
Total raped	8
Total untraced/lost	10
Total families compelled to leave the locality	50

to convert the area into shrimp cultivation compartments. The massive deforestation of 15-kilometre-radius Sonadia Island on the Bay of Bengal, where about 100 families live, has exposed the adjacent Moheshkhali to tidal waves, cyclones and other natural calamities (Daily Star, Vol. 3 No.1038, Friday August 09, 2002).

In Cox's Bazar, from the early 1990s to 2004, local organised resistance to shrimp farming was arranged through local movements and organisations such as the *Bargachasi Samity* or Sharecroppers' Society against 'outsider' control of public (khas) land for shrimp farming, the manipulation of shrimp fry prices by traders, and opposition to loss of common property resources. Attempts were made by the landless to gain access to public shrimp farming land via the *Kudal Bahini* (Spade Soldiers) organisation, and opposition to the use of 'outside' labour on shrimp farms (Gregow 1997). By 2004, the Sharecroppers' Society and the Spade Soldiers had ceased to operate and while there is still local resentment at the power of khas land lease-holders, the leader of the Sharecroppers' Society told the author he had gone back to sub-leasing several shrimp plots from a first-hand lease owner. Kudal Bahini was formed by landless labourers in Badarkhali to secure khas shrimp land. At its height in the mid-1990s it had more than 500 members. Most were also members of the Badarkhali Cooperative Society and part of their grievance was that the Forest Department had allocated public khas land to six members of the Society. Khas landholders in the region were well organised and well-connected politically, and were able to stop the movement.

In addition to the more vocal and organised protests against shrimp farming, there is some evidence of what James Scott refers to as everyday forms of resistance (Scott 1985). According to Scott, these include: 'foot-dragging, dissimulations, false compliance, feigned ignorance, desertion, pilfering, smuggling, poaching, arson, slander, sabotage, surreptitious assault and murder, [and] anonymous threats' (Scott 1989) The author's own field work in Chakoria, Southeast Bangladesh revealed some evidence of poaching and pilfering. One large shrimp farm owner told me of poaching taking place on his farm which required him to hire guards to protect shrimp stocks. Employees working on some of the larger shrimp farms were also searched as it was believed that shrimp were hidden in the men's *lungis* (tube-shaped and skirt-like wraps worn by men and boys) when they finished work. Guimaraes (1989) reports from Southwest Bangladesh that poaching was a common problem as shrimp attracted high prices on the 'black market' and was difficult to police. However, it is important to note that several of these tactics have been employed by the more powerful shrimp investors against opponents of shrimp farming, as noted earlier.

Conclusion

The emergence and growth of the Bangladesh shrimp export sector since the 1970s illustrates one aspect of the growing dominance of a global seafood industry aimed at producing high quality fish and fish products for a global market centred on the rich countries of the world and, more recently, growing middle class markets in Asia. It also illustrates Bangladesh's attempt to shift towards a more export-oriented development policy aimed at earning foreign exchange, generating employment and raising the living standards of the population through a shift domestically to a more market-based exchange and production system.

The incorporation of large areas of coastal Bangladesh into international circuits of shrimp production, distribution, exchange and consumption has brought many changes to communities and ecologies in shrimp farming areas of the country. From an initial private investor-led and government-backed yet poorly regulated expansion of shrimp farming into what had been traditional rice growing communities, shrimp farming has become firmly established as an important agro-industrial activity in which the economic and social benefits have spread unevenly across the rural populations. Shrimp farming is increasingly regarded, if not completely accepted, by local communities as a fact of rural life and financial benefits have extended beyond an initial core of urban-based investors to many medium- and small-scale farmers. However, it has also meant the physical and social displacement of many landless labourers and marginal farmers who lost access to public lands and work opportunities under the earlier rice-based rural regime. Particularly during the early years of the development of the sector, there was some organised opposition to shrimp farming and several landless labourers and marginal farmers were killed as they sought to protect their livelihoods from shrimp farm expansion. Ecological impacts have ranged from a destruction of some mangrove areas to increasing salinization of ground water and soil, which have affected, *inter alia*, rice production, grazing opportunities and availability of potable water. These impacts have, in turn, affected the capacity of many rural peoples to maintain their livelihoods, which has forced some of the poorest populations to exploit the fragile ecological systems such as the Sundarban mangrove forest in the country's southwest in order to survive.

More recently, there have been signs that the Bangladesh State along with NGOs and global governance agencies, is attempting to regulate the sector to protect Bangladesh's fragile coastal zone and its inhabitants from unregulated shrimp farming. One such initiative is the creation of zones for shrimp production to reduce competition with rice farming and to protect particular ecological environments. For example, in February 2012, the Bangladesh High Court ruled as illegal the use of salt water on forest and agricultural land (Daily Star 2012). The aim of this ruling is to protect such lands from forcible conversion to shrimp farming. Another initiative is the development of integrated shrimp farming approaches (Bostock et al. 2010). This includes mixed shrimp-fish-rice, prawn production systems, which seek to reduce the incidence of disease, to spread economic risks by diversifying production options and to enhance the compatibility of shrimp farming with the prevailing agro-ecological conditions (Ahmed et al.2010). One 'pro-poor' shrimp initiative is the Danish Danida-funded Greater Noakhali Aquaculture Extension Project, which from 2002 was designed to assist poor women to nurse freshwater prawn fry for sale to grow-out farmers who export the mature shrimp (Danida 2008). SIPPO (Swiss Import Promotion Program) began a pilot project in 2004 in South West Bangladesh to see if shrimp could be produced using traditional methods free of artificial compound feed, chemicals and fertilizers and recent evidence suggests improved environmental and economic outcomes for organice farmers (Paul and Vogl 2012, 2013). Shrimp farmers have moved towards obtaining shrimp fry from hatcheries in order to ensure a steady supply and reduce reliance on wild-caught fry, although some hatcheries have suffered from viral and other diseases, which can threaten the quality of shrimp as well as farmer livelihoods. The government ban on wild fry collection in 2000 has not stopped the practice and there is little evidence of government initiatives to assist fry collectors to shift to other work. Indeed, the criminalisation of such activity has exposed some of the poorest to official and unofficial intimidation. The FAO (FAO 2012) has initiated a program to organise selected shrimp farmers into cluster organisations to adopt best management practices in disease control, quality assurance standards, improved supply chains and cooperation among shrimp farmers. The idea is that these organisations will act as leaders in their field, setting an example for other shrimp farmers.

These initiatives indicate a growing awareness of the need for longer term strategies based around ecohydrological or ecosystem management systems which seek to control effluents, disease and salinity through more environmentally sensitive and polycultural techniques and mangrove and wetland restoration to absorb salt and other contaminants (Sohel and Ullah 2012). There have also been calls to integrated shrimp farming into Bangladesh's Integrated Coastal Zone Management Strategy (ICZMS), which has been under development since the late 1990s (Afroz and Alam 2012). However, the schemes mentioned focus largely on shrimp farms as economic and environmental units and only partially address wider community issues such as landlessness and poor governance.

In addition, concerns have been expressed that such initiatives are aimed predominantly at meeting certification and other standards set by richer countries and global agencies and do not engage local peoples sufficiently in setting such standards (Islam 2009; Vandergeest and Unno 2012). Shrimp farmers themselves play little if any role in the setting of such standards and remain dependent on distant market players and cross-governmental bodies who determine the buying price of shrimp and who continue to set global production and process standards. For example, globally there exists human rights and environmental group opposition to the new certification standards being developed by the World Wildlife Fund (WWF) for the ASC with support from some NGOs in Bangladesh linked to the more radical approach outlined earlier. However, it is unclear if shrimp farmers in Bangladesh are aware of these certification changes, which suggests the need for more research on the domestic political economy of Bangladesh to examine the degree to which domestic NGO opposition reflects the views of local shrimp farmers and other rural inhabitants.

The urgency of taking a more integrated approach to the place of shrimp farming in Bangladesh is highlighted by the growing threat of climate change to coastal dwellers in Bangladesh. The Government of Bangladesh has been at the forefront of global climate change governance and has embarked on an ambitious climate change adaptation strategy for the country supported by the World Bank, the British Government and several international and national NGOs and action research centres (Ministry of Environment and Forests 2009; GoB Climate Change Strategy and Action Plan 2009). Shrimp farming is included in new climate change adaptation programmes and it remains to be seen what changes will be required in the shrimp sector to assist in sustainable adaptation to the hazards of climate change.

References

Adnan S (1993) Shrimp-culture projects in coastal polders of Bangladesh: policy issues concerning socio-economic and environmental consequences. Shomabesh Institute, Dhaka

Afroz T, Alam S (2012) Sustainable shrimp farming in Bangladesh: a quest for an integrated coastal zone management. Ocean and coastal management, doi:10.1016/j.ocecoaman.2012.10.006.

Ahmed N (1967) Prawn and prawn fishery of East Pakistan. Government of East Pakistan, Directorate of Fisheries, Office on Special Duty, Services and General Administration Department In-charge, East Pakistan Government Press, Dacca

Ahmed N, Troell M (2010) Fishing for prawn larvae in Bangladesh: an important coastal livelihood causing negative effects on the environment. Ambio 39:20–29

Ahmed N, Garnett ST (2011) Integrated rice-fish farming in Bangladesh: meeting the challenges of food security. Food Secur 3(1):81–92

Ahmed N, Troell M, Allison EH, Muir JF (2010) Prawn postlarvae fishing in coastal Bangladesh: challenges for sustainable livelihoods. Mar Policy 34(2):218–227

Ahsan DA (2011) Farmers' motivations, risk perceptions and risk management strategies in a developing economy: Bangladesh experience. J Risk Res 14(3):325–349

Alam SM, Curtin University of Technology (2010) School of Social Sciences and Asian Languages. Faculty of Humanities, EU regulatory governance and the management of quality in the Bangladesh shrimp supply chain.

Alam SM, Pokrant B (2009) Re-organizing the shrimp supply chain: aftermath of the 1997 European union import ban on the Bangladesh shrimp. Aquacultur Econ Manag 13(1):53–69

Ali A (2006) Rice to shrimp: land use/land cover changes and soil degradation in southwestern Bangladesh. Land Use Policy 23(4):421–435

Azad AK, Lin CK, Jensen KR (2007) Wild shrimp larvae harvesting in the coastal zone of Bangladesh: socio-economic perspectives. Asian Fish Sci 20(4):339–357

Bagchi A, Jha P (2011) Fish and fisheries in Indian heritage and development of pisciculture in India. Rev Fish Sci 19(2):85–118

BBS (Bangladesh Bureau of Statistics) (2011) Statistical yearbook of Bangladesh-2010. Government of Bangladesh, Dhaka

Béné C (2005) The good, the bad and the ugly: discourse, policy controversies and the role of science in the politics of shrimp farming development. Dev Policy Rev 23(5):585–614

Berkes F, Colding J, Folke C (eds) (2003) Navigating social-ecological systems. Building resilience for complexity and change. Cambridge University Press, Cambridge

Beveridge MC, Little DC (2002) The history of aquaculture in traditional societies. Ecological aquaculture. The evolution of the Blue Revolution, 3–29

Bose S (1993) Peasant labour and colonial capital: rural Bengal since 1770, vol 3. The New Cambridge History of India. Cambridge University Press, Cambridge

Bostock J, McAndrew B, Richards R, Jauncey K, Telfer T, Lorenzen K, Little D, Ross L, Handisyde N, Gatward I, Corner R (2010) Aquaculture: global status and trends. Philos Trans R Soc B 365:2897–2912

Cato J, Lima dos Santos CA (1998) European Union 1997 seafood safety ban: the economic impact on Bangladesh shrimp processing. Mar Resour Econ 13(3):215–227

Choudhury JK, Hossain AA (2011) Bangladesh forestry outlook. Study working paper series, working paper no. APFSOS II/ WP/ 2011/ 33. Food and Agriculture Organization of The United Nations Regional Office For Asia And The Pacific, Bangkok.

Daily Star (2002) Shrimp Lords Destroy Coastal Mangroves. Vol 3 No. 1038, Fri. August 9. Accessed 27 May 2011.

Daily Star (2011a) Aila-hit people get relief to be barely alive, find no help to return to their normal life. May 1. http://www.thedailystar.net/newDesign/news-details.php?nid=183920. Accessed 27 May 2011

Daily Star (2011b) Shrimp farming deals major blow to South. Coastal land in Aila-hit areas turns barren due to salinity. May 26. http://www.thedailystar.net/newDesign/news-details. php?nid=187242. Accessed 27 May 2011

Daily Star (2012) Shrimp farming on forest, arable land illegal: HC. February 2. http://www.thedailystar.net/newDesign/news-details.php?nid=220894. Accessed 10 Feb 2012

Danida (2008) Impact evaluation of aquaculture interventions in Bangladesh. Orbicon Lamans, s.a. Management Services. www.evaluation.dk. Accessed 3 May 2012

Das TC (1931) The cultural significance of fish in Bengal. Man India 11:275–303

Das TC (1932) The cultural significance of fish in Bengal. Man India 12:96–115

Datta AK (2006) Who benefits and at what cost? Expanded shrimp culture in Bangladesh. In: Rahman AA, Quddus AHG, Pokrant B, Ali L (eds) Shrimp farming and industry: sustainability, trade and livelihoods. University Press Ltd, Dhaka

Delap E, Lugg R (1999) Not small fry children's work in Bangladesh's shrimp industry. Save the Children, Dhaka

DoF (Department of Fisheries) (2010) Fisheries statistical yearbook of Bangladesh (2008–2009), vol 26, no. 1. Fisheries Resources Survey System, Department of Fisheries, Bangladesh, Dhaka.

Duarte C, Marbá N, Holmer M (2007) Rapid domestication of marine species. Science 316:382–383

Eaton R (1990) Human settlement and colonization in the Sundarbans, 1200–1750. Agric Hum Values 7(2):6–16

Environmental Justice Foundation (2003) Smash and grab. Conflict, corruption and human rights abuses in the shrimp farming industry. EJF, London

Environmental Justice Foundation (2004) Desert in the Delta. A report on the environmental, human rights and social impacts of shrimp production in Bangladesh. EJF, London

FAO (2009) Situation Assessment Report in SW-Coastal Region of Bangladesh. Dhaka, FAO. (available at http://foris.fao.org/static/data/nrc/LACCII_situation_assessment_report.pdf.). Accessed 15 June 2011

FAO (2010) FishStat fishery statistical collections: aquaculture production (1950–2008; released March 2010). Food and Agriculture Organization of the United Nations, Rome. http://www.fao.org/fishery/statistics/software/fishstat/en. Accessed 20 Dec 2011

FAO (2012) Building trade capacity of small-scale shrimp and prawn farmers in Bangladesh – Investing in the Bottom of the Pyramid Approach. MTF/BGD/046/STF (STDF/PG/321). http://www.standardsfacility.org/Files/Project_documents/Project_Grants/STDF_PG_321_Application.pdf. Accessed 31 Oct 2012

Frankenberger TR (2002) A livelihood analysis of shrimp fry collectors in Bangladesh: future prospects in relation to a wild fry collection ban. DFID, Dhaka

Gain P (2005) Shrimp Fry Collection and Its Trade: A survey on shrimp fry collectors in Cox's Bazar, Bhola, Khulna, Bagerhat and Satkhira of Bangladesh. SEHD, Dhaka

Giasuddin M, Bari Md N, Rahman A (2003) Agricultural activities and their impacts on the ecology and biodiversity of the Sundarbans area of Bangladesh. J Natl Sci Found Sri Lanka 31(1–2):175–199

Gopal PB and Vogel, CR (2013) Key performance characteristics of organic shrimp aquaculture in southwest Bangladesh. Sustainability 4(5):995–1012

GoB (Government of Bangladesh) (2009) Climate Change staretgy and Action Plan. http://www.moef.gov.bd/climate_change_strategy2009.pdf

GoB (Government of Bangladesh) (2010) Bangladesh bureau of statistics. Government of Bangladesh, Dhaka

Gregow K (1997) When the trees are gone. Social and environmental costs of converting mangroves into shrimp fields. A case study from Chakoria Sunderbans, Cox's Bazar, Bangladesh. UBINIG (Unnayan Bikalper Niti Nirdharoni Gobeshona: Policy Research for Development Alternative). Dhaka

Guimaraes JP (1989) Shrimp culture and market incorporation: a study of shrimp culture in paddy fields in southwest Bangladesh. Development and Change 20(4):653–682

Gupta KG (1908) Reports on the results of enquiry into the fisheries of Bengal and into fishery matters in Europe and America. Calcutta: Bengal Secretariat Book Department

Gupta RK (1984) The economic life of a Bengal District: Birbhum 1770–1857. University of Burdwan Press, Calcutta

Habib E (n.d.) Legal aspects of shrimp cultivation. Bangladesh Environmental Lawyers Association (BELA), Dhaka

Hodder I (2012) Entangled. An archaeology of the relationships between humans and things. Wiley-Blackwell, Chichester

Hora SL (1948) Knowledge of the ancient Hindus concerning fish and fisheries of India: I. References to fish in Arthashastra (ca. 300 B.C.). J R Asiat Soc of Bengal Sc 14:7–10

Hoq ME (2007) An analysis of fisheries exploitation and management practices in Sundarbans mangrove ecosystem, Bangladesh. Ocean Coast Manag 50(5-6):411–427

Humphrey J (2006) Global value chains in the agrifood sector. UNIDO working papers, Vienna

Iqbal I (2010) The Bengal delta. Ecology, state and social change, 1840–1943. Cambridge imperial and post-colonial studies series. Palgrave Macmillan, London

Islam MS (2008) From pond to plate: towards a twin-driven commodity chain in Bangladesh shrimp aquaculture. Food Policy 33(3):209–233

Islam, MS (2009) In Search of "White Gold": Environmental and Agrarian Changes in RuralBangladesh. Society & Natural Resources. 22(1):66–78

Islam S, Wahab A (2005) A review on the present status and management of mangrove wetland habitat resources in Bangladesh with emphasis on mangrove fisheries and aquaculture. Hydrobiologia 542(1):165–190

Islam MS, Serajul Islam M, Wahab MA, Miah AA, Mustafa Kamal AHM (2003) Impacts of shrimp farming on the socioeconomic and environmental conditions in the coastal regions of Bangladesh. Pak J Biol Sci 6(24):2058–2067

Ito S (2002) From rice to prawns: economic transformation and agrarian structure in rural Bangladesh. The Journal of Peasant Studies 29(2):47–70

Ito S (2004) Globalization and agrarian change: a case of freshwater prawn farming in Bangladesh. Journal of International Development 16(7):1003–1013

Jack JC (1916) The economic life of a Bengal district. Oxford Clarendon Press, London

Jalais A (2010) Braving crocodiles with Kali: being a prawn seed collectors and a modern woman in the 21st century Sundarbans. Socio-Legal Rev 6:1–23

Karim R (2000) Shrimp culture and changing land use pattern in Rampal Thana, Bagerhat Zila: a spatial analysis. Unpublished PhD. Thesis, Department of Geography and Environmental Studies, Rajshahi University, Bangladesh

Karim M, Aftabuzzaman (1995) Brackish and marine water aquaculture: potential, constraints and management needs for sustainable development. A paper presented at the National Workshop on Fisheries Resources, Development and Management, Dhaka, October 1995

Kartiki K (2011) Climate change and migration: a case study from rural Bangladesh. Gend Dev 19(1):23–38

Kesteven GL, Ling SW (1951) Report on the fisheries of east Pakistan. Government of east Bengal, Department of Agriculture, Cooperation and Relief, Fisheries Branch, 1951. File No. 4E-4/51. National Archives, Bangladesh

Kumar U, Baten Md A, Al-Masud A, Osman K, Rahman MM (2010) Cyclone Aila: one year on natural disaster to human sufferings. Dhaka: Unnayan Onneshan-The Innovators. www.unnayan.org. Accessed 20 Dec 2012

Maniruzzaman M (1998) Intrusion of commercial shrimp farming in three rice-growing villages of southern Bangladesh: its effects on poverty, environment and selected aspects of culture. Unpublished PhD thesis, University of the Philippines, Diliman, Quezon City

Ministry of Environment and Forests (MoEF) (2009) Climate change strategy and action plan, 2009. Ministry of Environment and Forests, Peoples Republic of Bangladesh, Dhaka

Mukerjee R (1938) The changing face of Bengal. A study in riverine economy. University of Calcutta, Calcutta

O'Malley LSS (1925) Faridpur: Bengal district Gazetteers. Bengal Secretariat Book Department, Calcutta

Pakistan (1961) Pakistan 1960–1961. Pakistan Publications, Karachi

Paul BG, Vogl CR (2012) Key performance characteristics of organic shrimp aquaculture in southwest Bangladesh. Sustainability 4(5):995–1012

Paul BG, Vogl CR (2013) Organic shrimp aquaculture for sustainable household livelihoods in Bangladesh. Ocean and Coastal Management 71:1–12

Pokrant B (2006) The organisation and development of coastal brackish water export oriented shrimp production in Bangladesh: A critical review of the literature. Shrimp Farming and Industry: Sustainability, Trade and Livelihoods 299–320

Pokrant B (2009) Work, community, environment and the shrimp export industry in Bangladesh, India and Thailand. In: Gillan M, Pokrant B (eds) Trade labour & transformation of community in Asia. Palgrave Macmillan, London

Pokrant B, Reeves P (2003) Work and labour in the Bangladesh brackish-water shrimp export sector. South Asia 26(3):359–389

Pokrant B, Reeves P, McGuire J (1997) Riparian rights and the organisation of work and market relations among the inland fishers of colonial Bengal, 1793–1950. In: Ali MY, Chu-fa Tsai (eds) Openwater capture fisheries of Bangladesh. University Press Ltd. and the Bangladesh Centre for Advanced Studies, Dhaka

Pokrant B, Reeves P, McGuire J (2001) Bengal fishers and fisheries: an historiographical essay. In Bandhopadhyaya S (ed) Bengal: Rethinking History. Essays in Historiography. Manohar, Delhi 93–118

Rahman AA, Quddus AHG, Pokrant B, Ali ML (eds) (2006) Shrimp farming and industry: sustainability, trade and livelihoods. University Press Ltd, Dhaka

Rahman A, Chowdhury F, Nandy G, Rahman A (1997) The state of the coastal environment: a people's report based on their knowledge and opinions. Shamunnay, Dhaka

Rahman MH, Lund T, Bryceson I (2011) Salinity effects on food habits in three coastal, rural villages in Bangladesh. Renew Agr Food Syst, 26(3): 230–242

Rahman QM (1945) Tour note of Deputy Director of Fisheries (Administration), Bengal, for the month of July 1945. QM Rahman. Government of Bengal, Department of Agriculture, Fishing Branch, File No. 3M-78, 1945. Tour Note of the Deputy Director of Fisheries (Adm), Bengal

Ray N (1994) History of the Bengali People (Ancient Period). Orient Longman, Calcutta. (Translated with an introduction by JW Hood)

Reeves P (1995) Inland waters and freshwater fisheries: issues of control access and conservation in colonial India. In: Arnold D, Guha R (eds), Nature, culture and imperialism: essays on the environmental history of south Asia. Oxford, New Delhi

Richards JF, Flint EP (1990) Long-term transformations in the Sundarbans wetlands forests of Bengal. Agric Hum Values 7(2):17–33

Rivzi S (ed) (1970) East Pakistan district Gazeteers: Chittagong. Government of East Pakistan, Dacca

Saha M (2000) People's movement against shrimp farming in Bangladesh. Khulna, Bangladesh

Saha M (n.d.a) Movement against industrial shrimp in Harin Khola led by landless leader Korunamoyee. http://www.nijerakori.org/documents/Karunamoyee_case_study.pdf. Accessed 21 May 2012

Saha M (n.d.b) Socio-economic and emvironmental impacts of shrimp culture in the Khulna Region. Struggle against gher owners and terrorism by gher owners in South Khulna. An unpublished compilation of events (in Bengali)

Scott JC (1985) Weapons of the weak: everyday forms of peasant resistance. Yale University Press, New Haven

Scott JC (1989) Everyday forms of resistance. Cph J Asian Stud 4:33–62

Shahid A, Islam J (2003) Impact of denudation of mangrove forest due to shrimp farming on the coastal environment in Bangladesh. In: Wahab MA (ed) Technical Proceedings of BAU-NURAD Workshop on Environment and Socio-Economic Impacts of Shrimp Farming in Bangladesh

Sohel SI, Ullah H (2012) Ecohydrology: a framework for overcoming the environmental impacts of shrimp aquaculture on the coastal zone of Bangladesh. Ocean Coast Manag 63:67–78

Swapan MSH, Gavin M (2011) A desert in the delta: participatory assessment of changing livelihoods induced by commercial shrimp farming in Southwest Bangladesh. Ocean Coast Manag 54:45–54

Swedish Society for Nature Conservation (2011) Murky waters. The environmental and social impacts of shrimp farming in Bangladesh and Ecuador. Stockholm http://www.naturskyddsforeningen.se/sites/default/files/dokument-media/murky_waters.pdf. Accessed 21 May 2012

The Fish Site (2008) www.thefishsite.com/fishnews/vars/country/bd. Accessed 20 Dec 2012

Tutu A (2006) Grassroots perspectives on shrimp farming in southwest Bangladesh. In: Rahman AA, Quddus AHG, Pokrant B, Ali L (eds) Shrimp farming and industry: sustainability, trade and livelihoods. University Press Ltd, Dhaka

Uddin T (2008) A study on shrimp processing activities in Bangladesh. Journal of Fisheries International 3(1):34–38

USAID (2006) A pro poor analysis of the shrimp sector in Bangladesh. Office of women in development. U.S. Agency for International Development. GEW-1-00-02-00018-00, Task no. 02. pp. 29–33 (www.usaid.gov)

Van Schendel W (2009) A history of Bangladesh. Cambridge: Cambridge University Press

Vandergeest P, Unno A (2012) A new extraterritoriality? Aquaculture certification, sovereignty, and empire. Political Geography 31(6):358–367

Webster JE (1911) Noakhali: Eastern Bengal and Assam district Gazetteers. Pioneer Press, Allahbad

Zafar M, Ahsan, MN (2006) Marketing and Value Chain Analysis of Mud Crab (Scylla sp.) in the Coastal Communities of Bangladesh. In: Islam MA, Ahmed K, Akhteruzzaman, M (eds) Value Chain Analysis and Market Assessment of Coastal and Marine Aquatic Products of Bangladesh. Bangladesh Fisheries Research Forum, Dhaka 25–53

Chapter 7
Evolution and Development of the Taiwanese Offshore Tuna Fishery

Ta-Yuan Chen

Abstract The depletion of tuna stocks in the waters of Southeast Asia has presented a substantial challenge to the fishing communities of post Second World War Taiwan. The purpose of this chapter is to the link the impact of quantitative changes in offshore tuna resources to the growth and decline of Taiwan's fishing industry, and to trace the development of longline fishing techniques and their long-term impact on Taiwan's offshore tuna fisheries. The chapter focusses on the main centres of the offshore fishing industry in Taihoku (Taipei) State and Takao (Kaohsiung) State. It begins with the introduction of tuna fishing to Taihoku State in the early 1910s, and shows how the centre of tuna fishing gradually shifted to Takao State in the mid-1930s, demonstrating how the exploitation of tuna resources in Southeast Asian waters reshaped the fishing communities of pre WWII Taiwan. The chapter also analyses the interactions between the changes in tuna abundance, the development of onboard fishing facilities and the growth and decline of the tuna longlining industry in post-WWII Taiwan.

Keywords Taiwan history · Taiwanese fishing · Longline tuna · Distant-water tuna · Southeast Asia fishing

The depletion of tuna stocks in the waters of Southeast Asia has presented a substantial challenge to the fishing communities of post Second World War (WWII) Taiwan. A number of studies have been made on relevant issues from the perspective of marine ecosystem studies. However, there are still relatively few studies undertaken from the perspective of marine environmental history (e.g. Butcher 2004). In order to address this gap, the purpose of this chapter is to trace the development of longline fishing techniques and their long-term impact on Taiwan's offshore tuna fisheries, and to the link the impact of quantitative changes in offshore tuna resources to the growth and decline of Taiwan's fishing industry. I limit the temporal range of my study to the period since 1912, when tuna fishing was introduced in Taiwan. I limit the scope of my chapter to the main centres of the offshore fishing industry in Taihoku (Taipei) State and Takao (Kaohsiung) State; tuna fishing was

T.-Y. Chen (✉)
Asia Research Centre, Murdoch University, 90 South Street,
6150, Murdoch, WA, Australia
e-mail: dr.henry.chen@gmail.com

J. Christensen, M. Tull (eds.), *Historical Perspectives of Fisheries Exploitation in the Indo-Pacific*, MARE Publication Series 12, DOI 10.1007/978-94-017-8727-7_7,
© Springer Science+Business Media Dordrecht 2014

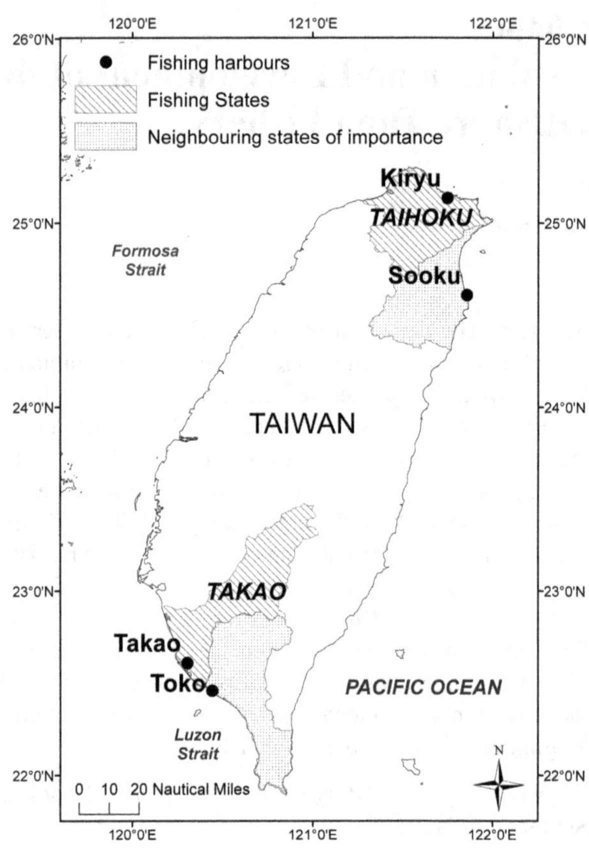

Fig. 7.1 Taihoku State and Takao State and their fishing ports: Kiryu, Sooku, Takao and Toko fishing ports

introduced to Taihoku State in the early 1910s, although the centre of tuna fishing gradually shifted to Takao State in the mid-1930s (Fig. 7.1). The history of Taiwan's distant-water tuna fisheries in Southeast Asia has been covered elsewhere at greater length (Chen 2009).

In the first part of the chapter, I will explain how the exploitation of tuna resources in Southeast Asian waters reshaped the fishing communities of pre-WWII Taiwan. In the second part of the chapter, I will describe and analyse the interactions between the changes in tuna abundance, the development of onboard fishing facilities and the growth and decline of the tuna longlining industry in post-WWII Taiwan. The main tuna species that the Taiwanese offshore fishers harvest in the waters of Southeast Asia include yellowfin tuna (*Thunnus albacares*), bigeye tuna (*T. obesus*) albacore tuna (*T. alalunga*) and southern bluefin tuna (*T. maccoyii*). To conduct research from a macro and comprehensive perspective, in this chapter 'tuna' refers to the above four species. Statistical data used in the chapter are mainly taken from *Taiwan Yuyeshi Ziliao Suanbian: Tongjipian (1) Mingjhi-Dajheng*, [Historical Data of Taiwan's Fishing Industries (1) Meiji-Taishō Period] and *Taiwan Suisan Tokei*

[Statistical Data of Taiwan's Fisheries], and fisheries yearbooks published by the fishing authorities in post-war Taiwan.[1]

The Development of Offshore Tuna Longlining Fishery in Pre-WWII Taiwan

Tuna longlining fishing was introduced to Taiwan in 1912. This new fishery developed separately in Taihoku State and Takao State. In Taihoku, fishers used Kiryu and Sooku as their home ports and fished in the nearby fishing grounds. The early development of Taihoku's tuna longlining fishery was very difficult, partly because fishers were not familiar with the fishing grounds of the tuna fishery, and partly because mackerel, the major source of tuna fishing bait, was expensive. Fortunately, the problem was soon solved after fishers began using mullets and milkfish as bait. Local fish farms provided longlinermen with large quantities of milkfish at reasonable prices. In 1918, people in Taihoku began to export tuna products to Japan (Taiwan Suisan Yōran 1928). The success of this trade between Japan and Taiwan greatly accelerated the development of the tuna fishery in northern Taiwan, because both the Sooku and Kiryu fishing ports were geographically close to the main markets for tuna products in Japan (Yoshinobu 1938). In 1923, the Japanese government spent ¥ 650,000 constructing the fishing-related infrastructure of the Sooku Fishing Port. The newly-built port facilities not only reduced the running costs of longliners, but also encouraged Japanese fishers from Oita and Ehime to use Sooku as their homeport. In the early 1930s, more than 100 fishing vessels were using Kiryu and Sooku as supply bases (Taiwan Suisan Yōran 1930).

When tuna longlining fishing was first introduced to the Toko and Takao fishing ports, local fishers only used sail-powered boats in coastal waters. They even believed, wrongly, that sail powered longliners were more suitable to harvest tuna than motored vessels. However, the situation gradually changed after the mid-1910s when Taiwan's tuna fishing industry gradually expanded its fishing territories to greater distances which only motored vessels could reach. The success of the motored tuna longlining industry in northern Taiwan greatly encouraged fishers in the south to motorise their boats (Taiwan Suisan Yōran 1928).

The development of the tuna fishing industry in southern Taiwan soon caught up with that of northern Taiwan for three reasons. Firstly, fish farms in southern Taiwan harvested large amounts of mullet and milkfish, which provided local longlinermen with sufficient fishing baits for their commercial fishing activities. Secondly, both the Takao and Toko fishing ports were adjacent to Southeast Asia, the most important tuna fishing ground in the pre-WWII years (Taiwan Suisan Yōran

---

[1] *Taiwan Yuyeshi Ziliao Suanbian* was edited by the Academia Sinica in post-WWII Taiwan; *Taiwan Suisan Tokei* were fisheries yearbooks published by the colonial authorities.

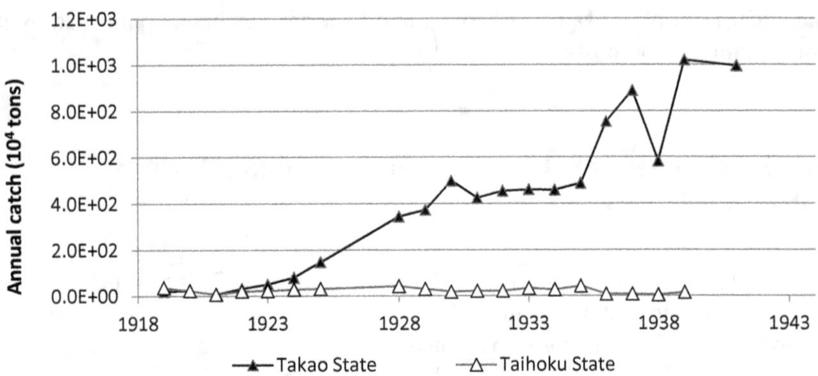

Fig. 7.2 Annual Tuna Catch of Taihoku State and Takao State, 1919–1941

1930). Thirdly, due to the availability of refrigeration facilities, tuna longlinermen in southern Taiwan also started to export their tuna products to Japan after 1923. By 1928, approximately 150 offshore longliners were installed with 30-horsepower engines in southern Taiwan. Some large-sized longliners were even installed with 120-horsepower engines (Taiwan Suisan Yōran 1928). The number of motored longliners in southern Taiwan increased at a great rate, which enabled fishers to expand their fishing territories from coastal waters to the heart of Southeast Asian waters (Taiwan Suisan Yōran 1940). The development of the tuna fishery in the south of the island soon exceeded that of the north (Yoshinobu 1938).

The view that the centre of Taiwan's tuna longline fishery shifted from the north to the south of the island is supported by the statistical data in pre-war fisheries yearbooks. Figure 7.2 indicates that during the period 1916–1923, the tuna landings in Taihoku were more than Takao's, although only by a small margin. The year 1924 is considered a milestone in the history of Taiwan's tuna fishery. In that year, the volume of tuna catch in Taihoku was only 226,009 kins (approximately 135.6 t; 1 kin is equivalent to 600 g); however, 788,045 kins of tuna were yielded in Takao State, so that the annual tuna catch of Takao State was about three times that of Taihoku's. After 1924, the annual tuna catch of Taihoku stagnated at between 200,000 and 400,000 kins for more than a decade. However, the annual tuna catch in southern Taiwan was making astonishing progress during the same period of time. The subsequent gap in the tuna catch between the south and north therefore widened rapidly. In 1939, 10,195,164 kins (6,117 t) of tuna were harvested; in that year the development of Takao's offshore tuna industry had reached its peak. However, at the same time, the annual catch of Taihoku's tuna fishery was at its lowest point, at only 28,000 kins. The output of Takao's tuna fishery was 364 times more than Taihoku's. Without a doubt, Takao State had become the centre of Taiwan's tuna fishery.

The geographic distribution of 'bonito resources' (or Skipjack tuna *Katsuwonus pelamis* and true bonitos *Sarda* sp.) overlapped with that of the tuna species to a large extent (Taiwan Suisan Yōran 1928). Considerable quantities of tuna were also

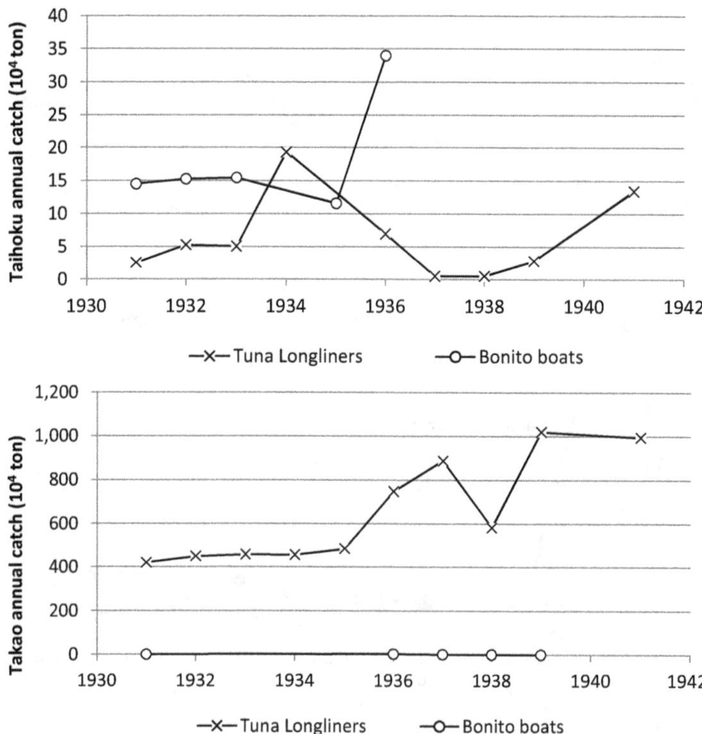

Fig. 7.3 Annual tuna catch of Taihoku state and Takao state per boat types (Bonito boats & tuna longliners), 1931–1941

harvested by bonito boats as by-products in northern Taiwan. Thus, further distinguishing the tuna output yielded by bonito boats enables us to observe the growth of Taiwan's tuna longlining fishery more precisely. I use the annual tuna catch in 1931 as an example. Figure 7.2 shows that the annual tuna catch of Taihoku was 169,706 kins that year. However, 4,241,712 kins (approximately 2,545 t) of tuna were harvested in Takao. It seems that the annual tuna catch of Takao State was 25 times more than that of Taihoku. In fact, the gap between tuna catches in the two places was much wider than 25 times. Figure 7.3 shows that only 25,000 kins of tuna were yielded by Taihoku's tuna longliners; the rest were harvested by bonito boats. In the same year, tuna longliners in southern Taiwan harvested 4,187,995 kins in total, which was 176 times more than that of their counterparts in the north. The island-wide tuna catch harvested by offshore tuna longliners was 4,262,928 kins (approximately 2,558 t). The tuna catch from the Takao State accounted for 98.2 % of the total catch.

Besides comparing the volume of tuna landing, the view that the centre of Taiwan's tuna fishery was gradually shifting from the north to the south of the island is also supported by other statistical comparisons (Fig. 7.4). In 1935, the number of Takao's tuna longliners had exceeded that of Taihoku. In 1933, the total tonnage of

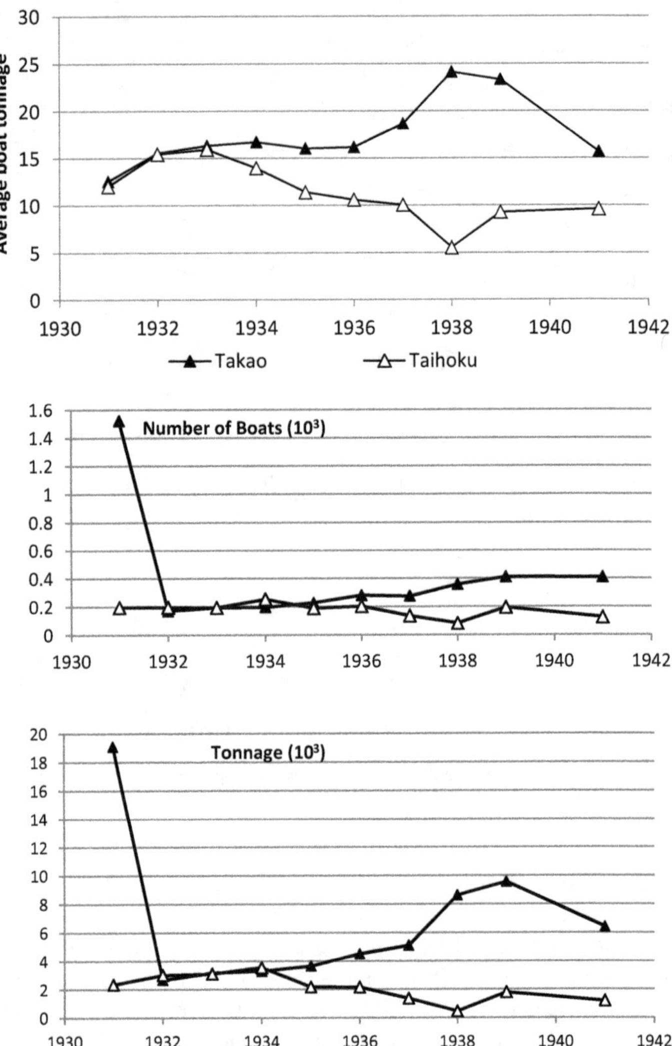

Fig. 7.4 Number and tonnage of Taihoku and Takao states' tuna longliners, 1931–1941

Takao's longliners exceeded that of their counterparts in the north, but 1 year later, Taihoku caught up again. However, since 1935, Takao's tuna longliners have maintained a distinct lead in total tonnage. The year 1935 was a turning point of Taiwan's tuna fishing industry. The number and total tonnage of Taihoku's tuna longliners and their tuna catch started to shrink rapidly thereafter.

Two factors lay behind the shift in longline fishing to the south of the island. Firstly, the tuna fishing grounds off northern Taiwan extended only 50 or 60 miles from the northeastern coast of Taiwan to the Senkaku Islands where the Japan

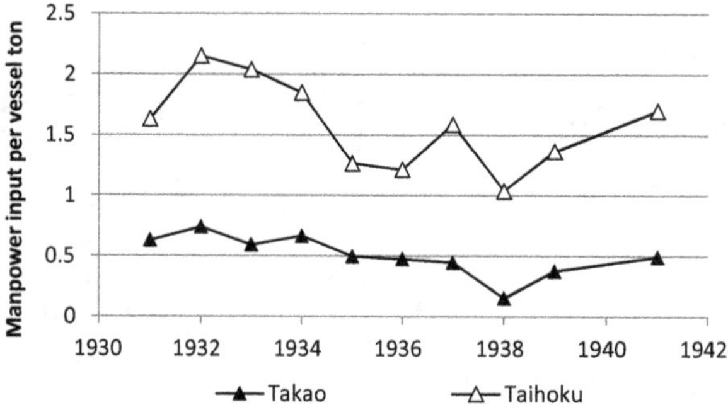

Fig. 7.5 Manpower input per vessel ton in Takao and Taihoku's tuna fishing industry, 1931–1941

Current passes by. However, the tuna fishing grounds off southern Taiwan were vast and productive. Secondly, the southward expansion of tuna fishing was also greatly encouraged by the fishing authorities (Taiwan Suisan Yōran 1927, 1928). Many fishing research vessels were built by the fishing authorities of Taiwan to conduct fishing experiments and research. Encouraged by the Government's fishing experiments, the tuna longliners from southern Taiwan continually expanded their fishing territories southward. In 1928, the most distant fishing grounds that they could reach were 250 miles from the coast (Taiwan Suisan Yōran 1928). In 1934, some tuna longliners had fished in the waters of the Sulu Sea, and even the Celebes Sea (Takao Shisei Jū Shunen Ryakushi 1934).

In addition to the availability of vast and productive tuna fishing grounds in Southeast Asia, the Southward Development Policy (Nanshin Seisaku) of the Japanese colonial government also had a profound influence on the southward expansion of Taiwan's tuna fishing territories. The southward presence and regional movement of the Japanese navy directly encouraged the southward development of Taiwan's fishing industry. The military expected fishing boats to collect intelligence information for the navy, and fishing vessels from Taiwan were also keen to offer strategic services (Kenichi 2001). Tuna longliners from southern Taiwan now became more active in the waters of Southeast Asia than before. They even entered the waters of the Philippines, Indochina and Borneo, and conducted fishing activities in the coastal waters of those countries.

The average tonnage of Takao's tuna longliners continuously increased along with the development of tuna fishing generally in Southeast Asian waters (Fig. 7.4). In the early 1930s, the average tonnage of tuna longliners from Taihoku and Takao states were almost on par with each other. However, the average tonnage of Taihoku's longliners continuously dropped off afterwards. It even fell to 5.5 t in 1938. In the same year, the average tonnage of Takao's tuna longliners reached its historic peak, at 24.1 t per longliner. This means that tuna longliners that fished in Southeast Asian waters were approximately five times larger than those in northern Taiwan.

The fishing facilities onboard Takao's tuna longliners also enabled fishers to work more efficiently than their counterparts in Taihoku. This view is supported by the comparison of the level of manpower input between the two places' tuna fishing (Fig. 7.5). In 1932, longliners only employed 0.7 fishers per registered ton in Takao

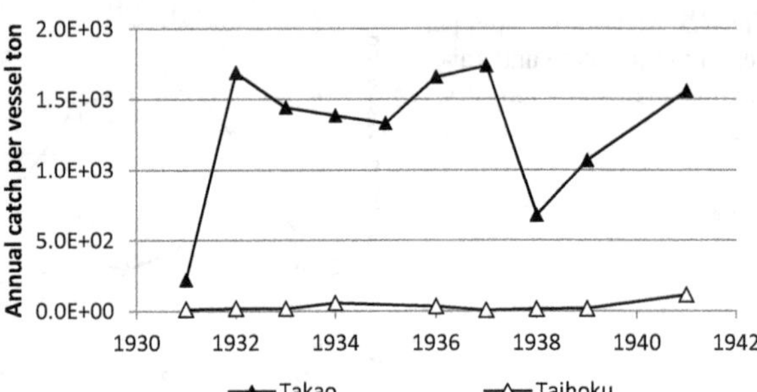

Fig. 7.6 Tuna catch per vessel ton in Takao and Taihoku state's tuna longlining industry, 1931–1941

State. However, in the same year, their counterparts in Taihoku had to hire 2.1 fishers per registered ton. It should be emphasised the lesser manpower input per vessel ton did not prevent longliners in southern Taiwan from harvesting more tuna. We can see this by comparing tuna catch per vessel ton yielded by Taihoku's and Takao's longliners. In 1941, every vessel ton in Takao State harvested 1557.3 kins of tuna; however, in Taihoku, only 112.7 kins were yielded (Fig. 7.6). Obviously, tuna longliners in southern Taiwan were much more efficient than their counterparts in the north.

Figure 7.6 shows that tuna catch per vessel ton remained stable from 1932 to 1941 even though the tuna longliner fleet in Takao State expanded rapidly during the same period. Therefore, despite the growth of Takao's longliner fleet and the improvement of onboard fishing facilities, which resulted in an increase in fishing power and expansion into new fishing grounds, there was no appreciable change in tuna catch rates during this period. In other words, overexploitation of tuna resources in the waters of Southeast Asia had not yet emerged as a problem in the pre-WWII years.

The Impact of the Tuna Fishery on Pre-WWII Taiwan

The exploitation of tuna resources in the waters of Southeast Asia had a profound influence on the society and industrial development off pre-WWII Taiwan. A large number of Japanese fishers, especially longlinermen, had migrated to Taiwan. Japanese fishers had introduced longlining techniques to Taiwan, and played a significant role in the development of Taiwan's offshore tuna fishery (Shinhat 1936). Some even remained in Taiwan and worked as fishing masters for Taiwanese distant-water fishing enterprises in the early post-WWII years (Interviews with Wu Tianrui, Kaohsiung, 5/4/2002; Cai Wun'yu, Kaohsiung, 13/5/2002). The exploitation of tuna resources in the waters off Taiwan also led directly to the rise of modern ancillary industries to fishing, including the ice-making industry, refrigeration industry, and tuna-canning industry (Shinhat 1936).

In the beginning of the industry, tuna harvested by Taiwan's fishing vessels was sold and consumed exclusively in Taiwan Island. The price of tuna products in

Taiwan was slightly cheaper than those in Japan. The situation began to change in 1918 when trade dealers in northern Taiwan started to export fish products to Japan (Taiwan Suisan Yōran 1927). At the outset, all the fish products were exported to Japan through Kiryu Port. In 1924, the opening of the direct sea route between Takao and Yokohama enabled longlining companies to export their tuna products directly from Takao port. Tuna landed in northern Taiwan was exported to Moji, Shimonoseki, Kobe and Yokohama. Tuna harvested in Takao State was mainly sold to Yokohama (Taiwan Suisan Yōran 1927). The birth of the tuna trade between Taiwan and Japan had left a profound impact on the tuna longlining industry of post-WWII Taiwan. A solid business partnership between Taiwanese longliner owners and Japanese fish dealers had been forged in the pre-WWII period, and these business links remain strong today.

Overexploitation of tuna did not emerge as a problem in the pre-war years, as catch rates were maintained by the expansion into new grounds and the increasing efficiency of vessels without noticeably impacting on the abundance of tuna. However, the exploitation of tuna resources had a profound influence on the society and economy of pre-war Taiwan. The industrial tuna fishery was introduced to Taiwan; businessmen started to form fishing companies; modern fishing ancillary industries grew up; business links between Taiwanese vessel owners and Japanese fish dealers were established, and numerous Japanese fishers migrated to Taiwan. This pre-war influence continued into the post-war years. Joint-ownership fishing companies in post-war Taiwan are the best example of this. The tuna fishing territories were further expanded after the end of WWII, but the running expenses of a tuna longliner became prohibitive in post-war Taiwan. Those who were interested in running a tuna fishery usually adopted the pattern of joint-ownership, and raised funds and built longliners together. The ways they ran the business and shared the profit were exactly the same as pre-war styles. The experience in the colonial period also enabled people to work comfortably with those from different ethnic backgrounds. Fishing migrants were forced to return to Japan after the end of WWII. The vacuum that the Japanese fishers left was soon filled by new fishing migrants from different parts of Taiwan, or even from China.

The Growth and Decline of the Offshore Tuna Fishery in Post-WWII Taiwan

The fishing authorities of post-WWII Taiwan also encouraged vessel owners to exploit tuna resources in the waters of Southeast Asia. The South China Sea, Sulu Sea and Celebes Sea were all considered ideal fishing grounds for fishers to restart their tuna longline fishery in the post-war years (Taiwan De Nonglin Jianshe 1950). Like the pre-war years, investigations into tuna resources were still being launched in the waters of Southeast Asia. Taiwan's fisheries experimental vessel, Haicing Hao, was frequently sent to the South China Sea for the investigation of marine resources in 1954, 1960, 1961 and 1962. From 1956 onwards, it was also tasked to investigate

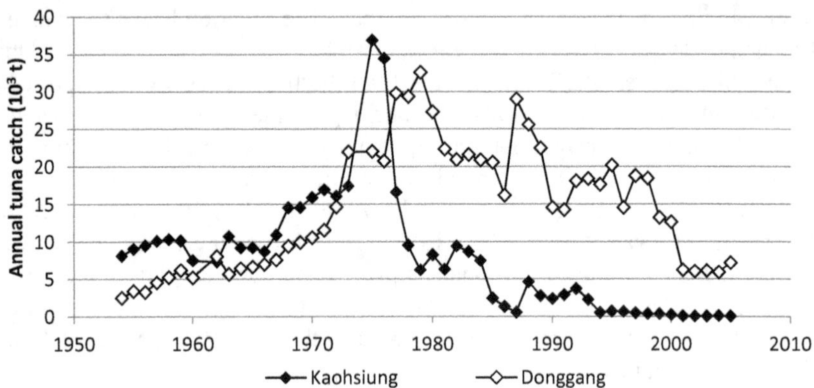

Fig. 7.7 Volume of tuna catch yielded by Kaohsiung and Donggang's offshore tuna longliners, 1954–2005 (No data for 1961 and 1974)

the tuna population in Southeast Asian waters, the Indian Ocean and the South Pacific Ocean (Yuye Fajhan 1971).

As we have seen, the centre of Taiwan's offshore tuna longlining fishery shifted from Taihoku (Taipei) to the Takao (Kaohsiung) State in the pre-war years. However, along with the development of the fishing industry in post-WWII Taiwan, the centre further shifted from Kaohsiung to Donggang (Toko). This is evident from a comparison of the annual volume of the tuna landed in each locality. Figure 7.7 shows the volume of tuna catch unloaded by Kaohsiung and Donggang's offshore tuna longliners from 1954 to 2005, and therefore provides a clear picture of the development of the offshore tuna longline fishery in Southern Taiwan over five decades. In 1954, the volume of tuna catch unloaded at the Donggang Fishing Port, now measured in tons instead of kins, was 2,459 t. In the same year, Kaohsiung's offshore tuna longliners harvested 8,115 t of tuna; the volume of Donggang's offshore tuna landing was approximately one-third of Kaohsiung's.

Kaohsiung had developed into a centre of distant water tuna longlining industry in the pre-war era, and its offshore tuna fishery still remained strong before the mid-1970s. Figure 7.7 indicates that from 1954 to 1975, the volume of tuna harvested by Kaohsiung's offshore longliners grew at a steady pace. The development of Kaohsiung's offshore tuna fishery eventually reached a peak in 1975, when the volume was as much as 36,873 t. However, from that year onwards, the annual tuna catch harvested by Kaohsiung's offshore longliners gradually declined.

The volume of tuna yielded by Donggang's offshore longliners also robustly grew from the mid-1950s to the late-1970s. The volume of tuna harvested by Donggang's offshore longliners exceeded that of their counterpart in Kaohsiung respectively in 1962, 1973 and 1977. The year 1977 was a milestone in the history of Donggang's offshore tuna fishery; from that year the offshore tuna landings in Kaohsiung would never be able to compete with Donggang's tuna catch again. Donggang Fishing Port officially replaced Kaohsiung, and became the centre of Taiwan's offshore tuna longlining fishery. In 1979, the development of Donggang's offshore tuna fishery

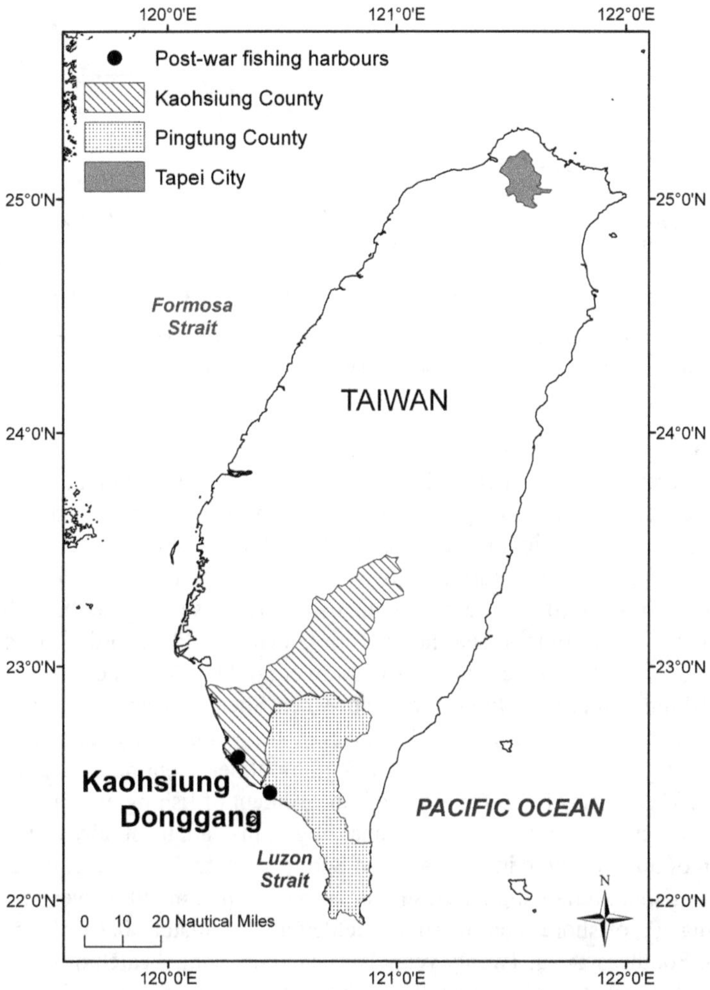

Fig. 7.8 The Kaohsiung (Takao) and Donggang (Toko) fishing ports in post-WWII Taiwan

culminated in a peak catch of 32,592 t. After that year, the volume of its tuna catch gradually dropped as well (Fig. 7.8).

The volumes of Kaohsiung and Donggang's tuna catch (offshore fishery) both dwindled over the next two decades. Figure 7.7 shows that in Donggang's case, the volume of tuna yielded by offshore tuna longliners decreased gradually. However, for some reason, the volume of Kaohsiung's tuna landing abruptly slumped respectively in 1994 and 2001. In 1994, its offshore tuna landing suddenly fell from 2,237 to 505 t. In 2001, it plummeted to 42 t. In 2005, the annual volume of Kaohsiung's offshore tuna catch was merely 26 t. It appeared that Kaohsiung's offshore tuna fishery in Southeast Asia had become moribund.

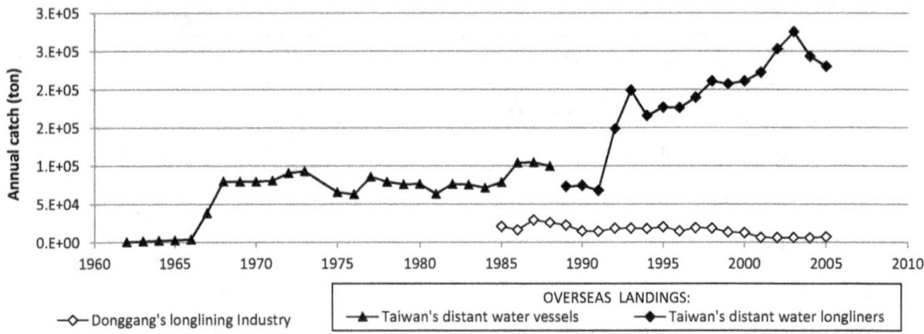

Fig. 7.9 Annual tuna catch evolution from 1961 to 2006 for 3 different Taiwanese fishing fleets: Donggans' longing industry, Taiwan's distant water vessels and Taiwan's distant water longliners

The rapid development of Taiwan's offshore tuna longline fishery had led to a massive decrease in tuna stock in the waters of Southeast Asia and the South-western Pacific Ocean. As an example, I use a 30-t offshore tuna longliner that operated in the waters of the Philippines. Before 1950, the hold could be easily filled to the brim with fish and would return to Taiwan in 3 weeks. In 1964, it took on average more than 2 months to fill the hold with fish. Taiwanese longlinermen therefore had to spend longer time at sea than before (Jicyuan n.d.). In order to fish more efficiently, Taiwanese vessel owners sought to equip their longliners with state-of-the-art fishing equipment (Interviews with Chen Shengli, Kaohsiung, 16/6/2002; Hong Fucai, Kaohsiung, 21/5/2002). However, the volume of tuna catch yielded by Kaohsiung and Donggang's offshore tuna longliners continued to shrink year after year. This clearly indicates that the improvement of fishing equipment did not necessarily increase the volume of the tuna caught. Instead, it actually worsened the depletion of tuna resource in the waters of Southeast Asia. Donggang, the centre of Taiwan's offshore tuna fishing industry, is as an example. In 1985, every vessel ton of Donggang's offshore tuna longliner yielded approximately 0.7 t of tuna in the waters of Southeast Asia. Twenty years later, in 2005, annual catch of tuna that per vessel ton harvested was less than 0.3 t (Fig. 7.9). Clearly, the intensive fishing of Taiwan's offshore tuna longliners has contributed significantly to the depletion of tuna resources in the waters of Southeast Asia.

The Improvement of Taiwan's Onboard Fishing Equipment and its Impact on the Tuna Fishery in Post-WWII Southeast Asia

The overexploitation of tuna presented a substantial challenge to longlinermen in the post-WWII years. In order to fish more efficiently, Taiwanese fishers regularly updated their onboard fishing facilities soon after the newest equipment appeared on the market. They achieved better fishing results in the short term. However, from a long-term point of view, their up-to-date fishing devices only accelerated the trend towards the overexploitation of tuna in Southeast Asian waters.

In the early post-war years, most offshore tuna longliners were installed with semi-diesel engines. There were many drawbacks with this old model of marine engine; they were heavy and highly fuel inefficient, which was a heavy financial burden to vessel owners or fishing companies in the long term. Moreover, semi-diesel engines were also large in size and occupied considerable space on board the vessel. As a result, the space for the fish hold and sleeping berths had to be heavily compressed, which not only brought about inconveniences to the daily lives of longlinermen in their long voyages, but also reduced the capacity to store fish onboard. In addition, the engine operation always caused severe hull vibrations, which shortened the ship's operating life (Nongye Yaolan Dishiji Yuyepian 1962).

To re-establish the business connections between the Japanese fish dealers and the offshore fishers of Taiwan, especially those in Donggang, the government encouraged the export of tuna from southern Taiwan in the mid-1970s. However, selling tuna products to the Japanese market was not as simple as the Taiwanese offshore fishers had expected, because the quality of the tuna meat required for Japan's sashimi market was extremely high. The meat had to be completely fresh, and its skin had to be free of scrape marks. Some longliners had installed traditional semi-diesel engines which produced constant vibrations during the voyage, which left fish bruised and with scrape marks on their skin. Less than one-tenth of the catch met the strict quality requirements of Japan's sashimi market. With a view to solving this problem, offshore longliner owners started to install diesel engines (Interview with Cai Wun'yu, Kaohsiung, 29/5/2002).[2] This new model of marine engines effectively reduced noise and vibrations, so damage to the fish was kept to a minimum on long voyages. The small size of these new engines also helped vessel owners minimise fuel expenses and opened considerable onboard space for fish holds. Thus, offshore longliners could harvest tuna more efficiently than before (Interview with Chen Shengli, Kaohsiung, 22/4/2002).

When the tuna longlining fishery was introduced to Taiwan at the beginning of the twentieth century, the longlines that fishers used were made from bark thread. Longlines made from bark thread were thick, and could scare off the tuna; therefore, the fishing returns were not satisfactory. Bark thread longlines were gradually replaced by factory-made cotton longlines in the 1950s. They were comparatively thinner than bark thread longlines, which helped fishers harvest tuna more efficiently. Furthermore, the prices were low and affordable (Interview with Cai Bian & female relative, Kaohsiung, 3/7/2002).

Both bark thread and cotton yarn easily rotted and became frayed in water. In order to lengthen the usage life of longlines, before mid-1950, people in fishing communities had to clean the longlines and provide a protective coating at least once

---

[2] As a matter of fact, the onset of the business relationships between the Japanese fish traders and Taiwanese distant water fishing companies had been established not long after the end of WWII, and this relationship was strengthened after overseas supply bases were developed by Taiwan's fishing companies.

a week. The maintenance of longlines was time-consuming and laborious. Some longlines were protected with mineral pitch. Longlines treated with mineral pitch could effectively prevent both the cotton yarn and bark string from being eroded by sea water. However, both bark longlines and cotton longlines became thick after being treated with mineral pitch.[3] Old offshore longlinermen said in interviews that the tuna were always frightened away by 'fat' longlines, especially in rough seas. In order to avoid this problem, fishers started to use nylon yarn as a substitute for cotton longlines. By the 1960s, nylon yarn was used throughout the fishing industry (Interview with Guo Shihfu, Kaohsiung, 27/6/2002).

The introduction of nylon longlines had a substantial impact on Taiwan's offshore tuna industry. For a start, people in the fishing communities no longer needed to spend so much time on the maintenance of longlines. To some extent, the use of this new product changed the traditional outlook and employment opportunities for workers of the offshore fishing communities of Taiwan. Furthermore, nylon longlines are pliable but extremely strong. The length of current nylon longlines is several times longer than that of the older cotton longlines, which enables fishers to harvest tuna much more efficiently than before. However, it also inevitably worsens the problem of over-exploitation of tuna resources in the waters of Southeast Asia. From 1967 onwards, an increasing number of offshore tuna longliners in Taiwan were equipped with auto fishing reels, which enabled onboard longlinermen to fish more efficiently during fishing voyages, and also helped longliner owners reduce labour costs, as they did not need to hire as many workers as before (Taiwan Yuye Jhi Yanjiu 1974).

Along with the development of Taiwan's offshore tuna industry, fishing territories were gradually extended to waters thousands of miles away from the homeports. To fix the vessel's location at sea and navigate to remote destinations, fishers could no longer simply rely on past fishing experiences. Instead, they had to use some onboard equipment. In the early post-WWII years, besides the compass, small-sized tuna longliners were equipped with radio; experienced fishers could fix the vessel's location at sea by judging the direction of radio sound waves sent from onshore stations in Kaohsiung and Donggang (Taiwan Yuye Jhi Yanjiu 1974; Interview with Cai Bian, Kaohsiung, 3/7/2002). Sextants were also commonly used in offshore tuna longliners heavier than 20 t. Those who knew how to use sextants would not easily impart their skills to other fishers, because in the early years, sextant users had a higher chance of being hired as fishing masters. In order to cultivate this skill, the fishing authorities held seminars at port areas for offshore fishers (Taiwan Yuye Jhi Yanjiu 1974; Interview with Jhen Sanbian, Kaohsiung, 2/2/2002). Nowadays, most of Taiwan's longliners are equipped with the Global Positioning System (GPS) and satellite phones. Telecommunication across oceans and continents no longer presents a problem. Fish finders had been commonly used by offshore tuna longliners heavier than 10 t since the mid-1970s. This revolutionary device enabled fishers to gauge water depth and mark the exact location of tuna shoals in the fishing grounds (Interview with Lin Changren, Kaohsiung, 8/5/2002). From a long-term point of

---

[3] During the 1940s and 1950s, longlines that were used in the waters of Southeast Asia were dyed with mineral pitch.

view, however, these increases in fishing efficiency only worsened the problem of depletion of tuna resources in the waters of Southeast Asia.

The Depletion of Tuna Resources in the Southeast Asian Waters and the Development of Taiwan's Global-Scale Fisheries

The depletion of Southeast Asian waters was a direct cause of the growth of Taiwan's global-scale fisheries. In 1950, Taiwanese offshore tuna longliners only fished in the Bashi Strait, South China Sea, and Sulu Sea. To prevent problems caused by overfishing in traditional fishing grounds close to the island, the fishing authorities of Taiwan began to encourage vessel owners to explore new fishing grounds in more remote seas. By 1954, Taiwanese longlinermen started to fish in the Banda Sea and Flores Sea. During the period from 1958 to 1960, some Taiwanese longliners also conducted fishing activities in the waters off Sumatra and Java and in the Bay of Bengal in the eastern part of the Indian Ocean (Interview with Lin Changren, Kaohsiung, 8/5/2002).[4] In 1953, when an American company, the Van Camp Sea Food Company, established a modern tuna canning factory in American Samoa, tuna longliner fleets from Taiwan, Japan and Korea were invited to fish there (Sen 1977).[5] Eleven Taiwanese longliners from the Kaohsiung Fishing Port substantially fished in Samoan waters as pioneers, completing an unprecedented expedition in 1964 (Tai'an 1973).[6] This was a milestone in the history of Taiwan's fisheries development; it was the first time that Taiwan's civilian fishing companies had undertaken such a long voyage and fished in the heart of the southwestern Pacific Ocean (Taiwan no Gyogyōgenkyō 1978).[7]

The success of the fishing venture in Samoan waters encouraged many Kaohsiung longliner owners to exploit tuna resources in the waters of remote oceans. Generally speaking, large-sized longliners fished in the waters of the Indian Ocean. However, for safety reasons, smaller longliners preferred to fish off Samoa in the Western Pacific. They could travel along the coastlines of the Philippine archipelago before

---

[4] In the 1960s, the Jhongguo Fishing Company (JFC), a state-run fishing company, had fished in the South Indian Ocean. However, by the JFC's pioneering fishing activities did not encouraged the Taiwanese tuna longliners to harvest tuna in the Indian Ocean.

[5] In 1963, another American company, Star Kist Food Inc., also set up a canning factory in Samoa.

[6] The year that Taiwanese civilian longliners first went to Samoa was 1964. The longliners which joined this expedition numbered as many as 11. This point is in accord with information contained in *Taiwan ni okeru Maguro Gyogyō no Genkyō* [The Present Situation of Taiwan's Tuna Longline Fishing, 1981].

[7] Before 1964, Taiwan's fisheries' circles knew little or nothing about the fishing grounds in the South Pacific Ocean. Taiwanese fishing expeditions were mainly conducted in the East Indian Oceans by a state-run fishing company, the Jhongguo Fishing Company (JFC), in 1955 and the Atlantic Ocean in 1960. However, the JFC's efforts did not contribute towards the establishment of the Taiwanese global-scale fisheries due to its constant mismanagement.

moving to the Indonesian archipelago and then the Solomon Islands (Interview with Jheng Sanbian, Kaohsiung, 13/5/2002). Through this expansion, Taiwan's distant water fishing industry gained a head start over all other Asian fishing nations apart from Japan (Taiwan no Gyogyōgenkyō 1978).[8]

In 1968, the fishing authorities in Taiwan implemented a new '5-Year Programme for the Acceleration of Taiwan Fisheries Development' (or 'Jiasu Fajhan Taiwan Yuye Wunian Jihua', 1968–1972) (Yuye Fajhan 1971; Taiwan no Gyogyōgenkyō 1978). Due to this new programme, substantial loans from the Asia Development Bank and the World Bank were allocated to Taiwanese fishing companies to build new distant water vessels. As a result, the fishing grounds of the Taiwanese fishers were continuously extended across Southeast Asia and beyond (Taiwan Yuye Jhi Yanjiu 1974; Sianchih 1984). By 1968, the aggregate tonnage of Taiwan's distant water tuna longliners had reached 85,300 t. By the end of 1972, the number of the Taiwanese tuna longliners operating in the waters of the Southern Pacific Ocean was as high as 238. The number that fished in the waters of the Atlantic Ocean to-talled 142 (Yuye Fajhan 1971). The Taiwanese tuna longlining fishing industry had emerged as a global scale fishery in the 1970s, harvesting tuna in the Pacific, Indian and the Atlantic Oceans. In 1974, the annual volume of Taiwan's distant water tuna landing was only behind only Japan and the United States; Taiwan had become a major international tuna fishing power (Taiwan Yuye Jhi Yanjiu 1974).

In the early years of post-WWII Taiwan, most civilian tuna longliners from Kaohsiung and Donggang had to return and unload their tuna catch at homeports after their fishing activities in the waters of Southeast Asia. From start to finish, the fishing voyage took 2–3 months (Interview with Pan San'guang, Kaohsiung, 30/5/2002). However, along with the rapid expansion of Taiwan's longline fishing territories, the problem of logistical support caused by the geographical distance between Kaohsiung and fishing grounds in the southern hemisphere presented a substantial challenge to Taiwan's fishing enterprises. Vessel owners removed this obstacle by cooperating with foreign fishery agencies on the condition that the en-tire catch must be sold to them at negotiated prices. Cooperation with American and Japanese fish dealers greatly benefited Taiwanese distant water fishing companies because they were now assured of continuous financial aid and the port services which fish dealers provided at their overseas supply bases (Nian Taiwan Dicu Yuye Ciyeti Jingji Diaocha Baogao 1991).

Recent Fishing Activity in the Waters of Southeast Asia

As mentioned, the gradual depletion of tuna resources in the waters of Southeast Asia drove Taiwanese tuna longliner fleets to harvest tuna in more remote oceans. Did it mean that the traditional fishing grounds were abandoned by Taiwanese

---

[8] The first fishing company to use overseas supply bases and fish in the Indian and Atlantic Oceans was the China Fishing Company (CFC). However, the CFC's efforts were not proven successful.

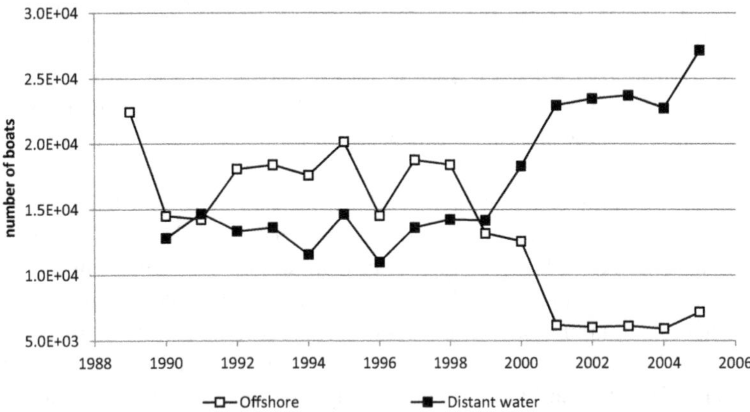

Fig. 7.10 Development of Donggang's tuna longlining industry, 1989–2005

fishers? Some Taiwanese longlinermen continue to work in the waters of Southeast Asia. In the wake of the development of Taiwan's tuna longline fishery, those who had been fishing in Southeast Asian waters now expanded their operations further into the Indian, South Pacific and Atlantic Oceans, and those who had been fishing in the coastal waters off Taiwan Bashi also began to operate in the Banda Sea or Celebes Sea (Taiwan Yuye Jhi Yanjiu 1974; Sianchih 1984).[9] In other words, those longliner owners whose vessels used to fish in the waters of Southeast Asia now undertake fishing activities in more remote oceans, with the vacuum in Southeast Asia filled by Taiwanese coastal fishers, especially those from Donggang.

In Kaohsiung, as the centre of Taiwan's fishing industry, distant water tuna longliners have fished in remote oceans like the Indian, Pacific and Atlantic Oceans. Encouraged by the success of distant water longliners, Kaohsiung's offshore tuna longliners left the waters of Southeast Asia earlier than their counterparts in Donggang Fishing Port. Presently, Kaohsiung's tuna longliner owners have totally given up their traditional fishing territories in the waters of Southeast Asia, to concentrate on the exploitation of tuna resources in other oceans. However, some of Donggang's tuna longlinermen still remain in the waters of Southeast Asia. This view is strongly supported by the fact that the annual volume of Kaohsiung's offshore tuna landing in 2005 has dropped to 26 t, but in the same year, Donggang's offshore longliners still harvested 7,168 t of tuna (See Fig. 7.7).

The thinning of tuna populations in nearby fishing territories has forced Donggang's longlinermen to harvest tuna resources in more remote waters. Figure 7.10 indicates that when the volume of Donggang's offshore tuna landing was shrinking, the volume of tuna catch landed at the Donggang Fishing Port by Donggang's distant water vessels was increasing at a steady pace. In 1999, the tuna catch harvested

---

[9] The Taiwanese fishing grounds continuously expanded in Southeast Asia, even while the distant water fishing vessels of Taiwan were operating all over the world since the 1960s.

by Donggang's distant water longliners exceeded that of the offshore longliners. This year was the milestone in the development of Donggang tuna fishery, when Donggang was no longer just an offshore fishing port.

There is final point to highlight. Donggang's longlinermen left their traditional fishing grounds in the waters of Southeast Asia much later than their counterparts in Kaohsiung Fishing Port. This was partly because traditionally Kaohsiung was the centre of Taiwan's distant water fisheries, so that its fishers felt confident in undertaking long voyages and conducting fishing activities in remote oceans, and partly because fishers in Donggang were much more conservative. They preferred to build small-sized offshore longliners, and fish in nearby waters. When they were forced to harvest tuna in more remote seas by the emerging problem of overexploitation, they still chose to fish in the waters of the southwest Pacific Ocean or the eastern part of the Indian Ocean, which, comparatively speaking, are not very far away from their traditional fishing territories in Southeast Asia. Only a select few in Donggang's fishing community have followed Kaohsiung fishermen's example and explore new fishing territories in the waters of the Atlantic Ocean or the Mediterranean Sea. Many longliner owners at Donggang fishing port stated in interviews that they preferred to fish in fishing grounds abandoned by their counterparts in Kaohsiung. The biggest benefit of this conservative approach is that, although fishing in depleted waters, they do not need to take the risk of poor catches resulting from unfamiliarity with more remote non-traditional fishing territories (Interview with Chen Shengli, Kaohsiung, 18/4/ 2002; Hong Fucai, Kaohsiung, 19/4/2002).

Conclusion

The exploitation of tuna resources in the waters of Southeast Asia had a significant influence on the economy and society of pre-WWII Taiwan. Numerous fishing migrants from Japan and other parts of Taiwan arrived at the major fishing ports of the island where they established fishing companies and the industrial tuna longline fishery. Modern fishing ancillary industries emerged alongside the development of tuna fishery, such as ice-makers, refrigeration storages and tuna canneries. The work of the fishing migrants and the rise of the tuna ancillary industries enriched the cultural, social and economic life of Taiwan's fishing communities, while the birth of the tuna trade between Taiwan and Japan laid a solid foundation for fisheries cooperation between the two nations in the post-WWII years.

The depletion of tuna resources in Southeast Asian waters in the 1970s posed a major threat to the sustainability of Taiwan's offshore tuna fishery. To cope with this problem, state-of-the-art fishing facilities were installed onboard tuna longliners soon after they appeared on the market. The increased fishing efficiency enabled fishers to harvest tuna more efficiently. Nevertheless, from a long term point of view, it worsened the problem of tuna depletion. This deadlock demonstrates how marine ecological changes and human fishing activities affected each other. The interaction between the improvement of fishing equipment and the decline of the marine ecosystem in Southeast Asian waters was a 'vicious circle' from which fishers could not extricate themselves unless they abandoned their traditional fishing territories and fished in other oceans.

The depletion of tuna resources in Southeast Asian waters in the 1970s did not devastate Taiwan's tuna fishery. Instead, it stimulated the industry to explore tuna resources in the waters of remote oceans. To cope with the ecological changes in the marine environment of Southeast Asian waters, the fishing authorities in Taiwan provided loans and encouraged fishing companies to build large distant-water longliners. To solve the problem of logistical support and open up overseas tuna markets, Taiwanese fishing enterprises cooperated with American and Japanese fish dealers. By the mid-1970s, the annual volume of Taiwan's distant water tuna landing was only behind those of Japan and the United States. Taiwan's tuna fishery had developed into a global scale fishing industry.

References

Butcher JG (2004) The closing of the frontier: a history of the marine fisheries of Southeast Asia, c.1850–2000. ISEAS Publications, Singapore

Chen HT (2009) Taiwanese distant-water fisheries in Southeast Asia, 1936–1977. International Maritime Economic History Association, St. John's, Newfoundland

Jicyuan Y (n.d.) Taiwan de Weidiao Yuye (Taiwan's Tuna longlining fishery). Zihyou Jhongguo Jhih Gongye (Industries of liberal China) 21(4):20

Kenichi G (2001) Kindai Nihon to Tōnan Ajia (Modern Japan and Southeast Asia). Iwanami Shoten, Tokyo

Nian Taiwan Dicu Yuye Ciyeti Jingji Diaocha Baogao (1991) Financial survey of Taiwan's fishery enterprises of 1989. Kaohsiung City Government, Kaohsiung

Nongye Yaolan Dishiji Yuyepian (1962) A brief introduction to the agricultural industries, vol 10: Fisheries. Department of Agriculture and Forestry, Taipei

Sen J (1977) Shuichan Jingying Sue (The management of fisheries). Sishi Foundation, Taipei

Shinhat L (1936) Taiwan Hattatsu Shi (The history of the development of Taiwan). Minsyū Kōronsya, Taihoku

Sianchih H (1984) Economic development and the structural adjustment of the production of Taiwan's fisheries. The Joint Annual Meeting of ROC Agricultural Groups. Unpublished, Taiwan

Tai'an S (1973) Taiwan Yuanyang Weidiao Yuye (The distant water longline fishing of Taiwan). Joint Commission on Rural Reconstruction, Taipei

Taiwan De Nonglin Jianshe (1950) The construction of the agriculture and forestry industries in Taiwan. Provincial Government of Taiwan, Nantou

Taiwan no Gyogyōgenkyō (1978) The present situation of Taiwan's fishing industry. The Interchange Association (Japan), Tokyo

Taiwan Suisan Yōran (1927) A brief guide to the fishing industry of Taiwan. Shosankyoku, Taihoku

Taiwan Suisan Yōran (1928) A brief guide to the fishing industry of Taiwan. Shosankyoku, Taihoku

Taiwan Suisan Yōran (1930) A brief guide to the fishing industry of Taiwan. Shosankyoku, Taihoku

Taiwan Suisan Yōran (1940) A brief guide to the fishing industry of Taiwan. Shosankyoku, Taihoku

Taiwan Yuye Jhi Yanjiu (1974) A study on the fishing industry of Taiwan. Bank of Taiwan, Taipei

Takao Shisei Jū Shunen Ryakushi (1934) A brief history of Takao City, 10-year anniversary. Takao City Government, Takao

Yoshinobu K (1938) Taiwan no Suisangyō (The fishing industry of Taiwan). In: Iichiro T (ed) Taiwan Keizai Sōsho, vol 6 (Series on the economy of Taiwan). Taiwan Keizai Kenkyukai, Taihoku, pp 105–154
Yuye Fajhan (1971) The development of the fishing industry. Provincial Government of Taiwan, Nantou

Chapter 8
History of Industrial Tuna Fishing in the Pacific Islands

Kate Barclay

Abstract The island countries and territories of the Pacific Ocean are relatively sparsely populated so there has historically been less fishing pressure on marine animal populations than in many other parts of the world. Industrial tuna fishing around Pacific Island countries began in the first half of the twentieth century, re-emerged after World War II in the 1950s and developed slowly until the 1980s when new fishing practices and new entrants increased catches steeply and steadily in a curve that continues to the present day. It has been estimated that industrial tuna catches are about ten times the volume and over seven times the value of all other fisheries in the Island Pacific combined—both commercial and artisanal. Furthermore, other fisheries that have been tried commercially in the region have not been resilient to industrial scale fishing pressure. Tuna fisheries may be the only potentially sustainable industrial wild-catch fisheries for the Island Pacific. Thus far fishing does not seem to have harmed the capacity of skipjack and albacore to maintain their populations, but it is having a deleterious effect on the biomass of yellowfin and bigeye. Industrial tuna fishing also incidentally kills other animals, but as yet there is not enough data collected to accurately gauge the ecosystem impacts of industrial tuna fisheries. Various attempts have been made to manage industrial tuna fisheries in the region. The main body responsible is the Western and Central Pacific Fisheries Commission (WCPFC), established in the 2004. Neither the WCPFC nor other bodies have thus far managed to reign in the overfishing of yellowfin and bigeye.

Keywords Industrial tuna fishing · Pacific Islands tuna history · WCPO · WCPFC · Fisheries management history

---

This chapter is based on existing research about tuna fisheries in the Pacific Islands region. The first part of the paper is based on various histories of tuna fishing in the Pacific and of national fleets within the fishery, and statistical data on changes to the fishery from the 1950s collated by stock assessing scientists at the Secretariat of the Pacific Community in Noumea. The second part of the paper on impacts to marine animal populations is based on scientific papers about fishing effects on stocks in academic publications and in the papers of the Western and Central

---

K. Barclay (✉)
Faculty of Arts and Social Sciences, University of Technology Sydney, PO Box 123, 2007 Broadway, NSW, Australia
e-mail: Kate.Barclay@uts.edu.au

J. Christensen, M. Tull (eds.), *Historical Perspectives of Fisheries Exploitation in the Indo-Pacific,* MARE Publication Series 12, DOI 10.1007/978-94-017-8727-7_8,

Pacific Fisheries Commission. Joe Hamby and Robert Gillett provided helpful information for the development of this chapter. Thanks to Antony Lewis and Thom van Dooren for constructive comments on an earlier version.

The island countries and territories of the Pacific Ocean are relatively sparsely populated so there has historically been less fishing pressure on marine animal populations than in many other parts of the world. Fishing effort has been steadily growing in recent decades, however, as overcapacity and overfishing in other places has sent fleets looking for new fishing grounds. Artisanal fishing has long been an important source of food and livelihoods for Pacific Islanders, but the focus of this chapter is industrial fisheries.[1] It has been estimated that industrial tuna catches are about ten times the volume and over seven times the value of all other fisheries combined in the Island Pacific, including prawn trawling, inshore commercial fisheries, and inshore subsistence fisheries (Gillett and Lightfoot 2002). Furthermore, while various other kinds of fisheries have been tried in a large-scale commercial manner (spearfishing, lobster fishing, live fish collection, giant clam fishing, bottom fishing) these stocks have not been resilient to fishing pressure. At this stage tuna fisheries are the only industrial wild-catch fisheries that may be sustainable in the Pacific Islands (Gillett 2007).

Industrial tuna fishing around Pacific Island countries began in the first half of the twentieth century, and re-emerged after World War II in the 1950s, developing slowly until the 1980s. Then tuna fisheries expanded greatly with new fishing practices and new entrants increasing catches steeply and steadily to the present day. The Japanese fleet was the most active in the region from the 1950s to the 1970s, and used the pole-and-line and longline methods to target fish mostly for the international cannery market. In the 1970s the Japanese fleet started also longlining for sashimi tunas. Taiwanese and Korean longliners also followed this pattern. In the 1980s the new, much more efficient method of purse seining came to be widespread in the fishery, and US purse seiners started operating in significant numbers in the Western Pacific region. Since the 1990s fleets from the Philippines, China, Europe and Pacific Island countries themselves have also come to be significant players, both in longlining and purse seining. The advent of purse seining is largely responsible for steadily rising tuna catches since the 1980s, although innovations in longlining have also led to a smaller increase.

The main tuna species targeted in the Western and Central Pacific Ocean (WCPO) are skipjack (*Katsuwonus pelamis*), yellowfin tuna (*Thunnus albacares*), bigeye tuna (*T. obesus*) and South Pacific albacore tuna (*T. alalunga*). Fishing does not seem to have harmed the capacity of two of these—skipjack and albacore—to maintain their populations. Fishing is, however, having a deleterious effect on the biomass of the other two species—yellowfin and bigeye—especially in terms of very large fish within populations. Yellowfin and bigeye have been targeted by

---

[1] There is no clear way to distinguish industrial from artisanal fisheries, especially for smaller scale industrial fisheries, but a working definition of industrial fisheries in the Island Pacific includes criteria such as: vessels greater than 15 m in length; fishing in offshore areas; selling catches for export and/or processing in a factory; and having data on the fishery collected for resource management purposes (Gillett 2007).

sashimi longline fleets, but more damage has been done by the much higher volume purse seine fishery. Purse seiners in the region usually target skipjack, although they may also target yellowfin. One of the main problems is that in some styles of purse seine fishing utilizing fish aggregating devices juvenile yellowfin and bigeye are mixed up in schools of skipjack. Fisheries regulation to date has not managed to significantly limit yellowfin and bigeye catch rates.

Industrial tuna fishing also incidentally kills other animals. Longlining in particular may catch seabirds, turtles and sharks. Because fisheries management has in the past been based on science about the target species in isolation from the rest of the ecosystem, the longitudinal data collected only covers the target species. With the move to ecosystem-based fisheries management the Western and Central Pacific Fisheries Commission (WCPFC) has commenced a more systematic collection of data about the mortality of other species in tuna fisheries, and techniques to minimize this mortality.

History of Industrial Tuna Fishing in the Island Pacific

In 1899 the *Albatross* surveyed the potential for industrial tuna fishing for US fleets around countries from French Polynesia in the southeast Pacific over to Guam in the northwest (Gillett 2007). There was no commercial tuna fishing, however, until after the Treaty of Versailles, when Japan gained control of many German colonial territories in the Pacific, including what is now Palau, the Federated States of Micronesia (FSM), Marshall Islands, and the Northern Mariana Islands. Japan invested substantial effort in developing industries in these islands, in line with its strategy to increase its food production base through colonial territories (Peattie 1984; Gillett 2007). The Japanese government offered subsidies for tuna fishing vessels, fishing gear and processing equipment (Fujinami 1987).

Three commercial fishing and processing operations were established in Palau in the 1920s. Activities accelerated in the 1930s with 116 pole-and-line vessels based in Japanese territories in the Island Pacific. Longline vessels based in southern Japan also fished the area. Most of the catch was processed on the islands into a dried product *katsuobushi*, and there were also two or more canneries processing fish for shipping back to Japan (Gillett 2007).

Japanese fishing in the Island Pacific ceased for some years after defeat in World War II, then recommenced in 1952. As with the pre-war fishing expansion, the Japanese government supported the post-war development of distant water fishing fleets, including large-scale pole-and-line and longline tuna vessels, to boost Japan's food production capacity (Fujinami 1987). Japanese government post-war support for fishing companies operating in the Pacific has included buyback schemes to assist with fleet restructuring, price support schemes, and low interest loans (Bergin and Haward 1996; Barclay and Koh 2008).

Although Japanese fishing vessels returned to Micronesian waters in the early 1950s, the USA had taken over control of most of the former Japanese colonial territories and restricted Japanese economic activity onshore in these places until the

Fig. 8.1 Map of exclusive economic zones in the Western and Central Pacific. (Courtesy ABARES)

mid 1970s (Gillett 2007). The Japanese fleet thus established bases in other parts of the Pacific. From the early 1950s through to the early 1960s Japanese longline bases were established in Fiji, American Samoa, Vanuatu, New Caledonia and French Polynesia. Developments in longline vessel technology improved such that even Japan based longliners increased their range throughout the Pacific to 40° south by the early 1960s (Matsuda 1987). Japanese longline vessels mostly caught albacore, which was exported to canneries in Hawai'i and the USA mainland (Gillett 2007; Fig. 8.2).

The USA government had commandeered California based pole-and-line vessels for service in the Pacific during WWII, and the several hundred fishermen serving on those vessels became aware of the potential of Western Pacific tuna fisheries (Felando 1987). Government and private fisheries surveys were carried out in the 1930s, 1940s and 1950s, but generally tended to concentrate on the Eastern Pacific Ocean (EPO, east of the black line in Fig. 8.1). In 1953 Van Camp Seafood Company bought a tuna cannery in Pago Pago in American Samoa. The cannery had been built to process catch from a Fiji based fishing company in the late 1940s but the fishing venture had failed. Van Camp also established a pole-and-line fishing base with 8–15 vessels and a freezing facility in Palau in 1964. Starkist joined Van Camp in Pago Pago, establishing a cannery in 1963 (Felando 1987; Gillett 2007).

Japanese pole-and-line vessels used live bait, which meant they could not operate far from a shore base, so this method was limited to waters close to Japan until the early 1960s. Then technological improvements allowed them to roam further

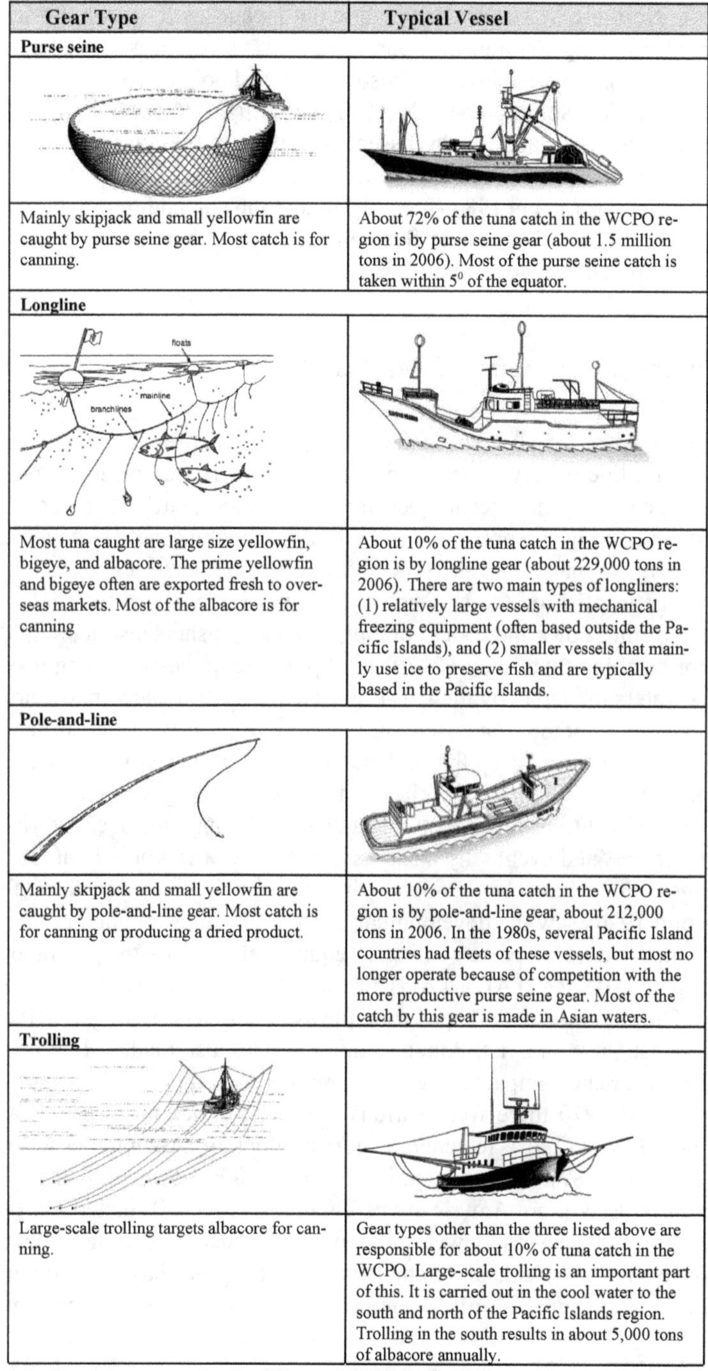

Gear Type	Typical Vessel
Purse seine	
Mainly skipjack and small yellowfin are caught by purse seine gear. Most catch is for canning.	About 72% of the tuna catch in the WCPO region is by purse seine gear (about 1.5 million tons in 2006). Most of the purse seine catch is taken within 5^0 of the equator.
Longline	
Most tuna caught are large size yellowfin, bigeye, and albacore. The prime yellowfin and bigeye often are exported fresh to overseas markets. Most of the albacore is for canning	About 10% of the tuna catch in the WCPO region is by longline gear (about 229,000 tons in 2006). There are two main types of longliners: (1) relatively large vessels with mechanical freezing equipment (often based outside the Pacific Islands), and (2) smaller vessels that mainly use ice to preserve fish and are typically based in the Pacific Islands.
Pole-and-line	
Mainly skipjack and small yellowfin are caught by pole-and-line gear. Most catch is for canning or producing a dried product.	About 10% of the tuna catch in the WCPO region is by pole-and-line gear, about 212,000 tons in 2006. In the 1980s, several Pacific Island countries had fleets of these vessels, but most no longer operate because of competition with the more productive purse seine gear. Most of the catch by this gear is made in Asian waters.
Trolling	
Large-scale trolling targets albacore for canning.	Gear types other than the three listed above are responsible for about 10% of tuna catch in the WCPO. Large-scale trolling is an important part of this. It is carried out in the cool water to the south and north of the Pacific Islands region. Trolling in the south results in about 5,000 tons of albacore annually.

Fig. 8.2 Main tuna fishing gear types. (Courtesy Robert Gillett)

afield to the Northern Marianas and Palau in the Japanese off-season. By the 1970s they were also fishing south of the equator in a sweep southeast of Japan as far as Fiji (Gillett 2007). Japanese investors also established pole-and-line bases in Papua New Guinea (1970), Solomon Islands (1971) and Fiji (1976), with around 300 vessels (both distant water and locally based) operating in the region till the peak of this fishery in the 1980s. The highest annual pole-and-line catch for the region was 380,000 metric tons (mt) in 1984 (Langley et al. 2006). In addition to Japanese and Japanese-established pole-and-line fleets in the region, there have been pole-and-line fleets in French Polynesia and Hawai'i.

Development of the Purse Seine Fishery

Expanding Japanese tuna catches in the 1950s put severe pressure on the California based pole-and-line fishery, so the California fleet reoriented to catching tuna with purse seine gear using new techniques and technologies, such as the power block to mechanize hauling the net, and synthetic fibres making nets stronger and less labour intensive to maintain. Like pole-and-line fishing purse seining is a 'surface' fishery—targeting fish on the surface of the water (as opposed to longlining that targets fish further down the water column). Japanese fishers also took up the new method for catching tuna, with 60–70 small purse seine vessels being used in the temperate waters off Japan by the late 1960s (Gillett 2007). New innovations were required, however, before purse seining could be used in the tropical Pacific, because the clear water and deep thermocline meant tuna schools were faster-moving and deeper-diving (Gillett 2007). The Japanese and USA governments sponsored experimental expeditions to develop methods for catching tuna in equatorial waters with purse seines, and eventually a successful method was worked out setting nets around logs (some tunas school underneath things floating on the surface of the water). The mid 1970s saw the development in the Philippines of another technology that facilitated industrial purse seining in equatorial waters—the *payaw* or *payao* fish aggregating device (FAD). FADs for tuna fishing in the Pacific are floating pontoons. They operate on the same principle as the log sets, aggregating the fish so they can be caught with a net. Pole-and-line fisheries also made use of this invention, but it had greatest impact in facilitating purse seining.

From around 1970 there were various efforts to entice more American purse seiners into the WCPO so as to generate more product for the canneries in American Samoa (Felando 1987).[2] It wasn't until the early 1980s, however, that a combination of factors drew more American vessels into the WCPO, including: (1) some American purse seine vessels were demonstrating success fishing around Papua New Guinea and New Zealand; (2) the American fleet was having difficulties getting access to fishing grounds off Mexico and Costa Rica; (3) a new licensing agree-

---

[2] Other quasi-government support for US tuna fisheries in the Pacific included the Reconstruction Finance Corporation (in the late 1940s), the Pacific Oceanic Fisheries Initiated (1949–1959) and the Pacific Tuna Development Foundation from 1974 (Gillett 2007).

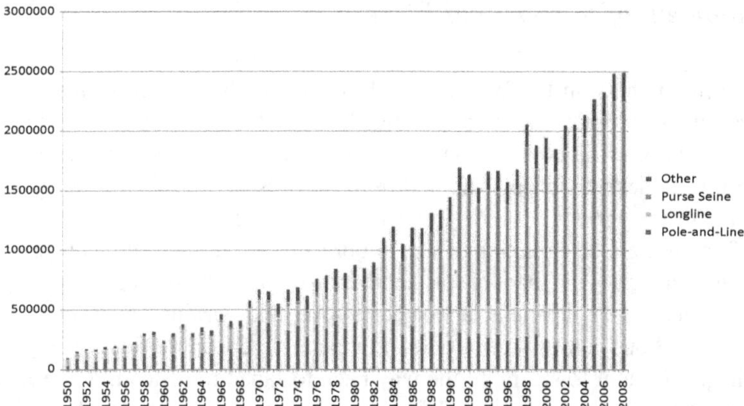

Fig. 8.3 Tuna catches (mt) in the WCPF-CA by Purse Seine, Longline, Pole-and-line and Other Gear Types, 1950–2008

ment was concluded between the American Tunaboat Association with US trust territories in the Pacific; (4) a strong El Niño event in 1982–1983 reduced fish availability in the EPO; and (5) campaigns against dolphin mortality in tuna fishing were encouraging fleets to find new grounds where dolphins do not school with tuna as they do in some parts of the EPO (Felando 1987).

The 1980s also saw the Japanese purse seine fleet grow in the WCPO. Many companies re-oriented from the pole-and-line to the purse seine method. New entrants also came at various stages from South Korea (hereafter referred to as Korea), Taiwan, China, New Zealand, Philippines, Indonesia, and most recently vessels based in various Pacific Island countries. The rise of the purse seine method marked a significant change in the nature of industrial fishing impacts on marine animal populations in the WCPO, with the steep and steady increase in catches from the 1980s attributable to expansion in the purse seine fishery (see Fig. 8.3).

An early attempt to limit purse seining in the region was made with the *Palau Arrangement* for the Management of the Purse Seine Fishery in the Western and Central Pacific in 1995. This agreement was made by a group of governments with the richest skipjack fishing grounds who since 1982 have worked cooperatively as the Parties to the Nauru Agreement Concerning Cooperation in the Management of Fisheries of Common Interest (PNA). Member countries are the Federated States of Micronesia, Kiribati, Marshall Islands, Nauru, Palau, Papua New Guinea, Solomon Islands and Tuvalu. The Palau Arrangement limited the number of purse seine vessels operating in their combined EEZs to 205. The input control measure used for purse seining in the PNA zone was changed in the mid 2000s from the vessel number cap to a Vessel Day Scheme (VDS), to enable flexibility through trading of fishing days across countries within the group and across vessels. As of 2010 around 220 large scale purse seiners operate in the WCPO.

Development of the Sashimi Fishery

The longline fishery in the WCPO mostly targeted albacore for cannery markets until various economic issues caused big shifts in the Japanese fleet in the 1970s (Matsuda 1987). During the 1960s the Japanese economy had consolidated its climb back from the devastation of war, and by the 1970s Japan was a wealthy society. The value of the yen rose, which affected profits for albacore exports to USA canneries. Increasing US regulation of fish imports due to public health concerns about the levels of mercury in albacore was another difficulty. The strengthening Japanese economy meant that Japanese labour costs rose, and various government restrictions on operations contributed to costs. For example, Japanese fleets were required by their government to steam all the way back to Japan to offload. Some Japanese investors went offshore, mostly to Taiwan and Korea, where they had business connections established during the Japanese empire. Costs for fleets also rose due to the oil price shocks in the 1970s, and the oil shocks caused recession in Japan, which also affected the Japanese fleet. Then there was the rollout of 200 nautical mile Exclusive Economic Zones (EEZs) around the coastlines of countries, meaning distant water fleets now had to pay for access to fishing grounds that were previously considered high seas and thus free access. The new fleets from Taiwan and Korea increased competition in longline fisheries, and also increased fish supply, which had a downward effect on prices. Finally, the growing affluence of Japanese consumers in the 1960s, along with supply increases related to Japanese export and import trends, gave rise to an increasing market for high value sashimi tuna (Bergin and Haward 1996, Issenberg 2007).

Japan-based longline fishing companies reacted to this business environment by reorienting their operations away from low value albacore for USA canneries, to high value sashimi tunas (bigeye and yellowfin) to sell in Japan. The Japanese longline fleet adapted new technologies in ultra low temperature (ULT, less than $-60\,°C$) freezing which enabled the preservation of tuna meat in a form suitable for sashimi markets (less cold freezing temperatures allow the flesh to go brown). With ULT vessels could stay fishing out at sea rather than having to steam frequently to the nearest port to airfreight chilled fish. This led to the building of larger longline vessels (>250 GRT) that could go out for months at a time and operate over large areas of the globe.

The Japanese longline fleet's shift to sashimi fishing did not mean that fishing pressure on albacore eased, because Taiwanese and Korean albacore fleets stepped in where Japanese fleets left the field. Taiwan had started distant water longline fishing for albacore in the Pacific in the 1960s, and by the mid 1970s the Taiwanese longline fleet was as significant as Japan's. Korea soon followed Taiwan. Taiwanese and Korean longline companies took over longline bases in the Pacific from the Japanese fleet. Taiwan and Korea also developed ULT vessels.

The development of ULT vessels did not mean the demise of smaller vessels using ice or other chilling methods. A portion of Taiwan's fleet has remained without ULT, and almost all Pacific Islands based longline fleets are non-ULT, especially those operating in tropical waters. There has also been a significant medium-scale non-ULT longline fleet from China operating in the EEZs of Pacific Islands countries since the late 1980s. This started in 1998 with 7 vessels, growing quickly to 457 vessels in 1994. Many of these vessels supplied a company called Ting Hong, which proved unsustainable and collapsed after a few years, causing the Chinese longline fleet to drop right back to around 120, where it has since remained (McCoy and Gillett 2005). Chinese longline vessels have been based in FSM and Marshall Islands (targeting sashimi) and Fiji (targeting albacore).

The development of sashimi fishing marked a change in the nature of impacts from industrial longline tuna fisheries on marine animal populations in the Pacific Ocean. In order to target bigeye Japanese vessels moved into different geographical locations (sashimi longlining has been concentrated in the tropical WCPO while albacore catches are greater in subtropical areas) and also extended their fishing effort deeper into the water column.[3] To catch bigeye the hooks on longlines are hung deeper in the water (100–300 m below the surface) than for catching albacore (Morgan and Staples 2006). The total number of vessels involved in the longline fishery has stayed relatively stable between 4,000–5,000 since the mid 1970s (Langley et al. 2006). Expansions in fish landings from the longline fleet from the 1970s (see Fig. 8.3), therefore, were due to the introduction of larger vessels and the extension of fishing effort into new species (see Fig. 8.4), new locations and deeper in the water column.

Prior to 1980 yellowfin was the preferred target in the WCPO sashimi fishery, but since 1980 bigeye has been preferred. Bigeye tunas sold on the sashimi market, chilled or frozen, fetch the highest prices of any tropical tuna. Bigeye from the WCPO has an estimated landed value of US$600 million annually, so although longline fishing has only accounted for around 11 % of the total catch in recent years, the landed value of the longline catch rivals that of the much larger purse seine catch (Langley et al. 2006).[4]

From the mid-2000s fuel prices once again became a major factor affecting longline fishing. From 2004 to 2008 steady fuel price increases meant some longliners were no longer economical to run, and many Japanese and Taiwanese vessels were tied up in port for long periods. The Global Financial Crisis reduced fuel prices in 2008, but fuel prices have risen again. Whether this results in an overall decrease in fishing effort depends on the balance between production costs (including fuel) and fish prices. This balance, in turn, is affected by factors such as demand for sashimi tunas and the extent to which vessel technology development (usually subsidized) can improve fuel efficiency for longliners. As Japanese seafood cuisine has global-

---

[3] For information on where in the Pacific the different tuna species are caught, see publications and reports of the Oceanic Fisheries Program (OFP) of the Secretariat of the Pacific Community (SPC) (http://www.spc.int/oceanfish/).

[4] It is worth noting that the highest value sashimi tunas are the bluefins caught in colder waters, fish from the tropical Pacific sit at the lower end of the sashimi price scale.

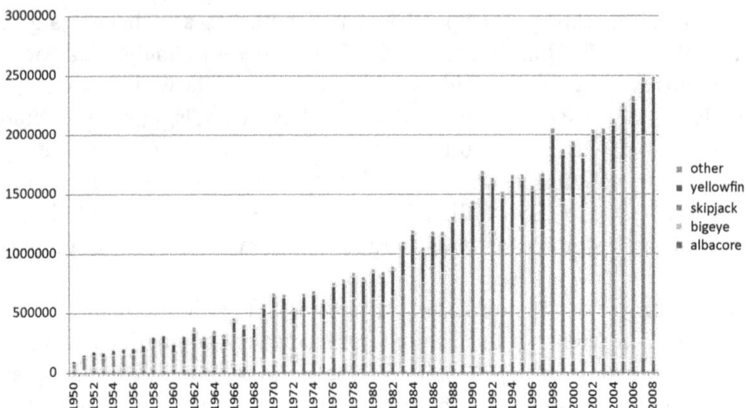

Fig. 8.4 Catches (mt) of yellowfin, skipjack, bigeye, albacore and other species in the WCPF-CA, 1950–2008

ized, demand growth in markets willing to pay high prices for sashimi, including in China, may sustain the industry even in the face of declining stocks and elevated fuel prices.

Development of Locally Based Fisheries

Exclusive Economic Zones of 200 nautical miles became a firm likelihood for the Pacific in the 1960s, and during the late 1970s they came into force. Both US and Japanese fleets resisted paying fishing access fees to Pacific Island countries for some time, although both capitulated in the 1980s (Schurman 1998). Fears that the USSR may gain a foothold in regional fisheries forced the USA government to negotiate access fees, resulting in the USA Multilateral Tuna Treaty in 1988 (Gillett 2007).

Many Japanese fishing companies responded to the advent of EEZs by establishing joint ventures with newly independent Pacific Island states. Support for Japanese tuna industries operating in the Pacific Islands was extended through the quasi-governmental Overseas Fisheries Cooperation Foundation (OFCF), among other avenues. OFCF soft loans could cover up to 70 % of the capital costs for joint ventures (Matsuda 1987). Around a dozen such joint ventures between Japanese fishing companies and Pacific Island governments were made from the end of the 1960s, the most significant being Solomon Taiyo in Solomon Islands and Pafco in Fiji, both of which established canneries that are still active today.[5]

---

[5] Because several import destinations, particularly the EU and USA, have protected domestic canning industries the demand is for frozen 'loins'—tuna meat prepared ready for canning—rather than cans. In recent years the Solomon Taiyo (now called SolTuna) and Pafco factories have exported loins more than cans.

Japanese interests in forming joint ventures coincided with another twist in Pacific Islands tuna fisheries in the 1970s; newly independent Pacific Islands states were interested in tuna as part of their economic development strategies. This was the era when former colonies, seeing the success of the OPEC countries in generating wealth from oil, hoped for a New International Economic Order, whereby wealth could be generated by raw materials (a reversal of the colonial economic order in which trade in primary commodities had benefitted the colonial powers) (Schurman 1998). The 1970s were also an era when government ownership of companies was in vogue, as it was believed that national ownership would guarantee that economic development benefits would be kept within the local economy rather than disappearing overseas. Eighteen state-owned tuna fishing companies were started in the Pacific Islands area in the 1970s, many as joint ventures with Japanese companies (Gillett 2007). During the 1980s and 1990s nearly all of the state-owned fishing ventures failed. None was profitable and some generated huge losses, although some also generated significant benefits for host economies in wealth distribution through employment and human resources development (trained and experienced fishing crews and technical staff have subsequently been used by private sector investors) (Barclay and Cartwright 2007).

Many of these government-owned companies engaged in medium-scale pole-and-line fishing since this method was relatively low-tech and labour-intensive, so it suited Pacific Island economic conditions and aspirations to have locally based and locally crewed vessels. This kind of fishing was carried out in several countries in the Western Pacific during the 1970s and 1980s, but since it required a combination of daily accessible live baitfish from reef and lagoon areas and the right kind of skipjack resources, it was most successful in Papua New Guinea and Solomon Islands, and seasonally in Fiji. Even there, however, competition with cheaper purse seine-caught fish caused problems and there was only one of these fleets still operating by 2000, in the Solomon Islands. The Solomon Taiyo fleet was subsidized, but even then it was only viable while the main buyers in the UK (such as Sainsbury's and Waitrose) were willing to pay a premium above the purse seine price for the positive social and environmental factors associated with the pole-and-line method. In 2000 changing retail conditions in the UK wiped out that premium, which made Solomon Taiyo's business unviable so the Japanese partner withdrew (Nakada 2005). By 2008 By 2008 Solomon Islands' fleet had dwindled to a couple of operable vessels. The increasing importance of Corporate Social Responsibility in food retail in Europe, however, means there is increased demand for environmentally 'friendly' tuna, so the pole-and-line method may yet be revived in Pacific Islands countries (Stone et al. 2009). Any developments in skipjack fisheries, however, need to be able to weather very volatile fish prices. In 1999 the world skipjack price dipped below US$ 400 per tonne, and stayed at record low levels till 2001, which meant some companies tied up their vessels because it was not worth fishing at that price. Prices recovered from 2004 then went the other way to record highs, sitting over US$ 1,500 per tonne for most of 2008 (FAO 2009).

In the 1990s Pacific Islands based small- and medium-scale (< 100 GRT) longlining took off. Pacific Island citizens have been significant players, and Japanese,

Taiwanese, and Chinese investors have been involved. There have been offshore subtropical albacore fisheries based in American Samoa, Samoa, Cook Islands, Fiji, French Polynesia, New Caledonia, Solomon Islands, Tonga and Vanuatu. There have been tropical offshore sashimi fisheries based in Palau, FSM, Marshall Islands, Guam, PNG, Solomon Islands, Cook Islands, and Fiji. These vessels have used ice or refrigerated sea water, rather than ULT capacity, so needed to return to port regularly to unload their catch for airfreight to market. By the mid 2000s dwindling catches of large yellowfin and bigeye suitable for sashimi markets, fuel price rises and difficulties with airfreight connections rendered some companies unviable and consolidated the rest into the albacore fishery (Barclay and Cartwright 2007). Pacific Island based companies caught 33 % of the total south Pacific albacore longline catch in 1998, and over 59 % in 2006 (Langley et al. 2006). Expanding domestic longline fleets remains a key aim for fisheries development in Pacific Island countries (Gillett 2008).

Pacific Island governments have often phrased their aspirations regarding the development of tuna fisheries and processing industries as 'domestication'. This overarching term means a range of things including: Pacific Islanders owning, managing and working in fisheries industries, turnover cycling through the domestic economy, and profits being reinvested locally. Purse seine fisheries take the largest portion of the regional tuna catch but purse seine vessels are high tech, large and expensive. PNA countries whose EEZs are used extensively by purse seine fleets have worked to domesticate purse seining. In 1994 the PNA signed the Federated States of Micronesia Arrangement for Regional Fisheries Access (FSMA). The FSMA gives multilateral fisheries access across the combined EEZs to purse seine fleets that meet the criteria for being 'locally based'. Not many purse seine vessels responded by basing themselves in Pacific Island countries, except in Papua New Guinea (PNG). In the mid 1990s the PNG government started using fisheries access to entice foreign fleets to invest in onshore processing (canneries and/or loining facilities). The first company to take up this offer was RD Tuna from the Philippines, with a cannery coming on line in 1997. Following the success of RD another Filipino company Frabelle and a US-Taiwanese venture South Seas Tuna Company established local operations in the mid 2000s. The PNG based purse seine fleet subsequently grew from two vessels before 1994 to around 40 vessels in 2006 (Lawson 2007). The Philippines is a major tuna player in the WCPO with its domestic catch and canning output one of the largest in the region, and with around 45 vessels currently based in PNG the Philippines has become a significant player in the Island Pacific. Growth in the PNG-based fleet accounts for most of the total increase in purse seine vessels operating in the region since the inception of the FSMA—from 147 in 1995—175 in 2006 (Langley et al. 2006), and over 200 from the late 2000s.

The PNA and its cooperative agreements the FSMA and Palau Arrangement are not the only instances of collective efforts for fisheries management and development by Pacific Island countries. The Secretariat for the Pacific Community (SPC) has been the umbrella organization for statistical data collection for fisheries management purposes since 1980, and has also long provided technical advice for development of small-scale commercial tuna fisheries. The Pacific Islands Forum Fisheries Agency (FFA) began in the late 1970s and has housed regional initiatives such as the US Multilateral Treaty, the Vessel Monitoring System for electronic surveillance of fishing vessels, a regional fishing vessel register, and newsletters on tuna trade and market issues.

Pacific Island development of fisheries and onshore facilities for canned tuna has been greatly affected by the international trade regime. The largest markets for canned tuna (the EU, Japan, the USA) have all had domestic canning industries, which are no longer competitive due to labour costs, but which have been protected by tariffs on processed fish imports. Some developing countries, especially those with past colonial relationships with importing countries, have had tariff exemptions. The relationship between the EU and former colonies in the Pacific has been particularly influential, contributing to the viability of processing facilities in PNG and Solomon Islands. These countries have higher production costs than competitors such as Thailand, so have survived due to tariff exemption. Pressure from the World Trade Organization to reduce tariffs and make preferential trade agreements WTO-compliant is causing changes in these relationships, possibly undermining the long-term viability of processing in Pacific Island countries (Campling et al. 2007). A collapse in regional processing may affect fishing practices, as fleets currently supplying those facilities may shift their fishing grounds.

In sum, the trends in Pacific Islands based tuna fisheries development started with a state-owned model in the 1970s, which led to disillusionment as most of these ventures failed, and the belief that maximizing access fees from distant water fleets was a more sound economic strategy. Then in the 1990s there was success in several countries with locally based longline fleets, and PNG successfully leveraged onshore development from distant water purse seine fleets, so these models of domestic development have been pursued by some countries, while access fees remain economically important for many of the PNA group (Gillett 2008). On the whole, however, Pacific Islands based fisheries have not had as much impact on marine animal populations as the distant water fleets, because distant water fleets have always taken much more fish.[6]

Trends Among Distant Water Fleets

The distant water fleets operating in the region have risen to prominence and declined or changed fishing practices for a range of reasons. The Japanese and USA fleets were the first big fleets and since the 1970s both have had relatively high production costs, in part due to home country regulations and economic conditions affecting their costs. For example, the fisheries sector has been an unpopular em-

---

[6] For information on catches by national fleets see Lawson (2007).

ployment option since the 1980s in Japan, so has shrunk dramatically. The Japanese and USA fleets have thus been undercut by fleets from Taiwan, Korea and China, which did not significantly innovate in fishing practice but have had lower production costs and 'aggressive' fishing practices (Gillett 2007). There are signs, however, that the Taiwanese and Korean fleets may be declining in the region. Taiwan dominated the albacore fishery from the mid 1980s, but by the mid 2000s Pacific Island fleets and others were taking a far larger proportion of the catch than Taiwan. Like the Japanese fleet before them, Taiwanese longliners have shifted focus to sashimi tunas, but this has not halted their decline. Since trade sanctions were imposed on Taiwanese sashimi imports to Japan in 2005 the Taiwanese government has been regulating its industry more strenuously. Skyrocketing fuel prices also affected longline fleets from 2004. Taiwanese and Korean sashimi catches dropped in the mid 2000s due to drops in active longline vessel numbers; from 133 (Taiwan) and 184 (Korea) vessels in 2002, to 117 (Taiwan) and 130 (Korea) in 2006 (Langley et al. 2006). According to an industry source, longlining is now less economically viable for Taiwanese companies than it was in the past (Wang 2009).

Impacts of Industrial Tuna Fishing on Marine Animal Populations in the Island Pacific

Fisheries have taken out more than 50 million tons of tuna and other predators from the Pacific Ocean since 1950. Some scientists calculate that this has meant there has been a catastrophic reduction in population biomass and collapses in oceanic food chains (Pauly et al. 1998, Myers and Worm 2003). Other scientists assert that those assessments are overly pessimistic, finding that some stocks are greater than 74% of their unexploited potential, and others are 36–49% of their unexploited potential (Sibert et al. 2006).[7] Time series data from 1950 indicates indeed that some stocks have increased, probably because decreasing stocks in other species are reducing competition. It may be that tuna fisheries' removal of slower growing predators from food chains are also causing population bulges in other fast growing predators for which statistics are not collected, such as mahi-mahi (*Coryphaena hippurus*) and wahoo (*Acanthocybium solandri*) (Sibert et al. 2006).

In addition to mahi-mahi and wahoo, other animals such as marlins, swordfish, sharks, turtles and sea birds are also killed by tuna fishing gear, especially longlines. For example, seabirds often try to take the bait from hooks on longlines as they float near the surface before they sink to the intended depth for catching tuna. Non-target fish and turtles may also be caught by longline hooks. Turtles may become entangled in fish aggregating devices used for purse seine fishing. Because fisheries science was conventionally based on single species, no longitudinal statistical information has been collected on mortality in non-target species caused by tuna fisher-

---

[7] For a refutation of Myers and Worm's (2003) analysis of Pacific tuna populations see Hampton et al. (2005).

ies in the WCPO. As ecosystem based fisheries management has become the norm, however, the WCPFC has for several years had a Working Party on Ecosystem and Bycatch Mitigation. This group is building knowledge about non-target species impacts, developing the basis for deciding which species need to be monitored and how best to collect and analyse data about these species, and investigating the effectiveness of measures such as circular hooks to reduce turtle catches, and 'tori-line' streamers to deter seabirds from attacking longline bait.[8]

Of the four types of tuna targeted by industrial fishing in the Island Pacific, skipjack has not suffered greatly. Skipjack are fast-growing and resilient to fishing. Fishing companies in the southern WCPO have since 2012 claimed that albacore is now showing signs of overfishing. Scientists agree, however, that there are stock conservation problems for bigeye and yellowfin in the WCPO, and fisheries managers have been attempting to decrease catches of these species since 2001.[9]

Impacts on Bigeye and Yellowfin

Yellowfin and bigeye are large fish. By 1970 the biomass of tunas larger than 175 cm from the tip of the snout to the centre of the fork in the tail had decreased by 40 % in the Pacific, and currently they are estimated to make up approximately 1 % of tuna populations, where they had made up around 5 % (Sibert et al. 2006).

The impact of fishing on the total yellowfin biomass in the WCPO was not significant before 1980, but has increased as catches have increased since then. Declines in stocks were first observed in the late 1990s (Hampton and Fournier 2001). Fishing is estimated to have reduced the yellowfin biomass by about 40 %, higher in the tropical zone (60 %) and lower in the subtropical zone. Most damage has been done by Indonesian and Philippines domestic fisheries and the purse seine fishery in the equatorial region, while the subtropical longline fishery does not appear to have affected stocks (Hampton and Fournier 2001, Langley et al. 2006). Yellowfin tuna is captured at different ages by different gear types—when juvenile it may be captured by purse seiners, when older it is caught by longliners and (under certain conditions) purse seiners. Yellowfin is targeted by longliners supplying sashimi markets. Purse seiners target yellowfin for cannery markets, and they also catch juvenile yellowfin incidentally when targeting skipjack.

Fishing impact on bigeye biomass has increased steadily from the mid 1970s with a sharp increase in the mid 1990s. The impact is highest in the equatorial Pacific, having reduced biomass by up to 80 % (which has been somewhat ameliorated by a high level of recruitment in stocks in the mid 2000s) (Langley et al. 2006).

---

[8] Papers on establishing data about and measures to limit bycatch are available in the Western and Central Pacific Fisheries Commission Scientific Committee meeting files (http://www.wcpfc.int/meetings/2).

[9] Debates about these efforts are in the meeting papers of the Standing Committee on Tuna and Billfish (http://spc.int/OceanFish/Html/SCTB/index.htm), and from 2005 meetings of the Western and Central Pacific Fisheries Commission (http://www.wcpfc.int/meetings/all).

Bigeye is targeted by sashimi longliners, and is also caught as juveniles by purse seiners—unintentionally because bigeye has no particular value as a canning fish.

The WCPO purse seine fishery is largely a skipjack fishery, with skipjack making up 70–85% of the catch, yellowfin 15–30%, bigeye less than 5%, and small amounts of albacore (only in the northern temperate Pacific) (Langley et al. 2006). Purse seiners have caught more or less bigeye and yellowfin depending on the fishing techniques employed. The main techniques are 'unassociated sets' (free-swimming schools of tuna) and 'associated sets'. Associated sets are when the net is set around something floating on the surface of the water that has aggregated tuna underneath: FADs fixed to the bottom of the sea; FADs drifting freely in the ocean; and drifting logs. Associated sets tend to have more juvenile yellowfin and bigeye mixed in with skipjack than do free swimming schools, which tend to be 'pure'—composed of one species.

A variety of factors influence the choice of fishing technique. Fixed FADs can only work in water shallow enough to tie the FAD to the ocean floor but deep enough to attract schools of pelagic fish. This method has been used around PNG and Solomon Islands by locally based fleets supplying canneries in those countries. The percentage of bigeye in catches is highest when drifting FADs are used, especially in eastern areas of the WCPO. Since 2000 the use of drifting FADs has declined and purse seine catches of bigeye have correspondingly declined. The choice to use drifting FADs is based on various factors. Some national fleets use them more than others. Drifting FADs have been used extensively by the USA fleet in the southern and eastern parts of the WCPO, close to the canneries in American Samoa those fleets supply. The Korean purse seine fleet, on the other hand has favoured fishing on free-swimming schools. Oceanographic effects also influence the use of drifting FADs. They are used more during El Niño events when the fish available to surface fisheries concentrate in the eastern part of the WCPO, as there are generally fewer free-swimming schools and logs available for fishing in the eastern than in the western part of the WCPO.[10]

Management of highly migratory tuna species must be done multilaterally so it is organized under intergovernmental Regional Fisheries Management Organizations (RFMOs). The WCPFC came into effect in 2004, after a decade of negotiations. The Parties to the Nauru Agreement (PNA) are the group of countries with the richest fishing grounds for purse seining, in which much of the damage to yellowfin and bigeye stocks is being done. The PNA first made an attempt to limit purse seine fishing effort through a cap on the numbers of distant water purse seiners operating in their combined zone in 1995, which has more recently been converted to the

---

[10] The majority of the information in this paragraph has been gleaned from Langley et. al. (2006). Notwithstanding the need to use more drifting FADs in the east, during El Niño there are more free-swimming schools available to surface fisheries in the east of the WCPO than there are during other parts of the oceanographic cycle, including free-swimming schools of very large yellowfin. Logs are washed down rivers from the larger land masses, so in the tropical Pacific they have drifted eastwards on the equatorial coun-tercurrent (ECC) from Southeast Asia and Papua New Guinea. They eventually sink, so are not usually available to use for fishing sets any further east than Kiribati. Antony Lewis, personal communication (email), 28 January 2010.

Vessel Day Scheme. As yet government measures, including restrictions on purse seine effort, have been unable to stem the overfishing of bigeye and yellowfin in the Western and Central Pacific Ocean. This mirrors the inability of other RFMOs to tackle overfishing, largely because member states either refuse to agree on effective measures or then fail to implement the agreed measures.

In light of continued overfishing, and because of the economic importance of tuna resources to PNA countries, the PNA group has pursued its own strategy for economic and environmental sustainability. In the 2008 3rd PNA Implementing Agreement it was announced that distant water vessels seeking to fish in PNA EEZs must sign an agreement not to fish on the high seas 'donut holes' between PNA EEZs and to accept 100% observer coverage and a ban on FAD fishing for the third quarter of each year (to relieve pressure on bigeye and yellowfin stocks). Since then the PNA has also entered into assessment of free-swimming skipjack for Marine Stewardship Council (MSC) eco-labelling certification.[11] There is great demand for MSC certified product by the large retailers in Europe and North America, and the PNA fishery includes close to half the world's skipjack stocks, so if success-ful this strategy could cause a major shift in global canned skipjack markets, and thereby in fishing practices. Improving purse seine fishing practices in the PNA zone, however, is not a total solution to the problem, because it does not address longline fishing, and because a great deal of damage to yellowfin and bigeye stocks is being done in the waters of Indonesia and the Philippines. The WCPFC and re-lated intergovernmental efforts remain the main avenues by which to fix those parts of the problem.

Conclusion

Tuna could be a vital renewable resource for the economies of the Island Pacific, if it is managed well. Most of the fisheries are still showing strong catches despite continual fishing increases since 1950, and accelerated effort from the 1980s, so it is imperative that effective resource management is implemented as soon as possible. It is important to learn more about the ecosystem effects of industrial tuna fisheries in the region and address any damage in bycatch species. Depletion in bigeye and yellowfin stocks is already known, and needs to be more effectively addressed than it has been to date. The Western and Central Pacific Fisheries Commission has been running now for several years so should be prepared to meet these challenges. The most dynamic actor on the scene at present is the PNA—a group of countries with the most productive EEZs in the WCPO. It is to be hoped that the PNA can leverage their position as key coastal states and take the lead in establishing more sustain-able fishing practices. The PNA, however, will be unable to fix all of the problems with bigeye and yellowfin stocks, as it has little leverage on longline fisheries or the

---

[11] The PNA group's collective purse seine fishing ground has been launched as 'Pacifical' (http://www.atuna.com/newsletter/pacifical.html).

crucially important fisheries in Indonesia and the Philippines. It will also be necessary for the membership of the WCPFC to move beyond their disparate individual interests and take decisive collective action.

References

Barclay K, Cartwright I (2007) Capturing wealth from tuna: case studies from the pacific, vol. Asia Pacific Press, Canberra

Barclay K, Koh S-H (2008) Neo-liberal reforms in Japan's tuna fisheries? A history of government-business relations in a food-producing sector. Japan Forum 20:139–170

Bergin A, Haward M (1996) Japan's tuna fishing industry: a setting sun or a new dawn? Nova Science Publishers, New York

Campling L, Havice E, Ram-Bidesi V, Grynberg R (2007) Pacific islands countries, the global tuna industry and the international trade regime—a guidebook. DEVFISH Project, Forum Fisheries Agency, Honiara

FAO (2009) The state of world fisheries and aquaculture, FAO fisheries and aquaculture department, food and agriculture organization of the United Nations, Rome

Felando A (1987) USA tuna fleet ventures in the Pacific islands. In: Doulman DJ (ed) Tuna issues and perspectives in the Pacific islands region. East-West Center, Hawai'i, pp 93–104

Fujinami N (1987) Development of Japan's tuna fisheries. In: Doulman DJ (ed) Tuna issues and perspectives in the Pacific islands region. East-West Center, Hawai'i, pp 57–70

Gillett R (2007) A short history of industrial fishing in the pacific islands, Asia-Pacific fishery commission, FAO regional office for Asia and the Pacific, Bangkok

Gillett R (2008) A study of tuna industry development aspirations of FFA member countries, Forum Fisheries Agency, Honiara

Gillett R, Lightfoot C (2002) Contribution of fisheries to the economies of pacific island countries, Pacific studies series. Asian Development Bank, Manila

Hampton J, Fournier D (2001) A spatially disaggregated, length based, age-structured population model of yellowfin tuna (Thunnus albacares) in the Western and Central Pacific ocean. Mar Freshw Res 52:937–963

Hampton J, Sibert J, Kleibert P, Maunder M, Harley S (2005) Decline of Pacific Tuna populations exaggerated? Nature 434 (28 April): E1–2

Issenberg S (2007) The sushi economy: globalization and the making of a modern delicacy, vol. Gotham Book, New York

Langley A, Williams P, Hampton. J (2006) The Western and Central Pacific tuna fishery: 2006 overview and status of stocks, Secretariat of the Pacific Community, Oceanic Fisheries Program

Lawson TA (ed) (2007) Tuna fishery yearbook. Western and Central Pacific Fisheries Commission, Pohnpei

Matsuda Y (1987) Postwar development and expansion of Japan's tuna industry. In: Doulman DJ (ed) Tuna issues and perspectives in the Pacific islands region. East-West Center, Hawai'i

McCoy MA, Gillett RD (2005) Tuna longlining by China in the pacific islands: a description and considerations for increasing benefits to FFA member countries, Forum Fisheries Agency, Honiara

Morgan G, Staples D (2006) The history of industrial marine fisheries in Southeast Asia, Asia-Pacific fishery commission, FAO Regional Office for Asia and the Pacific, Bangkok

Myers R, Worm B (2003) Rapid worldwide depletion of predator fish communities. Nature 423:280–283

Nakada M (2005) Personal communication with the author. Interview with Masao Nakada, former Operations Manager of Solomon Taiyo Ltd., July 2005, Honiara

Pauly D, Christensen V, Dalsgaard J, Froese R, Torres F (1998) Fishing Down Marine Food Webs. Science 279:860–863

Peattie MR (1984) The Nan'yo: Japan in the South Pacific, 1885–1945. In: Peattie RHMaMR (ed) The Japanese colonial empire, 1895–1945. Princeton University Press, Princeton, pp 172–210

Schurman RA (1998) Tuna dreams: resource nationalism and the Pacific islands' tuna industry. Dev Change 29:107–136

Sibert J, Hampton J, Kleiber P, Maunder M (2006) Biomass, size, and trophic status of top predators in the Pacific ocean. Science 314:1773–1776

Stone R, Toribau L, Tolvanen S (2009) Developing sustainable and equitable pole and line fisheries for skipjack, vol. Greenpeace International, Amsterdam

Wang W (2009) Personal communication with the author. Interview with Weichen Wang, third generation tuna fishing and trading manager. 18 November 2009, Sydney

Chapter 9
Southern Bluefin Tuna: A Contested History

Sid Adams

Abstract Southern Bluefin Tuna (SBT) is a threatened species of tuna. Harvested from the early 1950s the fishery provides an interesting case study of the interplay of technology and science. On the one hand, fishing effort has expanded on the resource. This has resulted in the significant reduction in the size of the stock. Indeed, by the early 1980s scientists were warning that the reduction in stock size had reached dangerously low levels. Paradoxically the stock's demise has occurred alongside a growing body of scientific research into the fishery. Indeed, the fishery remains one of the most researched fisheries in the world today. There has also been a regional fishing organization (RFO) created to achieve a more sustainable level of harvest between the fishing nations. By 2012 both initiatives have however not produced a significant improvement in the stock's biomass. Indeed, agreeing on a sustainable quota level has been at the centre of significant and abiding tensions between the parties. This chapter thus seeks to explain this conundrum. It will argue that the institutional setting of the fishing parties involved in the fishery is critical to understanding the tensions that have underpinned international management of the stock, the dispute over science and in explaining the precarious condition of the fishery today.

Keywords Southern Bluefin Tuna · Japanese tuna fishing · Australian tuna fishing · CCSBT history · IUU history

Southern Bluefin Tuna (SBT; *Thunnus maccoyii*) is a highly migratory stock that swim the waters of the Atlantic, Indian and western Pacific Oceans. Spawning in the region south of Java off Christmas Island, offspring migrate south, through the waters of Western Australia, before entering the Great Australian Bight. From here the fish either swim east across the Great Southern Ocean as far as New Zealand's territorial waters or west across the Indian Ocean, to Africa and the Atlantic Ocean. Figure 9.1 provides an overview of the stocks migratory pattern. Scientists have categorized the growth and maturity of the stock into two phases, a juvenile and adult stage; fish reach sexual maturity between 8 and 15 years, and live up to 40 years of age (Caton 1991; CCSBT 2011; Clean Seas 2011). Historically Australian fishermen have targeted juvenile fish in near-shore waters. As juveniles the fish swim at shallow depths and in

S. Adams (✉)
Asia Research Centre, Murdoch University, 90 South Street,
6150, Murdoch, WA, Australia
e-mail: s.adams@murdoch.edu.au

J. Christensen, M. Tull (eds.), *Historical Perspectives of Fisheries Exploitation in the Indo-Pacific*, MARE Publication Series 12, DOI 10.1007/978-94-017-8727-7_9,
© Springer Science+Business Media Dordrecht 2014

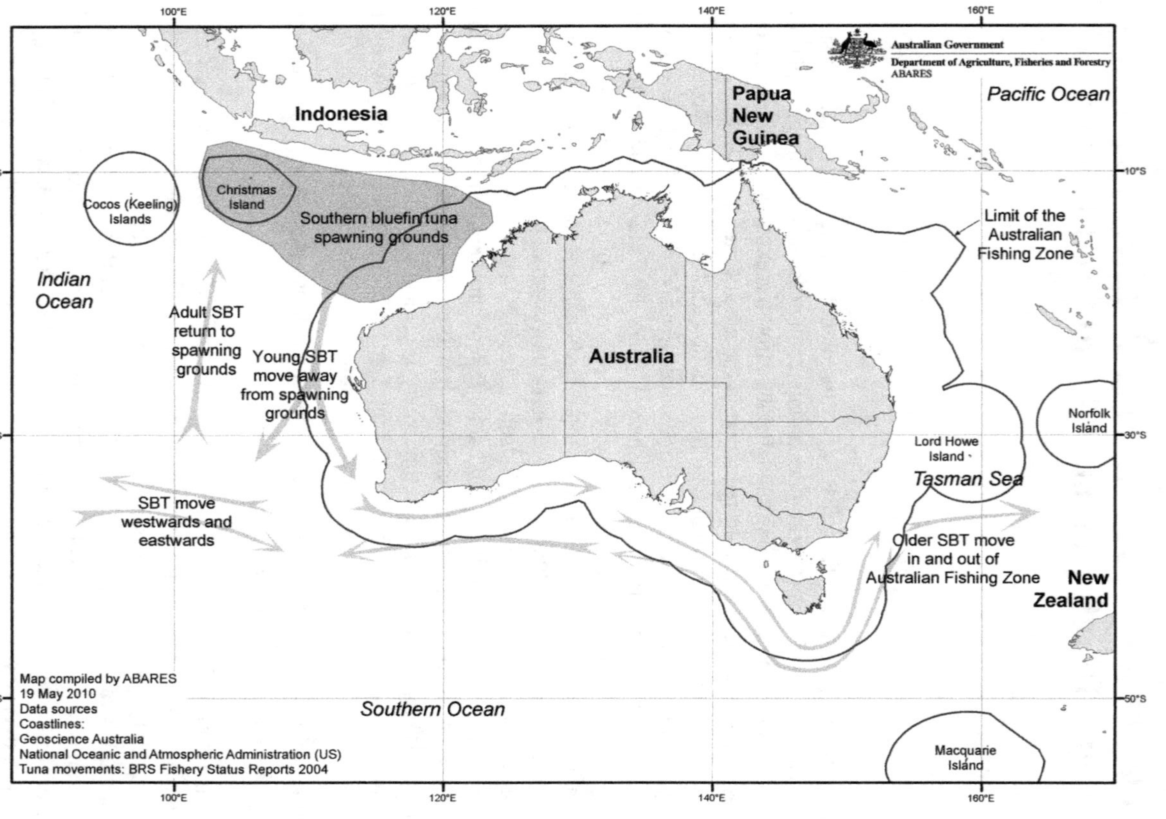

Fig. 9.1 Migratory movements of Southern Bluefin Tuna. (Courtesy ABARES)

larger schools before entering the high seas at between 3 and 4 years of age, while the majority of the adult stock is found in more remote offshore locations (Caton 1994).

The fishery has also attracted a diverse mix of fishing fleets both within Australia and across Asia (Hayes 1997). The Japanese were a pioneer in the harvesting of the resource. Beginning in the early 1950s, the fishing industry working in partnership with the Japanese government mapped the stock's distribution, discovered prized fishing grounds, and began to harvest the resource throughout the Indian and Great Southern Oceans through the development of a large scale distant water fleet targeting the stock with specialized longline vessels. An Australian SBT sector also developed during the 1950s. The Australian fleet concentrated on inshore catches, using a variety of surface gears to target the stock (Caton 1994). Despite these differences, both sectors did, however, rapidly expand their catch effort from the early 1950s. Indeed, evidence of overfishing soon became apparent, with declining hook rates in the Japanese catch by the early 1970s (Hayes 1997).

However, sustained fishing effort continued beyond the early 1970s. Fleets from New Zealand, Taiwan and Indonesia began targeting the stock from the 1970s and 1980s with South Korea emerging as an important player in the fishery by the early 1990s. More recently vessels from the Philippines, South Africa and the European Union have also begun to target the resource (CCSBT n.d.b). SBT has thus emerged as a genuinely international fishery that supports a range of international fleets. This reflects the prized commercial standing of the stock. SBT is one of the most lucrative commercial fisheries in the world today, able to attract premium prices as both fresh and frozen product on the sashimi market in Japan (Owen and Troedson 1993). Concerns that stock numbers are at dangerously low levels have, however, remained. Indeed, scientific estimates in 2009 and 2010 projected that the size of the stock was 5% or less (range 3–8%) of its virgin biomass (CCSBT Extended Scientific Committee 2009/2010).

The commercial and biological imperatives in the fishery are thus at the centre of management tensions in the fishery. Without question, fleet over-capitalization and illegal, unregulated and unreported (IUU) fishing effort has been central to the stock's demise. Technology has also been important in explaining the decline in stock numbers. The capacity to map and locate the stock, the vast distances and remote locations in which the fish are caught and brought to market and indeed the capacity to catch greater numbers of fish in often remote and inhospitable regions could not have been achieved without significant technological capability.

From today's perspective, however, the continued vulnerability of the stock requires explanation. During the early 1980s international efforts began with the purpose to move the fishery onto a more sustainable footing. Australia, Japan and New Zealand began informal tripartite negotiations in 1982 (Neave 1995). This resulted in significant restrictions on the catch during the 1980s.[1] By 1994 this more

---

[1] In 1983 Australia, Japan and New Zealand agree to a quotas of 21,000 t, 29,000 t and 1,000 t respectively. In 1985, Japan agree to further decrease its catch to 23,150 t. In 1988, major restrictions were introduced. A ceiling of 15,500 t was set for the fishery with 6.250 t to Australia, 8.800 t to Japan and 450 t to New Zealand. Global quotas were once again reduced in 1989. In that year global quota was set st 11.750–5,265 t (Australia) 6065 t (Japan) and New Zealand 420 t (Neave 1995).

informal arrangement evolved into the Commission for the Conservation of CCSBT (CCSBT) where the parties have continued to meet annually to decide quota levels between Commission members (Hayes 1997).

The critical question, then, is not the cause of stock decline, but why the stock continues to be in such a precarious condition after 30 years plus of international negotiations? More puzzling still is the situation wherein agreeing to sustainable quota levels has been at the centre of considerable conflict since the beginning of trilateral discussions. Failure to agree on quota levels has seen Japanese fishing vessels banned from accessing the Australian Fishing Zone (AFZ) and Australian ports. There have also been long periods of stalemate in the CCSBT where Commission members have been unable to agree on a Total Allowable Catch (TAC) with the dispute going outside the CCSBT in order arrive at a resolution between the parties. Revelations of illegal fishing by some Commission members further raises questions about the legitimacy of regulatory attempts amongst some CCSBT members. Clearly, this history requires explanation. Science, rather than the source of convergence around which management prescriptions proceed, is not only failing to arrest the stock's decline but critically has been at the centre of considerable conflict.

The Argument

It will be argued that the CCSBT is the intersection of actors who hold competing interests and agendas. Indeed, reduction in fish numbers has mobilized deeply conflicting priorities that have remained resolute on the international stage. The focus of the chapter will be on Japan and Australia. Both nations have historically been the main players harvesting the resource, and since the beginning of international negotiations the key interlocutors at both the trilateral and CCSBT meetings.

A first task then will be to account for the actors represented in the Australian and Japanese delegations. This will take us to consider the domestic contexts of the two protagonists in question as it is here the key players from both nations in the CCSBT have their origins. It will be suggested that while fishing policy is the purview of the state in both settings, management goals and priorities cannot be separated from the relationship between the state and key stakeholders in the development and formation of policy.

The relationship between the state and the fishing industry is, however, critical. Japan's position at the CCSBT reflects a corporatist alliance between the state and fishing industry while in Australia the national agenda (in direct contrast) has been forged independently of industry. Both positions reflect a distinct historical trajectory which will be explored in the next section of the chapter. Discussion of fishing technology will be central to this discussion. In response to the reduction in fish numbers technology has assumed a very different purpose and function within Australia and Japan. It will be argued that this not only reflects the structural relationship between state and fishing industry in the policy debate but an institutional setting where management goals and priorities have achieved a consensus amongst participants.

This chapter is also concerned with international negotiations since the early 1980s. Important to this discussion will be a summary of the CCSBT and its main institutional features. The discussion includes an overview of the controversies that have been a feature of international negotiations. This will be followed by consideration of the scientific debate. Science has been at the centre of the tensions that has characterized international meetings since the early 1980s. Explaining these tensions reveals the intersection of the very different institutional environments described in the pages that follow. The dispute, in the view of this chapter, is thus seen as having political origins rather than representing a dispute that can be understood purely in technical terms. A final section will then consider some broader implications from this argument.

Japan

Japan's SBT fleet has its origins in the 1940s (Caton 1994). The fleet did not, however, rapidly expand until the early 1950s. By the late 1950s fishing grounds had been discovered in the offshore waters of New Zealand and as far west as South Africa (Caton 1991); by the late 1960s the geographical range of the catch extended from the South Pacific west to the centre of the South Atlantic Ocean (Fig. 9.2 shows a statistical breakdown of the key fishing locations that been established by this time). Not surprisingly, the volume of catch rapidly increased with the expansion in geographical range. From 562 t harvested in 1952, the catch rose to 22,908 t in 1957 with a peak catch of 77,927 t in 1961 (Caton 1991).

The expansion of the SBT fleet was part of a broader opening up of Japan's distant water sector. Indeed, the 1950s and 1960s was a golden period for Japan's distant water fleet. With the lifting of the Macarthur Lines imposed at the end of the Second World War (WWII), Japan's distant water sector spread throughout the world's oceans in search of new catches for its fleet. Journalist Michael Wigan captures the energy, drive and resolve of the fleet during this period when he states,

> Japan …is the name the world's fish fear. It is the country that has caught more fish in the twentieth century than any other in more places, with the keenest and most dynamic, some would say the most unscrupulous pugnacity. The Japanese state has looked to the world's oceans as a whole and set about harvesting them with a single mindedness which was unprecedented. (Wigan 1998)

This passage underscores that for many decades Japan was the world's premier fishing nation. Wigan's observation also highlights the critical role of the state in the development of the fleet. Whether in partnership with industry or assuming responsibility for the sector, state support was critical in providing the infrastructure for the expanding fleet. Government scientific research is a case in point. It was vital in the search and discovery of fishing grounds as the distant water fleet expanded its reach during the 1950s and 1960s (Borgstrom 1964). State support was also crucial in finding new and innovative solutions to the myriad of challenges

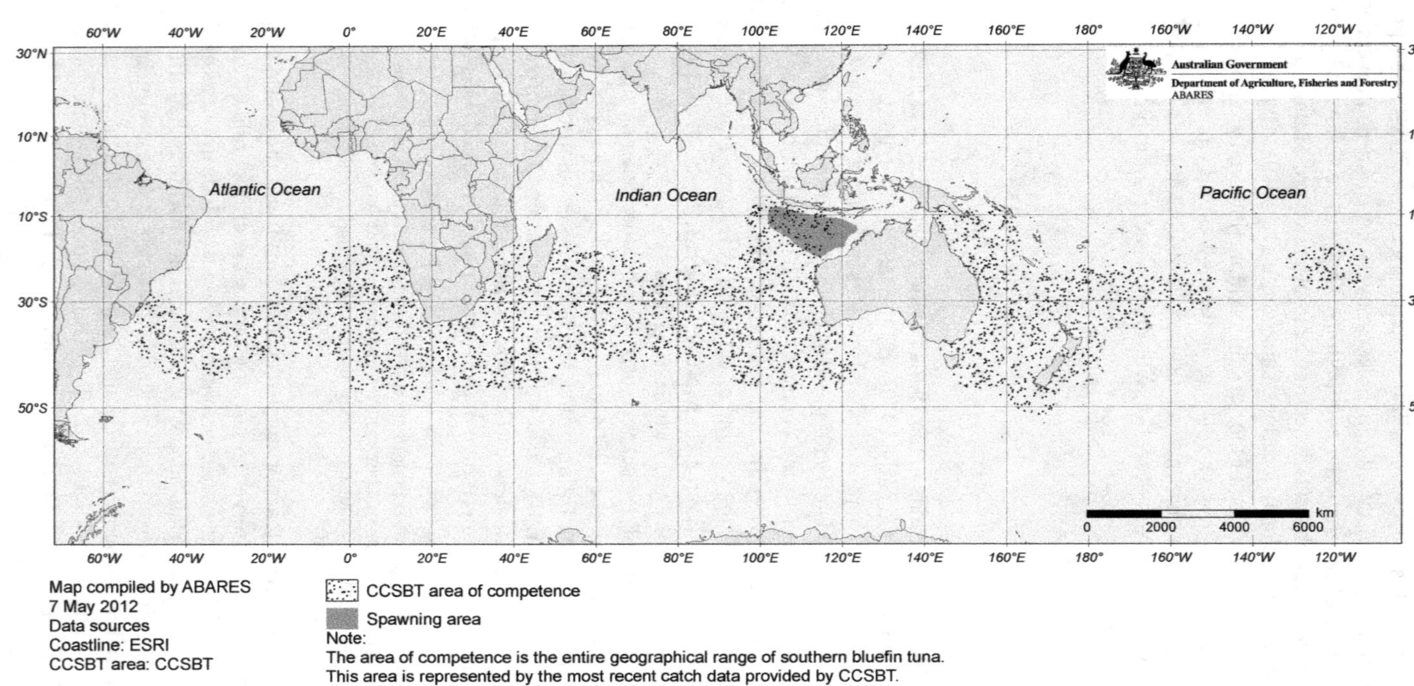

Fig. 9.2 Distribution Southern Bluefin Tuna. (Courtesy ABARES)

confronting industry[2] and was critical in providing technical solutions that would open up new markets and sources of income for industry. The development of ultra-low-temperature (ULT) freezing in the mid-1960s is a case in point. It allowed the reorientation of the SBT catch from canning markets to becoming a prime sashimi grade catch due to the longer storage time afforded to the catch without significant loss in the quality of the fish (Owen and Troedson 1993). Arguably the size and scale of Japanese fishing vessels was the most overt manifestation of the state's support of industry—Japanese fishing vessels were without peer in both size and sophistication in the two decades after WWII (Borgstrom 1964).

From the 1970s the activist role of the state in its SBT and its distant water fleet would continue. It would, however, take a somewhat different hue in response to the range of challenges confronting Japan's SBT fleet and more generally its distant water sector from this time. This included declining catch rates. In the SBT fishery global catch levels were contracting from a global peak of 81,750 t in 1961. Indeed during the 1970s and 1980s, the decline in catch levels was 50,000 and 45,000 t respectively, with reduction in catch effort continuing in subsequent decades (Caton 1991; CCSBT data 2011). This was in part a consequence of sustained fishing pressure on the resource. It was also, however, the consequence of the more complex international environment to which Japan was having to adjust from the mid-1970s (Bergin and Haward 1996). Indeed, the SBT fishery was indicative of Japan's distant water fleet where its fishing activities were becoming increasingly tied to rival coastal and distant water fleets resulting (as we saw) in formal reduction in catch effort from the early 1980s. There were, however, significant domestic challenges also confronting the sector from the early 1980s. These included rising labour and fuel costs and access fees to coastal waters all of which challenged the viability of the distant water tuna fleet from the early 1980s (Caton and Ward 1996). In response the state once again assumed responsibility for the sector. Critically, technology was to prove (once more) a central part of the government's response. A suite of government programs were implemented to address these challenges—subsidies to improve vessel design, subsidies to scrap and up-grade vessels, and subsidies to improve fuel efficiency (Owen and Troedson 1993). These initiatives combined to form part of a multifaceted response in support of its distant water fleet.[3]

The state has thus been a central player in the management and development of Japan's fishing fleet. Whether in periods of relative prosperity or decline, government played a dominant role in the management and development of policy for the industry. This is not to suggest that there was always harmony between government and industry in formulating policy; quite the contrary. However, in stark contrast

---

[2] This includes weather forecasting at sea, ability to detect surface and subsurface current fronts and thermoclines, strong thin longlines, snap-on hooks, fast line haulers, specialized rapid auctioning and sea and land fresh and frozen sashimi tuna supply chain.

[3] This includes payment of access fees to coastal waters, representation of industry interests at international meetings, and aid to coastal states. For a comprehensive discussion of these strategies see (Bergin and Haward 1996).

to Australia, commercial goals and priorities are assumed in the construction and development of policy.

In part this reflects the character of the business/government relations that have taken root in Japan. State management of its distant water fleet is emblematic of the guidance and strongly interventionist role of the state across the economy (Johnston 1982). Japan's pathway to economic modernity has fused commercial objectives into national plans, industry/government partnerships and critically state guidance of the economy (Johnston 1982). State support for the fishing industry thus needs to be understood within a context in which commercial aspirations, goals and objectives have been embedded in the state's regulatory structures and across political and economic life.

It would be somewhat misleading, however, to explain state involvement in the sector purely in these terms. Studies have revealed that across the sector government subsidies and programs to the distant water fleet far exceeded returns by industry during the 1980s (Owen and Troedson 1993). More puzzling still, there has been extensive collusion between state/industry officials in the SBT fishery, especially over the harbouring of illegal catch. This was revealed in 2006 where market research uncovered underreported catch levels of up to 178,000 t since the mid-1980s (Phillips et al. 2009). This, in turn, was the most dramatic expression of continuing resistance to reducing quota levels since the beginning of trilateral discussions in the early 1980s (see below). It also makes plain a political reality shaping Japan's national position. As the longevity of the stock appears at best a secondary concern and state support of industry has remained resolute long after the industry has remained profitable, it is clear that a viable commercial sector is not the main concern of policy. Rather, it is in essence the survival of the industry. Once again, an understanding of the past is important to appreciating this reality.

Since the Meiji Restoration of 1868, the agriculture and fishing industries have been important sectional interests, whose support was critical to achieving the political stability needed in order to attain rapid modernization of the Japanese economy (Johnston 1982). As a consequence both sectors have achieved strong representation in the decision making apparatus of the state. Indeed, in return for support from rural electorates, both sectors have been firmly entrenched in a strong corporatist alliance between the bureaucracy and industry (Barclay and Koh 2005). This has been a defining political reality that has shaped Japan's rapid economic development from 1868 and critically has endured (as we can see in the SBT fishery) to this day (Pempel and Tsunekawa 1979).

Shamed by the 2006 revelations, the Japanese delegation agreed to reduce its catch to 3,000 t from the previous level of 6,065 t in place between 2007 and 2011 (Findlay 2007). It remains doubtful whether this will see a long term substantive shift in fishing behaviour given the entrenched interests that construct national policy on this issue. Indeed, Japan's political economy will continue to caste a deep 'ecological shadow' over the SBT fishery. As we will see, the Australian context provides a significant contrast where the fishing fleet and technological development is secondary to a regulatory environment which places the protection of stock as the government's principle goal. It is to this issue that discussion will now turn.

Australia

The Australian SBT industry began in the early 1950s as a small inshore fishery, working off the coast of South Eastern Australia. In the 1950s two separate sectors emerged, off New South Wales (NSW), and South Australia (SA); in the late 1960s a third emerged in Western Australia (WA). Across all sectors, pole and line and purse seining (introduced in South Australia in the 1970s) have been the primary methods of harvest. Levels of production reached between 5,000 and 6,000 t in the NSW and SA sectors during the 1970s and 6,000 t in the WA fishery by the early 1980s (Caton 1994).

Despite more modest origins compared to its Japanese counterpart, the Australian industry had, to confront a significant crisis during the early 1980s. This was the result of the expansion in the number of operators across all sectors. The extent of the crisis was revealed in a Federal government inquiry published in 1984. It concluded that 45 vessels in the WA sector and 10 purse seiners (and a small number of pole boats) would more efficiently harvest the resource. This was a significant reduction from the 90 vessels in the WA sector and 35 vessels operating in the SA portion of the fishery at that time (IAC 1984). Declining incomes to operators were further exacerbated by a glut in domestic and international canning markets during these years (IAC 1984). Indeed, some 68% of the NSW and SA fishers and 89% of the WA fishers recorded significant financial losses during the early 1980s (Crough 1987).

The rising catch from the Australian industry was also causing growing scientific concern. While overall tonnage was significantly less than their Japanese counterparts, the expansion in catch effort by Australian operators was alleged to be preventing fish from reaching full maturity. Trilateral scientific meetings at this time concluded that 1 t of surface (Australian) catch had a commensurate impact of 2.25 t of longline catch on the parental biomass (Caton et al. 1990). These concerns appear to have been vindicated as by the 1980s there was a complete absence of SBT in NSW grounds with only small schools of fish being sighted in the early 1990s (Caton 1994).

The Australian SBT fishery was thus caught in a cycle of fleet overcapitalization and declining fish numbers; the fishery was beset by the 'Tragedy of the Commons'. In response to this crisis, there was a shift in management responsibility from the States to the Federal government and a management plan was implemented in October 1984. This would reveal a very different management environment than its Japanese counterpart, one which placed protection of the stock at the centre of its management priorities.

Central in driving this agenda forward was the Australian Federal Government. It was the architect of the 1984 management plan, which had a dramatic impact on the size of the fleet. By 1987 fishing capacity had been reduced by 50% (Wesney 1989). This reflected in part the significantly reduced quota level imposed on the sector at the start of the plan. In that year the quota ceiling was set at 14,500 t (Hayes 1997). However, for the WA and NSW sectors continued involvement in the fishery became highly problematic. Continuation in the fishery was now governed by Individual Transferable Quotas (ITQs). This property entitlement allowed operators to purchase a proportion of the TAC set by government. Small quota allocations to WA and NSW operators at the start of the plan, however, repositioned the fishery strongly in favour of the SA sector as the limited quota share made their long term involvement uneconomic. A significant transfer of wealth occurred from the WA and NSW sectors as quota entitlements were sold to the SA operators. Indeed, by the late 1980s the WA and NSW sectors had closed and the fishery centered on Port Lincoln in SA (Green and Nayer 1989).

The downsizing of fleet was just the first stage of a significant restructure confronting the sector. At issue was a drastic realignment of the fleet as industry confronted a radically new fishing environment by the late 1980s. Once again, the Australian government's response was in marked contrast to its Japanese counterpart.

The catalyst was the reduction in 'global quota' brokered in that year resulting in a reduction of 54% in the Australian quota from 11,500 t in 1988 to 5,265 t in 1989 (Neave 1995). This threw Australian industry into turmoil–'forcing' the industry to fish exclusively for the sashimi market in order to maximize returns. In theory, fishing for the Japanese sashimi market offered a more attractive and viable economic alternative to selling fish on the domestic canning market. As a premium high value market, financial returns were many multiples of traditional canning markets.[4] However, while some early attempts to fish for this market had been made it was still very much at a trial-and-error stage (Caton 1994). Targeting the sashimi market required extensive, hard-to-win knowledge of new fishing grounds, new fishing techniques and marketing knowledge and relationships to suit the nuances and peculiarities of this high value market. Large capital investments would also be needed to purchase new fishing vessels and/or to make current vessels more seaworthy in order to target the larger fish that swim in the deeper offshore waters. Training and experience in tuna longlining was also needed as local skills and experience in this fishery were not available at this time.

The conversion to high value SBT longline fishing would thus require time and much experimentation. Realistically, it would take a number of years to make these adjustments. The SA industry was in no position, however, to adopt this path, having incurred significant debt, the result in part of heavy borrowing to purchase quota from WA and NSW (Crough 1987). Instead, operators turned to skipjack fishing in the South Pacific in order to generate vitally needed income. This proved a disaster. The Australian industry could not compete with the highly subsidized tuna fleets of Japan, South Korea and USA, who could fish at a significantly lower cost base. With the failure of this initiative, a significant proportion of the SA tuna industry went into receivership in 1992 (TBOA 1996).

The road to high value fishing was thus proving to be a perilous journey for the Australian industry. The Australian SBT industry had undergone radical change as a consequence of Federal government management of the fishery. In part this reflected (as previously mentioned) Federal government policy in the fishery. However, the capacity to rapidly implement this change during these years requires explanation.

---

[4] Prices for SBT for canning reached A$1,200/mt, while for sashimi market prices reached A$ 30,000/mt (Franklin 1988).

As community backlash gained momentum over proposed changes to water access in the Murray Darling Basin or powerful interests challenge climate change policy within Australia, compromise and even reversal of government policy in these areas seemed almost to be inevitable, but this did not occur in the tuna fishery. Indeed Brian Jeffries, the elected President of the Tuna Boat Owners Association (TBOA), recognized that the extent of industry restructuring was without parallel in the Australian economy when he stated in 1992:

> It is hard to imagine any industry that has been through greater upheaval than the SBT industry. For example, secondary industries such as motor vehicles and textiles, and footwear have been allowed to change over a long period with substantial government assistance… In contrast, SBT has been persuaded to build up a big debt and then had the quota cut by over 60 % in one year. (Jeffries 1992)

Industry capacity to influence and shape management is therefore a crucial ingredient to appreciating developments in the Australian sector. In other words, while the Federal government has been an important player in driving this agenda forward it is the political capacity to implement this agenda which also requires explanation. In essence this reflects the position of industry within the Australian economy. The fishing industry within Australia has historically been a small scale cottage industry whose economic and social significance has been at best marginal within the Australian economy (Industry Commission 1991). The structural capacity of industry to influence policy has been further weakened by the fragmented nature of industry representation; the industry rarely speaks with a united voice to either government or the community (HRSC 1997).

The dominance of a precautionary approach in the management of the fishery thus reflects an industry where the interests of fishers (unlike other sectors in Australia) is of little political consequence, where the protection of fish stocks has strong support in the community, and, critically, where the government is determined to carry out its agenda. It is therefore remarkable that, considering the financial turmoil of the industry in the early 1990s and its limited political influence, by the mid-1990s the Australian SBT industry had been radically transformed. This was largely due to substantial cooperation between Australian industry and Japanese industry in successful sea ranching trials which quickly transformed the industry's fortunes (Bergin and Haward 1994). To this day, the Australian SBT fishery has largely continued in this way, operating from one location, in Port Lincoln, with some expansion into longlining on NSW grounds in the late 2000s (Hobsbawn et al. 2007).

However, the commercial success of industry does not belie the significant pressures that underpinned this change. Indeed, the political capacity to achieve these outcomes is central to explaining developments. The dramatic downsizing of the fleet and the government's resolve to make fleet adjustment the responsibility of industry could not have been achieved without the political capability to accomplish these ends. The limited capacity of industry to shape management outcomes is thus critical to understanding domestic developments in the fishery and critically appreciating the Australian position at the international negotiations—a position that is in direct contrast to Japan. How then does this explain the dynamics of international discussions and indeed the conflict and disagreements over science which has underpinned these discussions? It is to this issue that discussion will now turn.

International Management: The CCSBT

The signing of the Convention for the Conservation of Southern Bluefin Tuna in May 1993 created the CCSBT which came into force in May 1994. The Convention outlines the key objectives of the Commission and the key processes and procedures to achieve these ends. Article 3 of the Convention sets as a key objective 'the conservation and optimal utilisation of Southern Bluefin Tuna' (CCSBT n.d.a); Article 8 (3) for the Commission to set a TAC and to allocate this among members (CCSBT n.d.a), and for the Commission to meet on an annual basis to realize these objectives (Article 6[3]). A scientific committee was also created (Article (9) to help realize these objectives. Its function is to coordinate research and data, assess and analyse stock trends and report its findings on the stock to the Commission (CCSBT n.d.a). The scientific committee is thus central to the management of the fishery.

As we have seen, the CCSBT emerged from the more informal trilateral arrangements first established in the 1980s. Its scope, number of parties and sophistication has certainly widened with the passage of time. One constant, however, has been the significant tension that has characterized international management of the resource. In 1984, Japanese vessels were prohibited from fishing in Australian waters as a consequence of Japan's refusal to agree to the quota level demanded by Australia and New Zealand (Neave 1995). With Japan subsequently agreeing to catch reductions in 1985 this has been emblematic of negotiations moving from reluctant acceptance to long periods of stalemate where the Commission has been unable to reach a consensus on quota (Findlay 2007).

International negotiations over the stock have thus been marked by tension. More insidious have been long held fears of unreported catches occurring outside the ambit of the CCSBT (Polechek and Davies 2008). These were fully realized in 2006 with the discovery of the significant discrepancy between the declared global catch and the amount of product sold on the Japanese market, which as mentioned, was estimated as amounting to 178,000 t over a 20 year period. This suggests then a legitimacy crisis within the Commission particularly amongst successive Japanese delegations. Indeed, in the years when agreement has been reached suggesting greater unanimity between the parties this reflected the geopolitical realities of the fishery rather than a consensus based on science.[5] Certainly, science has been at the core of the tensions between the parties. It is to this issue discussion will now turn.

---

[5] Japan's acceptance of quota level during the 1980s was the result of a number of leverages the Australian government was able to successfully link to an agreement on quota. This includes access to Australian ports and Australia's Fishing Zone. The importance of both factors to Japan's fishing campaign in the southern hemisphere has been widely acknowledged and well documented in studies (Green 1991). In an interview by the author with a Australian industry official in March 1998, while acknowledging the importance of these leverages, he also added that in 1989 the

The Scientific Debate

At the centre of the dispute is assessment of the stock's recovery—in other words, how well the stock is rebuilding from its 1980 level (Ward et al. 1998). While science has expanded the knowledge of the stock, the stock's recovery, its resilience and capacity to rebuild from the low levels of the 1980s is the key point of dispute.

On this critical issue, Japan has typically leant towards more optimistic conclusions while Australia and New Zealand have been more circumspect. These differences have a long history. Virtual Population Analysis (VPA), a mathematical model used to project the recruitment potential of the stock, is a case in point (Caton et al. 1990). In the 1989 scientific meeting, a critical year in which global quotas were dramatically reduced, the difference in projections was stark: Australian VPA assessments predicted further long term decline in the stock at the same time as Japan projected more positive recruitment trends (Caton et al. 1990). This has been a typical outcome of VPA assessments in the CCSBT.

Similarly, Catch Per Unit Effort (CPUE) data—a measurement of stock abundance based on hook rates in the fishery (Sainsbury 1992)—has been a significant source of disagreement between the parties. While CPUE data was crucial in warning of the stock's demise this has been, like VPA assessments, a key point of dispute. The discrepancy centres on the interpretation of CPUE data. For example, when CPUE data indicates a positive return in the numbers of fish within a particular location (reflected in an increase in hook rates), Australia and New Zealand are more cautious in their overall assessment; that is, despite these positive trends, it is not assumed to be indicative of the fishery in its totality. Alternatively, Japan has historically held a contrary position; that positive recruitment in one locale is typical of the fishery overall. Critically, these differences are not largely self-evident from the data but reflect the preference towards two competing hypotheses: the 'variable square hypothesis' that the fish are not evenly spread across the fishery; and 'constant square hypothesis' which as the term suggests, assumes a more uniform spread of fish across the ocean.[6] Similar to VPA assessments, then, this variance in conclusions has a long history in the CCSBT.

These contrasts in VPA projections, and interpretation of CPUE data, thus underscore in part the 'opaque lens' from which stock assessments are constructed—science having to provide recommendations in a context of incomplete or developing knowledge. Indeed, James Crawford at the International Law of the Sea Tribunal captured these uncertainties when he stated:

---

acceptance by the Japanese delegation of the significant reduction in quota in that year, despite strong objections, reflected the perception by Japanese industry that Australian industry would decimate the resource in protest (through the use of purse-seiners in Port Lincoln) if Japan did not accept the quota level being proposed by the Australian and New Zealand governments.

[6] All scientific parties do however recognize that the 'constant' and 'variable' square hypotheses represent ideal types and that the 'reality' of the stock's abundance is somewhere between these models.

> The first point I want to make about the scientific disagreement is that it is not in essence a disagreement about the present state of affairs. It is a question of projection. Projections are just about predictions. They are based on the available data and series of assumptions. In this respect they are like weather forecasts. Weather forecasts require a lot of science and they require a lot of observation. They are based on a set of assumptions and yet we know… the weather is still uncertain even from day to day. With fish stocks the uncertainty is much worse because in our case we are trying to predict the state of fish stocks a considerable period in advance, something like twenty years. Such projections are difficult and may require very sensitive assumptions about a range of matters. (ITLOS 1999)

In other words, with the tightening of the science necessarily providing the basis for more 'objective' facts on which to base management decisions, the current differences in stock assessments reflect, as James Crawford states, assumptions and weightings that are at significant variance. Stock projections are not based on an immutable body of evidence. Indeed as differences have morphed into the management debate both sides have accused the other of practising politicized science—science that is less valid than their counterparts. Japan, in particular, has been quite explicit about what it sees as the more overtly 'political' nature of Australia's scientific position (CCSBT 1998). A final section will seek to account for the dispute where explanations, it will be argued, have institutional origins—their basis being in the relationship between industry and the state as discussed earlier in this chapter.

The Political Economy of Science

The scientific dispute, from the perspective of this chapter, is not about the 'objectivity' of a country's position. As each party has accused the other of practising 'politicized' science the implicit claim by each party is that the other is not adhering to the dictates of objectivity and impartial policy advice. The accusations themselves reflect how science has come to assume the mantle of objectivity able to dissolve sectional interest and provide a more objective basis from which management can proceed.

Ironically these taunts do hold an element of truth. However, rather than reflecting a cynical manipulation of science by powerful domestic constituencies, from the perspective of this chapter, science—its purpose, function and relationship to the other stakeholders—is steeped in the historical development of both countries, where the relationship between science, industry and the state has forged distinct roles, purpose and position in the domestic policy debate.

Writing in the 1960s, George Borgstrom provided a significant insight into the Japanese context when he observed:

> There is an unquestionable trend towards applied research, particularly such investigations which are of an immediate concern. There is no clear demarcation between scientific pursuits, control functions and routine observations. This is explained by the pressing needs of the rapidly expanding fishing fleet and in particular the tuna operation now spreading to all major operations…There is no question that the overwhelming demands of the active tuna fleets and fishing companies not only straight jackets research but also explains the fact the

research work is very patchy and diverge primarily into fields adjacent to those practical
areas which are given high priority. (Borgstrom 1964)

In others words, the purpose of scientific research is tied to the support and development of the fishing industry. Borgstrom goes on to illustrate how the rapid expansion of Japan's distant water fleet after WWII was underpinned by a vast research infrastructure, locating, identifying and collating data for the express purpose of discovering new sources of stock for the ever expanding distant water fleet (Borgstrom 1964). Indeed, this research network has also been mobilized as a key 'weapon' in supporting industry confronted with the significant challenges to its distant water fleet from the 1970s, not only in seeking out new sources of fish but new avenues (such as fish farming) in which industry can diversify its operations (Bergin and Haward 1996).

In contrast, industry's journey to becoming a high value fishery in Australia is emblematic of a management emphasis where the stock is the central locus of policy. In other words, management within the Australian context has foregrounded the 'precautionary principle' underpinning its management decisions. In somewhat pithy terms, in Japan scientific research has a distinctly commercial purpose; in Australia, scientific research is embedded within more conservationist parameters.

Critically, then, science within the respective policy communities holds a very different institutional purpose, function and role. This is a vital point. It suggests that the scientific conflict in the Commission is really about the role that science *should* play in the management of the stock. Both forms of science are perfectly 'rational' in their own terms—it is not at its core then a dispute that has 'technical' origins. Rather, it is a clash between two forms of science. Both *forms* of science have a long history—scientific research of a 'pure-basic' persuasion and 'applied scientific research'. In the case of the former its purpose is to discover general laws governing the natural world (the interactions that make up the ecosystem, the physical laws governing the universe); the latter, observation to achieve product innovation (Jasonoff and Wynne 1998). The key difference thus turns on the question of emphasis. Japan's leaning towards more optimistic VPA projections, and its leaning towards a 'constant square hypothesis, is indicative of a commercial milieu where 'finding fish' takes precedence over ecological considerations. Alternatively the more precautionary approach that underpins Australia's VPA projections and its more cautious interpretation of the CPUE data reflects a science embedded in the relationship between the stock and its natural surrounds.

In other words, the two forms of science, their goals and purpose are being reflected in the uncertainties of the stock assessments; in the different weightings given to the inputs that make up the VPA models and the interpretative frameworks applied to the CPUE data. However, in another important sense, the differences in science also have their institutional expression. In Japan, a dense network of private and state run institutes form a close nexus between industry and the state (Borgstrom 1964; Bergin and Haward 1996; Owen and Troedson 1993). In this respect the institutional configuration of scientific research mirrors that of an applied research environment—where a close relationship between science, industry, and focus groups is created to achieve product innovation. In Australia, the work of scientists is arguably quite independent from the policy setting, mirroring the autonomy typical of scientific endeavour of a more pure persuasion. Both settings, however, are not simply institutional configurations separate from social and political contexts. Indeed, the privileging of each 'scientific form' has deep historical roots rendering somewhat problematic the claim that science is simply captured by contemporary pressure group interests.

Indeed it is the contention of this chapter that the different scientific forms do not reflect some cultural predilection of the parties, nor as previously stated a primarily technical point of difference. Rather, it reflects the structural relationship between state and fishing industry as discussed earlier in this chapter. Science provides a functional role to state and industry in Japan while remaining more autonomous from industry within Australia. This reflects the different historical trajectories that have shaped state and industry relations in both settings.

Conclusion

This case study has argued that the problems confronting the fishery have political origins. While technology and fleet overcapitalization have often been identified as the cause of the reduction in the world's fish stocks, this chapter argues that the issue of governance is an important consideration. Indeed, the chapter argued that within the domestic policy debate in Australia and Japan, science has served a very different function and purpose in response to the decline in stock levels. These outcomes, it has been suggested, do not stand separately from the actors involved in the management debate but critically reflect the actors that shape and develop policy in both settings. Overfishing and fleet development is thus not an overarching/uniform response but critically reflects the domestic context in which fishing fleets are regulated.

The importance of the case study is to highlight the need to recognize the politics shaping the global decline in stock levels. This is an urgent task. In Southeast Asia for example evidence of overfishing has been well documented with fisheries throughout the region under significant stress. However, despite this evidence, fleet development continues largely unconstrained, and the region's fish stocks remain under severe pressure from overfishing (Williams 2007).

References

ABARE (2003) Australian fisheries statistics 2002. Australian Bureau of Agricultural and Resource Economics and Sciences, Canberra

ABARE (2005) Australian fisheries statistics 2004. Australian Bureau of Agricultural and Resource Economics and Sciences, Canberra

Barclay K, Koh S (2005) Neoliberalism in Japan's tuna fisheries? Government intervention and reform in the distant water longline industry. Asia Pacific School of Economics and Government, The Australian National University, Canberra

Bergin A, Haward M (1994) Recent developments in international management. Mar Policy 18(3):263–272

Bergin A, Haward M (1996) Japan's tuna industry: a setting sum or new dawn. Nova Science Publishers Inc, New York

Borgstrom G (1964) Japan's world success in fishing. Fishing News Books, London

Caton A (1991) Overview. In: Deriso RB, Bayliff WH (eds) World meeting on stock assessment of bluefin tunas: strengths and weaknesses. Inter-American Tropical Tuna Commission, La Jolla

Caton A (1994) Commercial and recreational components of the Southern Bluefin Tuna. In: Shomura RS, Majkowski J, Langi S (eds) Interactions of Pacific tuna fisheries: proceeding of the first FAO expert consultation on interactions of Pacific tuna fisheries, Noumea, 1991. Rome: FAO Technical Paper 336/2

Caton A, Ward PJ (1996) Access arrangements for Japanese longliners in Eastern Australian waters. In: Ward PJ (ed) Japanese longlining In Eastern Australian waters 1962–1990. Bureau of Resource Sciences, Canberra

Caton A, McLoughlin E, Williams MJ (1990) Southern Bluefin Tuna: scientific background to the debate. AGPS Bulletin No. 3, Canberra

CCSBT (n.d.a) Text of the convention for the conservation of Southern Bluefin Tuna. http://www.ccsbt.org/userfiles/file/docs_english/basic_documents/convention.pdf. Accessed 8 Feb 2012

CCSBT (n.d.b) Origins of the convention. http://www.ccsbt.org/site/origins_of_the_convention.php. Accessed 8 Feb 2012

CCSBT (1998) Report of the resumed fourth annual meeting of the commission for the conservation of Southern Bluefin Tuna, 19 to 21 February, Tokyo. http://www.ccsbt.org/site/reports_past_meetings.php. Accessed 8 Feb 2012

CCSBT (2008) Report of the Performance Review Working Group, 3–4 July 2008, Canberra, Australia. http://www.ccsbt.org/docs/pdf/meeting_reports/ccsbt_15/report_of_PRWG.pdf. Accessed 11 Nov 2010

CCSBT (Commission for the Conservation of Southern Bluefin Tuna) (2011) Website. Accessed 24 Sept 2011

CCSBT Extended Scientific Committee (2009/2010) Reports of the fourteenth and fifteen meetings of the scientific committee, CCSBT. http://www.ccsbt.org/site/reports_past_meetings.php. Accessed 8 Feb 2012

Clean Seas (2011) Clean seas website. http://www.cleanseas.com.au/. Accessed 27 Sept 2011

Crough CJ (1987) Australian tuna industry. Pacific Island Development Program, East-West Centre, Honolulu

FAO (2010) Regional fishery bodies summary description: commission for the conservation of Southern Bluefin Tuna (CCSBT). FAO, Rome

Findlay J (2007) Southern Bluefin Tuna fishery. In: Larcombe J, McLoughlin K (eds) Fishery status reports, 2006. Bureau of Rural Science, Canberra

Franklin P (1988) Australian Southern Bluefin Tuna fishery. Australian Fisheries Service, Canberra

Green G (1991) The value to Japan of access to the Australian fishing zone. ABARE, Canberra

Green G, Nayar M (1989) Individual transferable quotas and the Southern Bluefin Tuna fishery. ABARE Occasional Paper No 105, AGPS, Canberra

Haward M, Bergin A (2001) The political economy of the Japanese distant water fisheries. Mar Policy 25(2):91–101

Hayes E (1997) A review of the Southern Bluefin Tuna fishery: implications for ecologically sustainable management. TRAFFIC Oceania, Sydney

Hobsbawn PJ, Hender J, Rowcliffe S, Murphy R (2007) Australia's annual review of the Southern Bluefin Tuna fishery. In: Extended commission of the fourteenth meeting of the commission for the conservation of Southern Bluefin Tuna. http://www.ccsbt.org/docs/pdf/meeting_reports/ccsbt_14/report_of_CCSBT14.pdf. Accessed 23 Nov 2010

HRSC (House of Representatives Standing Committee on Primary Industries and Rural and Regional Affairs) (1997) Managing commonwealth fisheries: the last frontier. AGPS, Canberra

IAC (Industry Assistance Commission) (1984) Southern Bluefin Tuna. AGPS Report No. 349, Canberra

Industry Commission (1991) Cost recovery for managing fisheries (draft report). AGPS, Canberra

ITLOS (International Law of the Sea Tribunal) (1999) Press Release, 27 August. www.itlos.org/news/press_release/1999/press_release_28_en.doc. Accessed 4 Aug 2002 (pp 7–8)

Jasonoff S, Wynne B (1998) Science and decision making. In: Rayner S, Malone EL (eds) Human choice and climate change: the societal framework, vol 1. Batelle Press, Columbus

Jeffries B (1992) The tuna industry-future directions. In: Inaugural Southern Bluefin Tuna science-industry-management workshop, 24–25 June, Port Lincoln. CSIRO Division of Fisheries, Hobart

Johnston C (1982) MITI and the Japanese miracle: the growth of industrial policy 1925–1975. Stanford University Press, California

Neave P (ed) (1995) The Southern Bluefin Tuna fishery 1993. Fisheries assessment report compiled by the Southern Bluefin Tuna fishery stock assessment group. Australian Fisheries Management Authority, Canberra

Owen AD, Troedson DA (1993) The Japanese tuna industry and market: A.C.I.A.R project no. 8928. Australian Council for International Agricultural Research, University of New South Wales, Sydney

Pempel T, Tsunekawa K (1979) Corporatism without labor? The Japanese anomaly. In: Schmitter PC, Lehmbruch G (eds) Trends towards corporate intermediation. Sage Publications, Beverly Hills

Phillips K, Begg G, Curtotti R (2009) Southern Bluefin Tuna. In: Wilson D, Curtotti R, Begg G, Phillips K (eds) Fishery status reports 2008. Bureau of Rural Science and Australian Bureau of Agricultural and Resource Economics, Canberra

Polacheck T, Davies C (2008) Considerations of implications of large unreported catches of Southern Bluefin Tuna for assessments of tropical tunas, and the need for independent verification of catch and effort statistics. CSIRO Marine and Atmospheric Research, Hobart

Sainsbury K (1992) What is happening to the Southern Bluefin Tuna stock? In: Inaugural Southern Bluefin Tuna science-industry-management workshop, 24–25 June, Port Lincoln. CSIRO Division of Fisheries, Hobart

TBOA (Tuna Boat Owners of Australia) (1996) Submission no. 44. In: Joint standing committee on treaties, inquiry into the subsidiary agreement between the government of Australia and the government of Japan concerning Japanese long-line fishing, s318–s339. Parliament of Australia, Canberra

Ward P, Nektarios T, Kearney B (1998) Getting science into regional fishery management: a global overview. Fisheries Resources Branch, Bureau of Rural Science, Department of Agriculture, Fisheries and Forestry, Canberra

Wesney D (1989) Applied fisheries management plans: individual transferable quota's and input controls. In: Neher PA, Arnason R, Mollett N (eds) Rights based fishing. NATO ASI Series, Series E, Applied Sciences. Kluwer, Dordrecht

Wigan M (1998) The last of the hunter gatherers: fisheries crisis at sea. Swan Hill Press, Shrewsbury

Williams M (2007) Enmeshed. Australia and Southeast Asia fisheries. Lowy institute paper 20. Lowy Institute, Sydney

Chapter 10
The NSW Steam Trawl Fishery on the South-East Continental Shelf of Australia, 1915–1961

Anne Lif Lund Jacobsen

Abstract How was modern fishing methods, in the form of steam trawling, introduced in Australia? And what were the consequences for the fish stocks found on the South-East Continental Shelf? Through historical catch records and archival resources, the history of the NSW Steam Trawl Industry from 1915 to 1961 is unfolded. This reveals that government initiatives played a surprisingly decisive role in founding and sustaining the industry. Also that early signs of depletion of stocks and overfishing happened within the first decade of the fishery and, in the case of flathead, overfishing was so severe that flathead biomass on the South-East Continental shelf was permanently reduced. The study furthermore reveals how the trawl industry was influenced by government policy, market conditions, war and fishing effort with little understanding of the marine resources which they relied on.

Keywords South-east Australian trawl fishery · State trawling industry · New South Wales · Australian fishing history · Australian marine environmental history

Today commercial fishing in New South Wales (NSW), Australia, is one of the state's important food producing industries. It is estimated that landings of wild harvested fish from NSW's waters have an annual value of $ 94 million, of which ocean catches (by hauling, trap, line or trawl) comprised around half.[1] The earliest known ocean fishing grounds were located on the continental shelf at a depth of 200 m or less in waters mostly south of Sydney, but over time fishing spread out to also cover grounds in Bass Strait. Today the fishery includes several states, covering fishing grounds along New South Wales, Victoria, around Tasmania and South Australia, known as the Commonwealth Trawl Sector (previously the South East Trawl Fishery), and is managed by the Commonwealth under a single management plan with the exception of some inshore areas which are still managed by the relevant States.

---

[1] The value is based on NSW catch statistics for 2006/2007 by NSW Primary Industries, Fisheries and Aquaculture. See www.dpi.nsw.gov.au/fisheries/commercial/catch-statistics#NSW-reported-commercial-wild-harvest-for-2006-2007-by-fishery-including-gross-weight-%28tonnes%29-and-estimated-value-%28$%27000%29 (accessed 01.04.2011).

A. L. L. Jacobsen (✉)
Rigsarkivet, the Danish State Archives, Rigsdagsgården 9, 1218 København K, Denmark
e-mail: lif@lund-jacobsen.dk

J. Christensen, M. Tull (eds.), *Historical Perspectives of Fisheries Exploitation in the Indo-Pacific*, MARE Publication Series 12, DOI 10.1007/978-94-017-8727-7_10,
© Springer Science+Business Media Dordrecht 2014

However fishing, and especially on the open sea, has not always played a role in the NSW's economy. When looking up 'Fisheries' in the official government year book of New South Wales between 1888 and 1901 one would find this entry:

> The seas that wash the shores of New South Wales abound with fish, but this source of wealth to the State has been greatly neglected. (The Wealth and Progress of New South Wales 1887/88-1900/01)

The entry reflected a heated debate in NSW, fought in newspapers and in parliament, over why the State did not utilize the sea and its marine resources to develop a larger fishing industry. By 1888 the Australian sea fishing industry was mainly small and inshore-based and there had been no attempt to commercially exploit the species found on the continental shelf. Towards the end of the nineteenth century, as a pro-fisheries development position gained support, significant political pressure emerged for the State to support industry development. Although none had investigated the extent to which marine resources were available on the south-east continental shelf, it was taken for granted that rich fishing grounds existed and that they were suitable for large-scale commercial harvesting.

In 1914, an opportunity to develop offshore fisheries arose when high prices for meat made it attractive for the sitting Labor Government to expand its program of state-driven businesses to include a new fishing enterprise (Tyler 2006). This type of public development policy, with direct industry intervention, is known as 'colonial socialism' and has played a key role in Australia's economic development (Butlin et al. 1982).

The NSW State Trawling Industry—The Beginning of Trawling in Australia, 1915–1923

Australia's first demersal trawl fishing industry was established in June 1915 under the name of the 'NSW State Trawling Industry' (STI), when the NSW government founded an offshore, large-scale fishing industry, modelled on the British trawling industry in the North Atlantic. The aim of the government was to provide consumers, primarily in Sydney, with trawled fish. During 1909–1914 promising trawling grounds on the continental shelf had been identified by the Australian Commonwealth research vessel *Endeavour*. The NSW Government established the industry by investing in a fleet of three modern British built steam trawlers. The plan was to run the trawling industry on a commercial basis, and at the same time provide consumers with cheap trawled fish through a network of state-owned fish-shops located in the main population areas.

Initially the scheme was a success and during the next years six more vessels were built at the State Dockyard in Newcastle. To store catches central depots with large cold storage facilities was erected on No. 5 Wharf, Woolloomooloo Bay, Sydney, and in Newcastle. In order to distribute catches, an extensive network of fish shops serviced by lorries was established. In addition a network of coastal receiving stations (coastal depots), serviced by a transport vessel were initiated to boost local fishing communities. By 1923 trawled fish was responsible for 1,302 t or 14 % of the state's total fish landings (Jacobsen 2010). The species target was mainly tiger flathead (*Neoplatycephalus richardsoni*), but latchet (*Pterygotriglia polyommata*), chinaman leatherjacket (*Nelusetta ayraudi*), jackass morwong (*Nemadactylus macropterus*) and redfish (*Centroberyx affinis*) was also landed. In the beginning the species found on the trawling grounds were little known by the consumers but, as state trawled fish was sold below the price of privately caught fish from inshore waters, a market for these new species quickly developed.

However, the dual purpose of the STI, to develop a new fishing industry as well as provide cheap fish for NSW's urban populations on a commercial basis, made the STI rigid and expensive to run. As the ambition of the state-owned trawling industry increased after 1915, expenses spiralled out of control. As a result by 1922 there was no longer political support for the scheme, and it was decided to sell the industry.

The Rise of Private Steam Trawling

In February 1923 the STI's assets was announced for sale. In the sales advertisement it was specified that preference would be given to applicants planning to continue the industry (*The Argus* 14 April 1923, p. 6), as the Government was still keen to continue nurturing the development of an offshore fishing industry. During the next half-year all the company's seven trawlers,[2] as well as the Industry's buildings, gear and other assets were sold.

Despite the Government's obvious efforts to liquidate the STI as quickly as possible, buyers were not queuing up. No offers were received for the entire industry; and of the vessels, only the three British built trawlers initially attracted any interested buyers. Of all vessels only the British build *Koraaga,* bought by the locally based, newly formed Coastal Trawling Company,[3] was sold to a price similar to the vessels launch-cost; the costs associated with outfitting and delivery to Sydney was never recovered.[4] The two remaining British built trawlers were sold for £ 4,400

---

[2] A complete list of all NSW based steam trawlers participating in the south-east trawl fishery can be found in Jacobsen (2010) Appendix 2.

[3] Within 10 months of buying its first trawler, the company had paid off the trawler and was looking to expand its business. Over the next 2 years the company bought another former STI trawler and attempted to build a market for trawled fish in Melbourne, Victoria. On 22 October 1926 Coastal Trawling Company amalgamated with Red Funnel Fisheries Ltd.

[4] The original price of SS *Koraaga* from Smith's Dock Middleborough-in Tees, UK was £ 7,500. Additional costs of outfitting and delivering in Sydney amounted to £ 5,000. Sold in 1923 for £ 8,000.

and £ 3,400 respectively to New Zealand based Sanford Ltd, the largest fishing company in Auckland, where almost half of the New Zealand catches were landed. For about a year Sanford trawled the south east continental shelf before they sold the trawlers back to a Sydney company in 1925 because of low returns.

In hindsight the interlude with the New Zealand fishing company was symptomatic of some fundamental problems surrounding the trawl fishery on the south east continental shelf. As Sanford Ltd. was already a well-established trawling company in New Zealand it was in a position to realistically assess the situation of the fishery. Wanting to expand its business to Australia Sanford Ltd. had opened a depot in Sydney to handle the New Zealand imported fish as well as local species caught by its steam trawlers (*The Argus* 22 December 1924, p. 17). In 1925, as Sanford realised that 'while the distribution business [in NSW] was satisfactory, trawling was not' (Johnson 2004), the trawlers were sold off. What Sanford had grasped was that the productivity of the shelf was relatively low, especially compared to the New Zealand grounds. With an ample supply of New Zealand trawled fish, Sanford could eliminate the costly steam trawling operations for local species and concentrate on selling New Zealand caught fish which was in much more ample supply.

After the STI's British built trawlers were sold, the four vessels built at the NSW government's own dockyard were slowly sold during the second half of 1923. As an example the first NSW build trawler to be sold was *Goonambee,* which was handed over to a short-lived local syndicate called Tucabia Fisheries Ltd. for £ 3,500 on 12 months terms. When the STI had *Goonambee* delivered from the Government dockyard in May 1917 the total costs of delivery had been £ 23,725. As the rest of the vessels were sold at similar prices the Government's total write-off was substantial. As trawl fishing proved commercially viable under private management, the remaining vessels sold at slightly higher prices.

While it had taken about half a year for the NSW Government to sell all the vessels, the shops and other leases were more easily disposed of, although in most cases a cash deposit could not be made, and payment had to be made in instalments (Herlihy 1927). The new private trawling companies had bought most of the STI's vessels and storage facilities, but had refrained from taking over the shops and coastal depots, which generally found other uses.

By initially focusing on fishing and wholesale and not investing in retail facilities, the trawling companies had avoided some of the costly mistakes of the STI. By mid-1920's the private trawling industry had consolidated itself into three companies, which controlled steam trawling in NSW and until the 1950s were able to dominant the market for sea-fish (Jacobsen 2010). The three companies were known as Cam and Sons, Red Funnel and A. A. Murrell.

Cam and Sons

The oldest of the company to dominate the NSW steam trawling industry was Cam and Sons, also early known as Cam Brothers, Mess. Cam and Sons, and later Cam and Sons Ltd. Owned by a family of Italian descendants Charles Caminitti (or Cam) began selling fish from his own shop in 1913 and in 1918 became a wholesale fish-agent at the Commonwealth Co-operation Fish Exchange in Redfern (State Fisheries 1922). By 1920 he had expanded the business further by becoming sole agent of the State Trawling Industry, which by then had closed the wholesale part of its business (State Trawling Industry 1920). Trading must have been profitable, because in 1923 Charles Cam was able to buy his first trawler from the STI. During the 1920s business was thriving and Cam expanded the trawling business until the fishing fleet consisted of eight vessels by the end of the decade.[5] Of the eight trawlers, three were built by governments for commercial or defence purposes. In 1934 Cam and Sons consolidated the company by incorporating it under the name Cam and Sons Ltd. and transferring all assets previously owned by the founder Charles Cam to the new company. All the shares were distributed to family members, and Charles Cam remained in control of the company until his death in 1947, where he was succeed by his son Rocco (Cooke 2006).

Red Funnel Fisheries Ltd.

The second largest steam trawl company was Red Funnel Fisheries Ltd. The company was established in December 1925 and consolidated itself by absorbing several smaller companies. During its existence, the company went through several changes of ownership. Red Funnel Fisheries was incorporated on the 2 December 1925 with a capital of £ 75,000, taking over two former STI trawlers from Douglas Paul Hann as well as the New State Fish & Ice Company (Red Funnel Fisheries 1925). In its setup Red Funnel was ambitious and committed to rapid development. In January 1926 Red Funnel incorporated the business of Carlyon Ltd., which included the former State trawlers *Goonambee,* and a cool store in Newcastle (Red Funnel Fisheries 1926). Red Funnel not only expanded by absorbing other companies; in June 1926 Red Funnel also bought the former navy British steam trawler *Gunner.* To further fuel the company's growth an extraordinary general meeting of shareholders was held in October 1926, and it was decided to increase the nominal capital by £ 100,000, making the total capital £ 175,000. According to Red Funnel's second progress report, the company had a profit of 13.3 % in the first 10 months and it had increased its profit earning assets 200%. By 1929 the fleet consisted of eight trawlers,[6] all of which had been built for various governments to serve commercial or naval purposes and the company was operating from the former State Trawler Depot at No. 5 Wharf, Woolloomooloo. The growth of Red Funnel continued until 1928/1929 when the company, along with the rest of the trawling industry, was hit by falling catches and economic recession. As a consequence the

---

[5] SS *Goorangai* (1923), SS *Charlie Cam* (1925), SS *Beryl II* (1926), SS *Camro* (1927), SS *Oliver Cam* (1928), SS *Alfie Cam* (around 1929) and SS *Mary Cam* (around 1929).

[6] SS *Bar-ea-mul* (1925), SS *Dureenbee* (1925), SS *Goonambee* (1926), SS *Gunundaal* (1926), SS *Koraaga* (1926), SS *Millimumul* (1926), SS *Durraween* (1928), and SS *Goolgawai* (1928).

company was reconstructed in 1935 and began operating under the name Red Funnel Trawling. After 1935 only sketchy information about the company's ownership and financial arrangements is available.

A. A. Murrell

Little is known about the third of the main steam trawling company operating in NSW. The only surviving records from the company are two logbooks kept at the National Archive in Canberra. What is known is that Arthur A. Murrell, a former clerk, began his business as a fish retailer and later became a licensed fish agent and fish merchant. In 1926 Murrell bought his first vessel (Lorimer 1984), the Scottish built trawler *David Blake*. The *David Blake* was originally built for the British Admiralty as a minesweeping vessel in 1918, but was decommissioned in 1921 and later sold to A.A. Murrell in 1926. The vessel was typical of the type of trawlers that had been bought in the 1920s by the private steam trawling companies, which favoured second hand steam trawlers of 'Castle' class, built before 1920. A. A. Murrell was successful in his venture; his business setup put him in control of all steps from capture to consumers, and produced an annual turnover of between £ 70,000 and £ 89,000 (*Sydney Morning Herald* 1 February 1929, p. 13). In January 1929 he expanded the business further by adding *Samuel Benbow* to his fleet. *Samuel Benbow* was sister ship to *David Blake*, also built to the British Admiralty in Aberdeen (www.aberdeenships.com, accessed 10.03.2010). The last vessel to be acquired by A. A. Murrell in the 1920s was the trawler *Tongkol* built for the Fisheries Authorities in British Malaya. Unlike the other steam trawling companies, trawl fishing was only a minor part of A. A. Murrell's fish merchant business and catches were mostly traded through his own retail outlets or wholesale customers.

As a consequence of the progress made by the three trawling companies by the end of 1929 the total Sydney based trawling fleet consisted of 18 vessels, seven which had been added during 1929. Most of the trawlers were of Castle class, from 220 to 278 t gross register, with crews of 13, who were paid fixed rate wage and catch bonuses. The vessels had cruising speeds of 9–11 knots and were equipped with wireless telephone so they could be in contact with their shore offices and the other vessels. Before each voyage the trawlers would load 15 t of ice used to cool the fish in specially insulated chambers. Fishing was done with otter trawl of about 140 ft long with a mesh size of 6 in in the wings decreasing to 3½ in in the cod end (bottom end of the trawl). The upper edge of the net was 90 ft long and the lower 140 ft. The rope at the edges was attached in each side to two 'otter boards' or 'doors' of 10 ft long, 4 ft 6 in high which purpose was to kept the net open during trawling. Attached to each otter boards was 400 fathoms of wire rope which was wounded up on two drums placed aft pulled by steam winches. As shown on Fig. 10.1 gas lamps were placed with regular intervals on the deck so fishing could take place during nigh (Roughley 1916). After 1925 the modified Vigneron-Dahl otter trawl became common. The trawl differenced from earlier models by the way the

Fig. 10.1 Crew on aft deck with net and stem winches on a State Trawler, 1914–1921. Courtesy State Library of New South Wales. (David G. Stead; [ML MSS 5715 11(25)])

doors was connected to the net by long bridles, allowing the net to spread wider and increased turbulence in the water to herded more fish into the net. As a consequence fishing efficiency increased by about a third (Klaer 2006).

As shown above, private companies were able to buy the assets of the former STI at much reduced prices, making establishment costs low and affordable for even small scale investors. The private trawling industry benefited from the fact that the NSW Government had introduced trawl fish to the consumers and developed a market that could absorb large landings of sea fish like flathead. The result of this support was that the private industry grew rapidly and expanded to the point where overfishing became possible. In particular, the valuable flathead stocks were beginning to be negatively affected by the fishing pressure, which in the 1930s would throw the private industry into deep crisis.

Location of Trawling Grounds

When trawling began in 1915, the main fishing ground was located close to Sydney, just south-east of Cronulla (called the Botany Ground), but when the initially high catches began to decline the trawlers began to search further away. Within a year the trawlers began to sail as far south as Green Cape to trawl for fish, on ground found between 1–8 miles or about 1.5–13 km off the coast (Roughley 1916). The long distances sailed by the trawlers in the early years can partly be explained by the lack of knowledge about the habits and migration patterns of target species. Although some of the fishing grounds had been mapped by the *Endeavour*, there was little knowledge of what species the grounds contained, so particularly in the earlier years the trawlers had to undertake a lot of exploration work to identify the best fishing grounds (Horn 1916). As the captains became better at finding the fish, catch per unit of effort (CPUE) rose on the grounds.

Research using trawling logbooks[7] has shown the gradual expansion of the waters fished by the fleet. During the time of the STI the main fishing ground was primarily found outside Botany Bay on what was called the Home Ground, and secondly in the waters between Eden and Merimbula. A little fishing also took place in waters north of Sydney and around Montagu Island. Nearly all trawling took place more than three nautical miles offshore, beyond the limit of territorial seas, but throughout the existence of the STI the most important fishing grounds were those closest to Sydney. No haul data has survived from the private trawling industry before 1937, so the early movements of the privately owned steam-trawling fleets are not known. From 1937 to 1943 the fishing effort was distributed over most of the fishing grounds, although most effort was expended south off the coast, near Cape Everard. As WWII progressed, the fishing effort decreased south of Eden and intensified in the grounds closer to Sydney. Haul data covering the final years of steam trawling in NSW from 1952 to 1957 shows that the main fishing effort was again concentrated in the most southerly grounds and relatively less fishing took place on the grounds closest to Sydney (Klaer 2001).

The average depth of fishing between 1918 and 1923 during the time of STI was 75–100 m and the period after 1937 between 110 and 130 m, indicating a transition to progressively deeper waters as CPUE declined (Klaer 2001). Given that the continental shelf is quite narrow off NSW (maximum of about 40 km) and the continental slope begins at 150–190 m depth, in the 1930s the private steam trawling industry had expanded the fishing areas to the very edge of the shelf.

The expansion of the fishing areas meant more days at sea. The STI trawlers were usually at sea for 3–4 days, after which they returned to Sydney with their catches, but by 1930 the length of voyages varied between 9 to as much as 25 days (Commissioner of Taxation v. Cam and Sons 1936 1942).

Catches and the Effect on Fish Stock

Fish landings by the NSW steam trawlers from the east continental shelf show a progressive and pronounced increase from 1923 to 1929. In 1929 landings reached an all-time record of 6,665 t. During the 1930s landings declined steadily, until they suddenly plummeted between 1939 and 1946, when most of the steam trawlers were taken over by the Royal Australian Navy to be used as minesweepers. The post-WWII period saw a brief return to the pre-war level of landings, but landings soon continued their previous rapid decline, particularly after 1954 when the trawling company Cam and Sons ceased to operate.

Figure 10.2 also shows the gradual emergence of a non-steam trawling industry, gaining strength from 1936 when Danish seiners were introduced to the fishery.

---

[7] As part of the HMAP-South East Australia a comprehensive study of data extracted from historic trawling logbooks have been done Neil Klaer to estimate absolute biomass trends. See Klaer (2006).

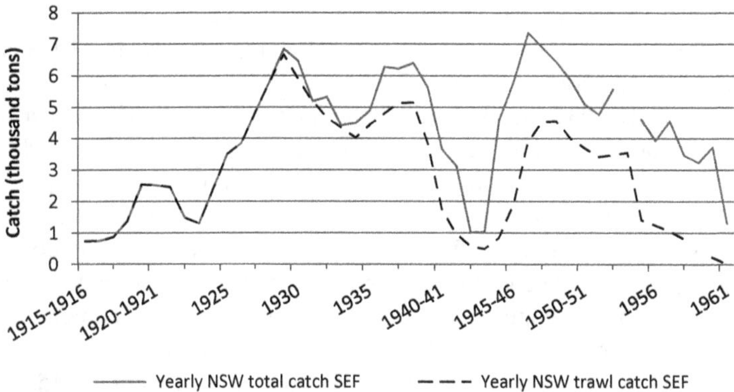

Fig. 10.2 Total catches in tonnes by NSW steam trawlers and other fishing vessels from fishery at the south east continental shelf (SEF). (Source: Klaer 2006)

Their number increased after WWII, culminating in 1947–1948 when 134 licenses were issued (Houston 1955). By the mid-1950s ocean landings from the seining and small vessel fishing industry were greater than landings from the steam trawlers.

In 1920 the trawl fishery was well-established and many of the fishing grounds on the continental shelf had been identified and were regularly trawled, affecting fish abundance. As the private industry took over in 1923 and increased the fishing effort, the stocks of the main target catch, flathead but also leatherjacket and latchet, began to be affected by the pressure. By the second half of the 1920s the first early signs that trawling was detrimentally affecting and changing the marine environment on the continental shelf can be detected (Klaer 2006). In his research on the historical catch records, Klaer has calculated the CPUE and how the structure of the main demersal fish communities was affected by fishing effort. For the first 5 years of steam trawling in NSW, CPUE continued to increase, and did not level off before 1920. This suggests that, in the early years of operation, the industry was still developing and not operating with maximum effect. However, over time, an increasing level of knowledge allowed trawlers to improve catch rates. This 'period of learning', as Klaer calls it, continued until about 1920 at which time CPUE levelled off (Klaer 2006). Dealing specifically with the economically important Botany Ground, Klaer's calculations show the same trend of increasing CPUE from 150 to 300 kg per hour from 1918 to 1923. After the STI left the fishery and the private industry took over in 1923 CPUE appears to level off, although the lack of data makes it impossible to fix the exact time when CPUE began to fall. However, archival research has shown that by 1928 the valuable flathead fishery on the Botany Ground had collapsed, throwing the trawling companies into crisis (Jacobsen 2010). Similar patterns of briefly increasing and then declining CPUE were found on other major fishing grounds although slightly later than on the ground near Sydney (Klaer 2006).

Without a complete time series of catch data it is impossible to estimate the precise impact of fishing on the marine environment during the second half of the

1920s and first half of the 1930s. Based upon the data available, it is obvious that CPUE dropped dramatically for all the known fishing grounds after 1923, suggesting that the fishing effort increased and/or the fish stocks were reacting negatively to exploitation. By 1937 the average CPUE for the steam trawling fleet was less than 50% of that achieved in the early 1920s (Klaer 2006). Time series data available for the post-war period show a significant decline in CPUE compared to the earliest data and radical changes in species composition; this finding has been confirmed by Klaer's modelling of fish stocks, which shows a steady decline in flathead, leatherjacket and latchet biomass and abundance indices between 1915 and 1961 (Klaer 2006). His population modelling also indicates a permanent biomass reduction in three out of the four studied fish species. In the case of tiger flathead the population was fished down to a low level of about 20% of its pre-1915 stock size in the 1950s and 1960s, but has today recovered to its current level of 40% (Scandol et al. 2008). Klaer judged fishing the most likely cause of the reduction in biomass. From this information it is clear that the steam trawl fishery was unsustainable, and in decline from the late 1920s.

Decline in Landings and Economic Recession, 1928–1938

Two events halted the progress of the steam trawling companies by the end of the 1920s; the downturn of the Australian economy during the Great Depression, and the collapse of the valuable flathead fishery at the Botany Ground in 1928.

The 1920s had been characterised by significant public spending on infrastructure funded by overseas borrowing (mainly English capital) to stimulate economic growth. The State Government's investments had heavily favoured activities in urban and metropolitan areas (Butlin et al. 1982). The collapse of the American stock market in 1929 triggered what is known as the Great Depression, a global economic downturn that lasted until the late 1930s. Australia's economy was hard hit by the Depression, because the nation relied on the export of agricultural products. When the prices of wool and wheat fell in early 1929, and exports dropped along with the withdrawal of English capital that had fuelled many of the 1920s public projects, Australians faced a severe financial crisis (Clark 2006/1963). Of all the states NSW was worst hit by the economic downturn. Unemployment in the industrial sector rose to a point where one in three unionists in NSW was unemployed in 1932. The highest rate of unemployment was found in the State's industrial districts and throughout the cities (Kingston 2006).

A stagnant economy and reduced economic activity hit the trawling companies hard. With the trawling companies selling their catches nearly exclusively at the Municipal Market they were vulnerable to changes in urban demand for their product. Due to its relatively high price fish was often considered a luxury food item for the average household and only when landings were abundant, and prices were lower than normal, could fish be considered a basic staple on a par with meat (Fisheries Branch 1929). Falling demand, low prices and increased costs strained the finances of the companies, a situation that lasted for most of the 1930s. Furthermore, the flathead fishery at the Botany Ground collapsed during the season 1927/1928 and overall catches started to decline. Between 1928 and 1932 catches declined by 30 % (Jacobsen 2010). The Fisheries Department estimated that trawl landings in general had declined from an average of 400 boxes (12.7 t) of fish per vessel per week in 1926 to 360 boxes (11.4 t) in 1927, to only 270 boxes (8.6 t) of fish per week in 1928 (Fisheries Branch 1929).

Decreasing productivity had a serious effect on the financial situation of the companies. Estimates prepared by Colin W. Mulvey, in February 1928 for the Australian Fisheries Conference, calculated costs and earnings of a trawler engaged in the Sydney trawl fishery in 1927 and 1928. Based upon an average catch in 1926 of 400 boxes per week sold at 21 shillings, total costs including commissions and fees (organisation expenses not included) amounted to £ 16,070 making a profit of £ 4,930 before taxes per year per vessel. Because those vessels built in NSW were about 50 % more costly in insurance and depreciation than their British counterparts, and generally more expensive to run, they only earned a profit of £ 3,760 before tax (Mulvey 1928). Using Mulvey's information about costs and earnings, it is possible to roughly estimate the economic impact of reduced landings. If landings in 1928 dropped to 270 boxes per week per vessel and were sold for the same price (21 shillings), the total cost per vessel would be £ 14,800 but the profit would instead turn to an annual loss of £ 623, and even more for the trawlers built in NSW.

As the Botany Grounds yielded less fish, fishing efforts on the more distant grounds intensified, raising production costs. In addition the companies had continued to invest in second-hand trawlers to keep up production (the trawling fleet increased from 11 to 18 vessels in 1929) which had to be paid off. The economic downturn beginning in 1929 would also have contributed to reduced earnings. Based upon the above estimates, there is no doubt that from about 1928 the industry experienced growth in operational costs and decline in earnings causing significant financial stress, especially when the catch from the Botany Ground did not recover. The collapse of this important, near-city fishing ground was symptomatic of a general depletion of the NSW trawling grounds.

When the catch declined continued in the 1930s (see Fig. 10.2) the trawl companies tried several strategies to improve their financial situation. Several times they used their market dominance to improve wholesale prices by withholding landings from the market by laying up vessels (Fisheries Branch 1929). Also there were attempts to land in Melbourne to build up a market for trawled fish there, but the venture was discontinued (Jacobsen 2010). To increase catches and make their trawlers pay both Red Funnel and Cam and Sons turned their attention to trawling grounds in New Zealand waters. From 1934 Cam trawlers were regularly fishing New Zealand grounds, landing their catch in Sydney (Johnson 2004). In 1938 they were joined by a trawler from Red Funnel. Trawling in New Zealand waters was not the only alternative fishing strategy Red Funnel explored in the 1920s. Danish seine boats, had been introduced into the shelf fishery in 1936, and by 1937 Red Funnel had acquired its own large seiner which it operated until WWII. Being the most versatile of the three trawling companies Cam and Sons also turned to coal mining during the 1930s to reduce costs, since coal was the source of fuel used to power the trawlers' steam engine and made up a large part of the trawlers' working expenses (Cooke 2006).

The Trawling Companies' Situation by 1939

By 1939 all the known trawling grounds on the Australian South East continental shelf were fully exploited, and the trawlers had increased their trawling depth and extended the fishery to the edge of the shelf. The companies' situation had improved somewhat as the Depression eased off, but the general economic outlook for the industry was still not promising.

Cam's venture into trawling in New Zealand and branching out to coal had given Cam a competitive edge over the other trawl owners and allowed the company to remain viable. However the company was left with a reduced and aging fleet compared to its situation in 1929.

Red Funnel Fisheries had also recovered somewhat from economic hardship and the decline in landings caused by overfishing of the continental shelf. After 1929 its shares had plummeted and by 1933 the company sold of some of its trawlers and was reconstituted into a new enterprise. This gave the company some much needed economic stability.

While the other two trawling companies had extended their operations to include other types of gear or fishing grounds, A. A. Murrell had continued to rely exclusively on fishing with steam trawlers on the continental shelf, supplying primarily its own shops and wholesale business. By the end of 1939 Murrell had virtually left the fishing industry, having sold all of his trawlers, and relied on its fish merchant business.

Second World War, 1939–1945

The outbreak of WWII radically changed the situation of the trawling companies, and brought them out of the economic deadlock they were in. The Royal Australian Navy (RAN) needed fishing trawlers for minesweeping in Australian waters and had to rely on the private companies to supply them. Between September 1939 and June 1942 the RAN requisitioned 12 out of 14 trawlers leaving only *Bareamull* from Red Funnel and *Dur-een-bee* from Cam and Sons to supply the market. The situation was initially not without appeal to the trawl owners. The Navy's requisitions reduced the companies' burden of maintaining and running aging vessels and the charter hire paid by the Navy provided the companies with a stable source of income, instead of having to rely on the return from fishing the declining resources on the continental shelf.

Steam trawlers were an essential part of Australia's naval defence and the owners, controlling the only fleet of large trawlers in Australia, used the situation to their advantage. First they pressed for a significantly higher rate than the one proposed by the RAN, which was based upon rates given by the British Admiralty, but with only limited success (Navy Office 1939–1946 [674/201/3230]). Later in 1941 the trawl owners suggested that the Navy buy the vessels outright instead of hiring them; they assured the Navy that all the trawl owners were prepared to repurchase the vessels after WWII (Navy Office 1939–1946 [674/203/442]). From June 1943 the RAN began to buy all the requisitioned trawlers. As the value of fishing vessels had increased dramatically since the outbreak of WWII, the trawling companies gained a considerable profit they could use to reinvest in newer trawlers after the war.

Just how significant the profit was can be illustrated by the sale of Cam and Sons trawlers. In June 1943 the RAN purchased all six steam trawlers[8] requisitioned from Cam and Sons for a total sum of £ 65,000. All of Cam's trawlers were between 23 and 30 years old at the time of purchase. In 1934 five Cam trawlers, which were of similar type and age to the ones purchased by the RAN, had been valued at £ 12,500 total. Even if the Cam trawlers' value in 1934 might then have been estimated below market prices for taxation purposes, the rise in value between 1934 and 1943 was impressive. Given that the trawlers were old and poorly maintained at the outbreak of the war Cam made a generous return on their initial investment.

Fishing During the Second World War and the Immediate Post-War Era

As WWII continued the RAN requisitioned more steam trawlers along with most of the independently owned seiners, and fish landings dropped dramatically while market prices rose. The fishing effort was severely affected, as by 1943 the fishing fleet was down to one steam trawler Red Funnel's *Bareamull*, which the Navy due to the deteriorated state of the wooden hull, had decided was unfit for service. Consequently the annual steam trawl landing was as low as 1,032 t, but on the other hand prices had more than doubled compared to pre-war levels (Klaer 2006). Although lacking a fishing fleet the other companies did not totally miss out on the favourable market conditions, as they instead increased their agent activities and began selling fish from the inshore fishing industry. In second half of 1944 the RAN also released a trawler to each of the three trawling companies, to increase productivity.

After WWII the conversion of the trawlers back to fishing vessels and their return to the owners was slow due to a labour shortage, congestion of dry docks and slipways, and industrial trouble. Most of the seven remaining trawlers were not

---

[8] SS Alfie Cam, SS Beryl II, SS Goonambie, SS Mary Cam, SS Oliver Cam and SS Samuel Benbow.

returned before the second half of 1946. When the trawlers were surveyed before being refitted, it was discovered that many were in great need of repair. In the case of Red Funnels *Goolgwai* and *Durraween*, Lloyd's surveyor in May 1946 found that the all of the steel decking, and a large part of the superstructure and engine room casing had to be replaced for the vessels to be in class. The officer in charge assessed that the cost of the repairs and necessary reconversion amounted to about £ 22,000 for *Goolgwai* and £ 20,000 for *Durraween*. Despite the damage being due to normal wear and tear mostly sustained before the RAN took over the ships, the Navy had very little luck with getting Red Funnel to share the costs (Navy Office 1939–1946 [674/201/3230]). In July 1946 Red Funnel successfully argued that the charter rate had been too low compared to what the company could have earned if the trawlers had been engaged in fishing during the time the vessels were charted, and since the trawlers had been in class when they were taken over by the navy it was the RAN's responsibility to bring them up to standard. In the end the Navy bore the brunt of the costs, acknowledging that they returned the vessels: 'in a condition vastly superior to that in which they were taken from the owners' (Deputy Super-intendent to the Department of the Navy, 10 May 1946, in Navy Office 1939–1946 [674/201/3230]). Although all three major trawling companies complained that they had lost money by chartering their vessels to the Navy rather than using them for fishing, there is no doubt that the companies ultimately benefited financially from WWII.

With a newly reconditioned fleet, high wholesale prices, a comfortable net worth and relatively high catch rates, the trawling companies were ready to renew their in-terests in steam trawling. Accordingly both Red Funnel and A. A. Murrell invested in newer trawlers from New Zealand. Red Funnel, which had lost one of its trawlers during WWII, bought in 1946 three newly built, former Royal New Zealand Navy minesweepers, of a similar design to those already fishing from Sydney, for less than a fourth of the original delivery price.[9] A. A. Murrell bought the two sister vessels under similar conditions the same year (Waters 1956). The very reasonable price for ex-navy vessels allowed the companies to invest in several trawlers at the same time, instead of having to expand their fleet more gradually. It also boosted the activity of the trawling companies in similar ways to that which occurred after the sale of the state trawlers in 1923.

Changes in Species Composition

Despite the capital injection and rejuvenation of the fleet that had taken place after the war, the optimism was short-lived and during the 1950s and early 1960s the economics of the trawling industry deteriorated further. The reason for the collapse was found in the changes in catch composition caused by over-fishing and the sub-sequent fall in earnings.

---

[9] A forth conventional trawler SS *Mulloka* was bought by Red Funnel around the same time.

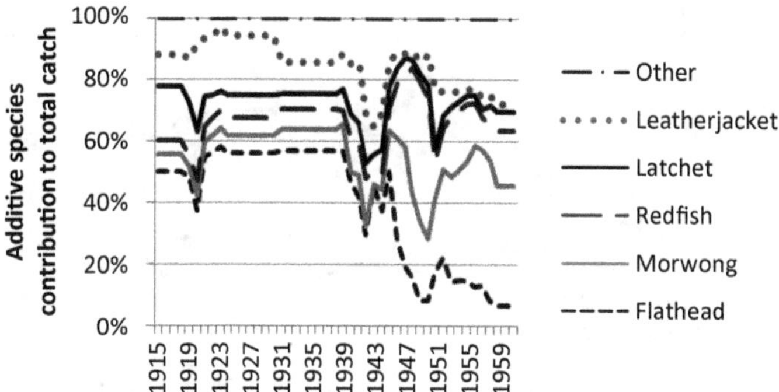

Fig. 10.3 Catch proportions by species for steam trawlers, 1915–1961. (Source: Klaer 2006)

Throughout its history the NSW steam trawling industry targeted several species, but flathead was the main target species and the most valuable in terms of price per kilogram. Klaer's estimate of the changes in the species composition of the catch reveals that after 1939 there was a clear trend towards decline in catches of flathead and an increase in catches of less valuable species such as morwong and redfish (Fig. 10.3).

Until 1939, flathead comprised about half the catch of the steam trawlers, but after 1939 the proportion declined rapidly until flathead made up only 6.8 % of the catch in 1961. Leatherjacket and latchet followed a similar trend, their proportion of the catch being reduced slowly since 1930 and nearly disappearing in the post-war period, although latchet in the final years increased to post-war levels. Catches of morwong and redfish increased dramatically, especially in the post-war period. In 1915 morwong only made up 5.6 % of the catch but in 1961 it accounted for as much as 38.7 % of all catches. The increase in redfish was more modest, from 4.5 % in 1915 to 17.9 % in 1961. By the end of the period 'other species' made up nearly one third of landings. Changes in species composition were most likely due to the impact of previous decade's unrestricted exploitation of the continental shelf.

Fish Prices and Earnings

The changes in species composition documented after 1939 had a profound impact on the earnings of the trawling companies. Recordings of monthly and annual prices received for fish by Red Funnel Trawlers is available from 1939 to 1952 (Klaer 2006). The data can be used to estimate a pattern for value of fish in NSW during the period, as well indicate overall trends in trawling companies' earnings.

Figure 10.4 clearly shows a significant increase in prices from 1941 and a gradual fall during the post-war period, but by 1952 prices were still higher than before

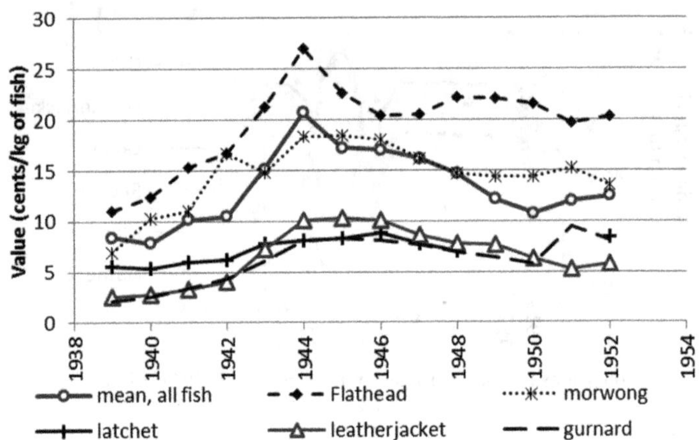

Fig. 10.4 Average price received by Red Funnel of main species per financial year, 1939–1952. Converted by Klaer to cents per kilogram. (Source: Klaer 2006)

the WWII. A system of fixed prices for fish had been introduced during the war,[10] to avoid overpricing and black marketing. Still the prices received by Red Funnel during WWII and in the immediate post war period were significantly higher than peace-time prices. However, the changes in species composition reduced the financial benefit of the high prices. Despite the fact that landings increased until 1948/1949 (Fig. 10.2) because of increased fishing effort, the overall value of landings began to decrease, due to the changes in species composition.

The trend in diminishing value of the catch, become even clearer when allowance is made for the number of trawlers involved in the industry. There was a dramatic increase in earnings per vessels from 1941 to 1944, but thereafter earnings per vessel began steadily to decrease (Table 10.1). By 1949 earnings per vessel were lower than in 1939. It was the valuable flathead catch that kept the total earning per steam trawler up. In 1939 flathead comprised 57% of total catch in weight by the NSW fleet, but 74% of the total value of annually landings, calculated from the prices recorded by Red Funnel. By 1952 flathead comprised only 22% of the catch, but totalled 36% of the value of annual landings. As the downward trend in flathead catches continued after 1952 so did earnings per vessel, throwing the companies into a rapid economic decline.

The experience of post-war decline was not unique to the NSW steam trawl industry. Similar experiences were had in the British near- and middle-water fishery, where many of the fishing vessels were steam trawlers. After being released from naval service British trawlers were experiencing high catches because the stocks had enjoyed a respite from fishing during the war, but catches began to decline again around 1946. At the time fish prices were high because of food shortages, but in November 1949 prices collapsed after the Government removed

---

[10] In December 1951 fish was removed from price control.

Table 10.1 Annual earning per steam trawler, based upon total landings and prices received by Red Funnel Fisheries, 1939–1952. (Source: Klaer 2006)

Cents/kg[a]	Flathead	Gurnard	Latchet	Leatherjacket	Morwong	Mean, all fish
1939	10.99	2.17	5.60	2.56	6.91	8.44
1940	12.39	2.56	5.35	2.75	10.28	7.87
1941	15.26	3.39	5.99	3.26	10.98	10.16
1942	16.71	4.30	6.25	4.06	16.71	10.49
1943	21.29	6.15	7.77	7.20	14.76	15.13
1944	27.01	8.07	8.08	10.09	18.31	20.71
1945	22.64	8.31	8.30	10.24	18.42	17.19
1946	20.40	8.05	8.84	10.06	18.07	17.01
1947	20.56	7.72	7.73	8.58	16.18	16.18
1948	22.21	6.99	7.00	7.78	14.66	14.66
1949	22.11	6.39		7.70	14.39	12.13
1950	21.60	5.78		6.26	14.32	10.74
1951	19.74	9.50		5.25	15.22	11.97
1952	20.37	8.09	8.43	5.80	13.54	12.45

[a] Index 100=1945

price controls. Because of the combination of declining catches caused by long-term overfishing and low post-war prices, the rate of vessel replacement was slow (Starkey et al. 2000). By 1952, 78 % of the British near- and middle-water fleet were built before 1921—the figure for the NSW steam trawling industry was very similar at 77 %.

By the early 1950s only Cam and Sons and Red Funnel remained in business, after A. A. Murrell had sold off its three trawlers around 1947. Cam was fishing with five trawlers[11] and Red Funnel with seven trawlers.[12] In September 1950 the Secretary of Red Funnel, Gordon Francis Thomson and the Managing Director of Cam and Sons, Rocco Edmund Cam testified to an Industrial Commission that both companies had sustained substantial losses during the last 2 years (*Sydney Morning Herald* 12 September 1950, p. 10). In October 1954 Cam and Sons announced that they were discontinuing business (*Fisheries Newsletter*, November 1954, 13[11], p. 21) Two of its former trawlers were taken over by family members (*Fisheries Newsletter* January 1955 14[1], p. 15), who for some years continued to operate the trawlers as individual owners, but in 1965 the trawlers were finally scrapped after having been laid up for years (Cooke 2006).

Due to financial difficulties, Red Funnel was also gradually forced to lay up its fleet, with the last trawlers ceasing fishing in 1958. The reason why Red Funnel was able to continue longer than Cam and Sons was likely due to the fact that the four

---

[11] SS *Alfie Cam* (built 1920), SS *Beryl* (built 1914), SS *Goonambee* (built 1917) SS *Mary Cam* (built 1918) and SS *Olive Cam* (built 1920).

[12] SS *Durraween* (built 1918), SS *Goolgwai* (built 1918), SS *Korowa* (built 1919), SS *Maldaana* (built 1942), SS *Matong* (built 1944), SS *Moona* (built 1943), and SS *Mulloka* might not have been added to the fleet before 1955. SS *Bar-ea-mul* was scuttled in December 1950 after having been laid up for several years.

active trawlers had an average age of about 11 years in 1955, while Cam's vessels were much older. Another reason for Cam and Sons decision to wind up its operations was probably the fact that the shareholding members of the Cam family were near retirement age. In July 1959 Captain Products Ltd. of Sydney,[13] which was a producer and distributor of canned fish products, acquired shares in Red Funnel Trawlers and set out to resume the fishery operating from No 5 Wharf at Woolloomooloo (*Fisheries Newsletter* July 1959 18[7], p. 9) In February 1960, after an extensive overhaul the fishing was resumed by one trawler (Roughley 1961). The plan was to gradually expand the activities to include all the trawlers, but it never happened and by 1961 the company had withdrawn its trawler and the NSW steam trawling industry was no more.[14] While the activities of the trawlers had been particularly destructive to the large fish inhabiting the ecosystem, the closure of the industry did not end the exploitation of marine resources; instead new industries emerged. New technologies enabled smaller boats to continue the fishery on previously inaccessible grounds, and fisheries for unexploited species and on deeper grounds were developed with the aid of the Commonwealth Government.

Conclusion

Steam trawling in NSW has a unique history because it was first promoted by the State before it became a fishery operated by private companies. Whilst the NSW Government's attempt to run a commercial steam trawling industry during 1915–1923 was an economic failure, it did succeed in proving it was possible to catch large quantities of fish on the continental shelf, and to establish a market for these fish in Sydney. Thus failed state entrepreneurship or 'colonial socialism' helped paved the way for private success.

The success of the private steam trawling industry was based upon access to affordable equipment, resources and capabilities left over from the NSW Government State Trawling Industry as well as easy access to reasonably productive fishing grounds. The 1920s was the golden era of steam trawling and the size of the fleet increased significantly during this time. The decline of the industry began around 1928 when the fishery at the Botany Ground collapsed and catches stated to fall. The economic problems were later aggravated when the depression reduced consumer demand and increased costs. The general reduction and ageing of the trawling fleet by 1939 illustrate the financial stress the companies were under during the 1930s.

The companies' financial circumstances improved during WWII when the Royal Australian Navy leased or bought most of the trawlers for minesweeping at very profitable rates. At the end of the war the industry had somewhat recovered from the long decline during the 1930s. The recovery was not achieved because of a wiser

---

[13] Formerly Downs Holdings Ltd.

[14] Red Funnel Trawlers Ltd. continued to hold the lease to the premises on No 5 Wharf until the 1980s where the buildings were demolished as part of the Woolloomooloo harbour-front development. The company still exists but has ceased trading.

use of marine resources, but was caused by improved market conditions which continued into the post-war period. The final collapse of the private trawling industry over the period from 1954 to 1961 was caused by the combination of overcapacity in the trawling fleet, the return to a free-market price system in December 1951, combined with long-term decline in landings of the most valuable species (flathead). Together these three factors undermined the economy of the industry which resulted in the demise of the steam trawl industry in NSW.

The era of steam trawling also had a long lasting effect on the marine environment. Klaer's modelling of fish stocks shows a steady decline in flathead, leatherjacket and latchet biomass and abundance indices since 1915 (Klaer 2006) His population modelling indicates a long-term biomass reduction in three out of the four studied fish species. In the case of tiger flathead the population was fished down to a low level of about 20 % of its pre-1915 stock size by the 1950s, but has today recovered to its current level of 40 %. The recovery is due to the introduction of mesh size regulations following extensive biological research and some reduction in fishing effort (Scandol et al. 2008).

Acknowledgments   The author wishes to thanks the participants at the HMAP-Asia workshop 2009, under the auspice of the Census of Marine life. Credit for funding my research goes to school of Geography and Environment, University of Tasmania. I am very grateful to Malcolm Tull and Joseph Christensen, Murdoch University, for valuable comments and help with the manuscript and to Kevin Rowling, Cronulla Fisheries Centre, DPI NSW, for his observations on fisheries research. Thanks also go to the following people for their advice and assistance: Elaine Stratford, University of Tasmania, Neil Klaer, CSIRO, René Taudal Poulsen, Copenhagen Business School, and Bo Poulsen, Aalborg University.

References

Butlin NG, Barnard A, Pincus JJ (1982) Government and capitalism. Public and private choice in twentieth century Australia. George Allen & Unwin, Sydney

Clark M (2006/1963) A short history of Australia, 4th rev. Penguin Books, Ringwood

Commissioner of Taxation v. Cam and Sons 1936 (1942) 14 ALJ 162

Cooke S (2006) Oceans, rivers and dreams. The story of the Cams and Cam & Sons. Stuart Cooke, Sydney

Fisheries Branch (1929) Annual report on the fisheries of NSW for the year 1929. Sydney: Legislative Assembly of NSW

Herlihy FJ (1927) Chief Secretary's Department, September 1927 [p. 36], NSW State Trawling Industry Historical Records 1915–1923, Lif Jacobsen Private Collection

Horn C (1916) Statement of Captain Charles Horn, 14 June 1916. Public Service Board: inquiry in the State Trawling Industry, 1916–1917 [8/513.3]. State Records of New South Wales

Houston TW (1955) The New South Wales trawl fishery: review of past course and examination of present condition. Aust J Mar Fresh Water Res 6(2):164–208

Jacobsen ALL (2010) Steam trawling on the south-east continental shelf of Australia. An environmental history of fishing, management and science in NSW, 1862–1961. PhD thesis, University of Tasmania, Hobart

Johnson D (2004) Hooked: the story of the New Zealand fishing industry. Hazard Press Limited for the Fishing Industry Association, Christchurch (Edited and completed by Haworth J)

Kingston B (2006) A history of New South Wales. University of New South Wales, Sydney

Klaer NL (2001) Steam trawl catches from south-eastern Australia from 1918 to 1957: trends in catch rates and species composition. Mar Fresh Water Res 52:399–410

Klaer NL (2006) Changes in the structure of demersal fish communities of the south east Australian continental shelf from 1915 to 1961. University of Canberra

Lorimer M (1984) The technology and practices of the New South Wales fishing industry 1850–1930. MA thesis, University of Sydney

Mulvey CW (1928) Letter from Colin W. Mulvey to S. Fowler, Secretary to the Australian Fisheries Conference, 28 February 1928. Documents relating to the Australian Fisheries Conference 1927 and 1929, Lif Jacobsen private collection

Navy Office (1939–1946) Correspondence. MP150/1/0 [674/201/3230; 674/203/442]. National Archives of Australia, Canberra

Red Funnel Fisheries (1925) Item 9920, Red Funnel Fisheries Ltd., Memorandum and Article of Association of the Red Funnel Fisheries Ltd., 2 December 1925 [17/5605]. State Records of New South Wales

Red Funnel Fisheries (1926) Item 9922, Red Funnel Fisheries Ltd., Memorandum of satisfaction of mortgage of change July 1926 [17/5605]. State Records of New South Wales

Roughley TC (1916) Fishes of Australia and their technology. Technical education series no. 21, R. T. Bake edn. Technological Museum Sydney, Sydney

Roughley TC (1961) Fish and fisheries of Australia, 4th edn. Angus & Robertson, Sydney

Scandol J, Rowling K, Graham K (2008) Status of fisheries resources in NSW 2006/07, vol 334. NSW Department of Primary Industries, Cronulla

Starkey DJ, Reid C, Ashcroft N (eds) (2000) England's sea fisheries. The commercial sea fisheries of England and Wales since C.1300. Chatham Press, London

State Fisheries (1922) Letter, 22 February 1922 [4/6636.2]. State records of New South Wales

State Trawling Industry (1920) Papers relating to the 1920 inquiry into the State trawling Industry [5/5366]. State Records of New South Wales

Tyler PJ (2006) Humble and obedient servants vol 2: 1901–1960. The Administration of New South Wales, UNSW Press, Sydney

Waters SD (1956) The Royal New Zealand Navy, Historical Publications Branch, Wellington

The Wealth and Progress of New South Wales (1887–1902) Compiled by T.A. Coghlan for New South Wales Statistician's Office. 2nd issue (1887/88)–13th issue (1900/01). Government Statistician's Office, New South Wales, Sydney

Chapter 11
Exploiting Green and Hawksbill Turtles in Western Australia: The Commercial Marine Turtle Fishery

Brooke Halkyard

Abstract Many attempts were made to exploit both the green and hawksbill turtle commercially from the mid-1800s. The first commercial export of hawksbill tortoiseshell appeared in the Western Australian trade tables in 1869 and the green turtle fishing industry operated intermittently between 1870 and 1961 prior to the industry becoming successfully established in the 1960s. Historical evidence suggests that up to 55,125 (archival records) and 69,000 (oral histories) green turtles were potentially harvested from Western Australian waters prior to the industry being closed down in 1973. Upper estimates indicate that 20,445 hawksbill turtles were harvested from northern Western Australia over the course of 84 years. It is argued that the exploitation of green turtles led to an observable decline in the numbers of these animals and it is likely that the fishing effort for the tortoise shell industry had an adverse impact on hawksbill turtle populations in the State's north-west. In a global context, the exploitation of the green and hawksbill turtles in Western Australia occurred at a time when there was an extensive international harvest of marine turtles. The relatively small-scale harvest that took place in Western Australia is likely to have been a factor contributing to the green and hawksbill populations of Western Australia being some of the largest populations remaining in the world. This research provides a detailed historical account of the commercial exploitation of marine turtles in Western Australia, including empirical accounts of the total number of animals harvested from turtle populations throughout the State.

Keywords Australian marine environmental history · Marine turtle fishing ·Ningaloo Reef · Green turtle (*Chelonia mydas*) · Hawksbill turtle (Eretmochelys imbricata)

Marine turtles, like many large marine vertebrates, have been commercially exploited for centuries. In many parts of the world, once abundant marine turtle populations have been decimated by the intensity of historical harvests (Bjorndal and

B. Halkyard (✉)
Department of Parks and Wildlife, PO Box 201, 6707, Exmouth, WA, Australia
e-mail: Brooke.Halkyard@DPaW.wa.gov.au

J. Christensen, M. Tull (eds.), *Historical Perspectives of Fisheries Exploitation in the Indo-Pacific*, MARE Publication Series 12, DOI 10.1007/978-94-017-8727-7_11,

Jackson 2003; Lutcavage et al. 1997). Six of the seven extant species of marine turtles occur in Western Australia: green (*Chelonia mydas*), hawksbill (*Eretmochelys imbricata*), loggerhead (*Caretta caretta*), flatback (*Natator depressus*), olive ridley (*Lepidochelys olivacea*) and leatherback (*Dermochelys coriacea*) (Limpus 2002). All marine turtles that occur in Western Australian waters are threatened species declared to be specially protected under the *Wildlife Conservation Act 1950*, though there are provisions for harvest by people of Aboriginal descent (CALM 2005) All species of marine turtle are also listed under the Commonwealth *Environment Protection and Biodiversity Conservation Act 1999* as 'threatened fauna' (DEC 2008). The hawksbill and leatherback are listed as critically endangered on the International Union for Conservation of Nature (IUCN) Red List, while the green and loggerhead as endangered, the olive ridley as vulnerable and the flatback as data deficient (IUCN 2009).

The purpose of this chapter is to recount the historical commercial marine turtle fishery of Western Australia which, at present, has only been partially documented (Limpus 2002; Halkyard 2005). Similar to the work conducted by Daley et al. (2008) on the Queensland turtle fishery, this chapter aims to provide an indication of the ecological impact of historical European practices on marine turtle populations in Western Australia based on documentary and oral history research.

Documentary evidence of the fishery was accessed primarily through the official records of the Western Australian State Records Office, chiefly the corporate files of what is today known as the Department of Fisheries. Other records were accessed through the Department's library, including annual reports produced by various Chief Inspectors of Fisheries. Export statistics were obtained from the annual editions of the *Colony of Western Australia* (1837–1869), the *Blue Book* (1870–1898) and the *Statistical Register of Western Australia* (1898–1968). Further documentary evidence was obtained from the journals of early explorers and other publications such as local and regional histories. Oral history sources included interviews with three fishermen who were engaged in the turtle fishing industry during its peak in the 1960s and early 1970s (interview transcripts in Halkyard 2005), and transcripts of earlier interviews. Interviews were conducted in 2005 using a methodology adapted from Weaver (1998).

As with Daley et al. (2008) and their synopsis of the Queensland marine turtle industry, a qualifying statement is needed for the reconstruction of the Western Australian marine turtle fishery. An exhaustive search of the records has been conducted in an attempt to obtain as much detail as possible regarding the operation of the industry. However, the documentary evidence is incomplete and many catch returns are missing, possibly because figures for green turtles were treated as confidential (Department of Fisheries and Fauna 1966a). Furthermore, export and catch figures were often inconsistent and open to interpretation, a point which was noted by the Director of Fisheries in 1973 with respect to export weight production figures:

> Even these figures are open to some question as the specification of the export has changed from time to time so that one is not sure that the figures are comparable. (17 May 1973, in Department of Fisheries and Fauna 1950)

Additionally, turtle research and monitoring has only been undertaken in Western Australia since the early 1980s (DEC 2008), meaning the marine turtle fishery in operated with no scientific baseline data on marine turtle populations. As a consequence of these factors, the extent of depletion of marine turtles cannot be estimated with confident precision and thus, estimates of population sizes prior to exploitation, based on historical evidence, has not been attempted in this chapter.

Commercial Marine Turtle Fishery in Western Australia 1869–1973

European exploitation of at least two marine turtle species has occurred in Western Australian waters since the earliest European exploration. William Dampier regularly recorded and harvested marine turtles during his visits to Western Australia in 1688 and 1689 (Limpus 2002) as did Philip Parker King during his surveys of the WA coast from 1818 to 1822 (King 1827). John Lort Stokes and his crew aboard the *Beagle* caught green and hawksbill turtles for personal consumption and for delivery to the Swan River Settlement during their surveys of the north-west coast from 1837–1843 (Stokes 1846). In 1864, Captain John T. Jarman and his passengers conducted a turtle hunt on the east coast of Barrow Island where 'the bays were swarming with them' (Cox 1977).

The abundance of green and hawksbill turtles and the potential for a large-scale turtle fishing industry on the north-west coast was recognised by the late 1800s, and frequently reported on by the Western Australian Government:

> Turtle of the most valuable qualities, including the aldermanic green turtle, Chelone Mydas, and the tortoiseshell producing hawksbill, Chelone imbricata, abound on the Western Australian coastline on Houtman's Abrolhos, and from Sharks Bay northwards. Excepting for local consumption no attempts have hitherto been made to turn these abundant natural supplies to practical account. There can be no doubt that there are numbers of locations on the Nor'-West coast, such as the Lacepede Islands, whereat extensive and profitable stations might be established for the wholesale export of the living animals, and for the curing of preparation of those commercial products of the turtle which have hitherto been mainly obtained from the West Indies and the Island of Ascension. (Fraser 1896)

The first commercial export of hawksbill tortoise shell appeared in the WA trade tables in 1869 and intent to commercially exploit green (or 'edible') turtle surfaces around the same time. The hawksbill turtle was primarily targeted for its tortoise shell, whereas the trade in green turtle products included turtle meat and extract for turtle soup (calipash and calipee), turtle oil (which was thought to have medicinal properties) and skin for leather. Its shell was considered far inferior to that of the hawksbill.

Applications to the government for exclusive turtle fishing rights over particular areas began to appear in the 1870s and several attempts were made to commercialise these animals without any marked degree of success. Applications to lease north-west islands for the purposes of turtle fishing appear in the records from 1871, though it is not clear to what extent turtle fishing was carried out (Cox 1977; Lands and Surveys Department 1881). It is possible that these were the first applications received by the Government for lease of an island other than for pastoral purposes (Cox 1977).

Prior to 1911, it was not possible to obtain an exclusive license (i.e. only one licensee with exclusive rights to fish for turtle within a given area) to take turtles under Section 30 of the *Fisheries Act 1905*. In recognition of the need for security of tenure to encourage investment in the turtle fishing industry, the Chief Inspector of Fisheries attempted to bring about a change in the *Fisheries Act 1905* to allow the granting of exclusive licenses for taking of turtles.[1] In 1911 the Act was amended by Parliament, which permitted the granting of exclusive licenses for the taking of green turtles only (Department of Aborigines and Fisheries 1919). The Fisheries Department instigated a further amendment in 1921 (*Fisheries Act Amendment Act 1921*) which allowed the granting of exclusive licenses to take hawksbill turtles (Fisheries Department 1920). Exclusive licenses were limited to 75 miles of foreshore.

In 1900, a factory was established at Beagle Bay for processing turtles caught at the Lacepede Islands, otherwise known as the 'Home of the Green Turtle'. However, the factory closed down 1 year later due to 'bad management' (Fisheries Department 1901b; Gale 1901). In 1901, a factory was also established at Cossack (Duckett 1990) and in 1903 a small soup factory was erected at Point Peron, Rockingham, where the turtles from Onslow were received (Gale 1904). There is no further evidence in these records as to how many turtles were processed at these factories, although the *Statistical Register of Western Australia* confirms export of turtle soup and dried turtle to a value of £ 237 in 1901 and £ 65 in 1902.

The Chief Inspector of Fisheries was keen to see this branch of the fishery successfully commercialised and in 1909 the Fisheries Department took steps to introduce calipash and calipee to hotels and clubs in Perth and other populated areas. A small consignment was also sent to London to be valued. The calipash was valued at 1/6 per lb and the calipee, if removed from the bone before drying, was valued between 1/– to 1/3 per lb and as high as 1/6 per lb (Department of Aborigines and Fisheries 1909).

There were a few notable instances where the attempts to establish an industry were fruitful, albeit short-lived. At various times between 1910 and 1934, H. Barron Rodway held exclusive licenses to take green turtle between North West Cape and Cape Lambert including waters and foreshores of numerous adjacent islands (Department of Aborigines and Fisheries 1919). The license operated under the business name of *Chelonia Co. Ltd* and it would appear that William Benstead was employed to manage the operations. Due to the World War I, the company experienced significant difficulty in securing finance and commencing operations until 1922, when the building of a turtle soup factory commenced at Point Peron, although whether this was the same factory as the one established in 1903 is uncertain

---

[1] In this context, an exclusive license refers to a licensee having exclusive rights over an area for the taking of turtle. Licenses would not be granted for an area of foreshore greater than 75 miles.

(Fisheries Department 1919). This represented the most committed attempt to develop an export industry for the green turtle up to this time.

Turtle fishing vessels operated out of Cossack and the turtles were shipped live to the Rockingham facility. This method of transportation met with extremely limited success. The first shipment of turtles died because they were not regularly hosed down with seawater and while the second shipment fared better, many turtles were still lost. More animals perished in the company's attempts to pen the turtles in water considerably cooler than northern waters (Durant 2004). In 1923, it is believed that a consignment of green turtles (potentially no larger than 50 lb each) escaped through openings in the factory's holding pens (Taggart 1984; Durant 2004). It would appear that around this time the company ceased its operations at Rockingham and did not recommence elsewhere. The company went into liquidation in 1934 after experiencing difficulty in raising fresh capital. At least 169 turtles were harvested during the company's operations (Fisheries Department 1919).

Meanwhile, Benstead relocated to Cossack and in 1924, he secured an exclusive license for the foreshores and coastal waters of Depuch, Forrestiere and Turtle islands. In collaboration with the *Roebourne Produce Company*, a long-term lease of the abandoned Customs house at Cossack was secured for the establishment of a turtle soup factory (Fisheries Department 1924). Due to the Great Depression, the company experienced problems finding a market for its turtle products, with unfavorable reports that the soup was diluted and mixed with beef (Durant 2004). The company went into liquidation in 1931 after harvesting at least 834 turtles from the waters around Cossack and Onslow and exporting 408 cases (19,464 lb) of finished turtle soup. The *Roebourne Produce Company* also produced detailed notes on the preparation of turtle soup, the recipe of which stipulated that 900 lb of turtle (live weight) combined with 20 lb of beef would produce 560 lb of turtle soup (Carse 1925).

In 1922, the *Broome Turtle Preserving Company* preserved at least 50 turtles in various forms, both for the export trade (London markets) and local consumption. The company held an exclusive license for the waters surrounding the Lacepede and adjoining Islands up to and including Callience Reef and Adele Island. The license was transferred to *Lacepede Products Limited* in 1924 but the company forfeited its license in 1928 following difficulties in securing capital for ongoing operations (Fisheries Department 1922).

Between August and December 1933, the *Montebello Sea Products Ltd* (license transferred to *Australian Canning Company* in 1934) processed at least 334 turtles caught at the Montebello Islands to produce 12,840 lb of extract for turtle soup at the Cossack factors (Fisheries Department 1921a, b, 1934). The company attempted to expand its capacity for operations with the installation of new factory equipment; however it ran into financial difficulty and operations ceased in 1935 when the company went bankrupt. During its peak operating phase, the factory was capable of treating 40 turtles per week, with an adjacent pen capable of holding 50–60 turtles (Fisheries Department 1934).

In 1939, *Westella Canning Company* obtained an exclusive license to take turtles from the coastal waters surrounding the Montebello Islands. (Fisheries Department

1939). The company processed its turtles at a factory in Belmont and for a few months, its turtle soup production was a financial success. However, due to a shortage of tin-plate and labour created during the Second World War (WWII), the company cancelled its license (Fisheries Department 1934). During its turtle soup making operations, the company received several shipments of turtles as a sideline to its other canning operations. At 30 June 1939, these shipments were valued at £ 77.5.8 (Department of Industrial Development 1938). Assuming values of 10/– per animal (as per the value received by *Montebello Sea Products Ltd.* in 1934), the *Westella Canning Company* would have received approximately 155 turtles in the space of 6 months.

In 1958 James Antonio Mazza and Alfred Robert Eric Russell of *North West Enterprises* conducted extensive market research in relation to green turtle products. They established that there was a market in the UK, USA and probably Europe for frozen turtles and dehydrated meats. They estimated the market value for 10 t of dried meat or 100 t of frozen meat (approximately 1600 turtles) at just over £ 22 (Fisheries Department 1958). This positive forecast contrasted with examinations by an Advisory Committee for the Minister for Industrial Development which stipulated that no overseas market had been proven and that there was no definite prospect of a continuing market (Department of Industrial Development 1959). After processing a sample of 25 turtles, the venture appeared to lapse after a request for financial assistance from the Western Australian government was rejected (Fisheries Department 1958).

1960–1973: The Advancement of the Industry

Prior to 1960, turtle fishing had been conducted on a sporadic basis but never to a significant commercial extent. Industry speculation suggested that the only way to establish an economical industry was to operate from a freezer vessel and return the catch to Fremantle for treatment. By 1961, *West Coast Enterprises* was using freezer boats to harvest turtles at a level unprecedented in the history of the industry. The company reportedly processed 40 t of turtle meat in its first 6 weeks of operation, from turtles caught at the islands of the Dampier Archipelago. The turtles were caught by 20 men using a fleet of six small catcher boats and processed aboard two large freezer boats (*Will Succeed* and *Collier*) prior to transportation to Robb's Jetty aboard State ships. Turtles up to 500 lb were harpooned at sea, trapped in throw nets or caught as they returned to the water following nesting (Department of Fisheries and Fauna 1950).

During its operation, *West Coast Enterprises* was accused of indiscriminate killing and concerns were expressed by industry personnel about the potential consequences to turtle populations. As a consequence of harvesting without an established export market, *West Coast Enterprises* requested financial assistance from the Western Australian government to pay an advance on its turtle products when the company was unable to find a buyer (Department of Industrial Development

1959). In addition, 128,000 turtle eggs remained in storage for 2 years before being advertised for tender (the outcome of which is unknown). The quantity of eggs in storage was equivalent to the annual egg production of roughly 240 females (Limpus 2002). The Department of Fisheries issued warnings for other operators not to engage in similar fishing practices, otherwise action would be taken to control their activities (Department of Fisheries and Fauna 1950). The company went into receivership in 1963 after harvesting at least 91,483 lb of turtle (live weight) (Department of Fisheries and Fauna 1950, Department of Industrial Development 1959).

From 1963 to the conclusion of the commercial turtle trade in 1973, all commercial take of green turtles was conducted under two exclusive licenses. The number of exclusive licenses was restricted to two due to concerns about the sustainability of the fishery (Department of Fisheries and Fauna 1969a). One of these was held by *Tropical Traders* (1963–1972) and the other was held by Strahinja (Stan) Stojanovic (1964–1969) and was later transferred to *West Coast Traders* (1969–1973). The operations of *Tropical Traders*, Stan Stojanovic and *West Coast Traders* comprised the bulk of the green turtle fishery in Western Australia.

Each licensee operated a freezer boat in adjoining areas from the North West Cape to the Montebello Islands. *Tropical Traders* utilised the processing vessels *Ngardee Mar* (operated by the Plug family from 1963 to 1969) and *Tringa* (operated by Andy Cassidy from 1970 to 1972). Stojanovic/*West Coast Traders* ran the processing vessel *East Winds*. Each license operated on a quota system, which was based on a competitive and limited European export market. The fishing season generally commenced mid-winter (June or July) and terminated in September or October, depending on when the quota for the available market was obtained (Department of Fisheries and Fauna 1969a).

Contracts for *Tropical Traders* stipulated a harvest of 4,000 turtles per season and that all turtles produced had to weigh 120 pounds dressed (gutted) (Plug 2005a; Weaver 1998). *Ngardee Mar's* capacity of 350 turtles was usually filled within three and a half days, at which point the vessel would travel back to Shark Bay to unload its consignment (Plug 2005b). In later years, a weekly delivery of 250 turtles was sent by Cassidy to Robb's Jetty for processing (Weaver 1998).

Sporadic catch records specific to *Tropical Traders* reveal that at least 8,549 turtles (roughly 756 t live weight) were harvested from the company's licensed fishing grounds (Limpus 2002; Department of Fisheries and Fauna 1950, 1963, 1966b). The total actual harvest would have been considerably higher but the exact quantity is unknown because explicit catch records for the company are incomplete and only statewide catch statistics are available. Commercial fishermen who worked on the fishing vessels reported a harvest of approximately 3,500–4,000 turtles for each year that the company operated its exclusive license, with the exception of the first two seasons in which approximately 1,000 turtles were harvested per year (Plug 2005b; Weaver 1998).

Prior to obtaining its exclusive license in 1963, *Tropical Traders* harvested an additional 487 turtles (49,303 lb live weight) from the Onslow area during October and November 1961 (Department of Fisheries and Fauna 1950). Fishing practices included nightly shore excursions where any turtles on the beach at the time (typically two to three) were turned onto their back to be collected by dinghies the following day (Plug 2005a). In 1964 Stan Stojanovic was issued an exclusive license to take turtles from an area extending from the North West Cape to islands of the Exmouth Gulf (Fisheries Department 1964). In 1967 the Muiron Islands (within the Exmouth Gulf) were excluded from the exclusive license following claims that turtle fishing would be incompatible with potential tourism ventures proposed for the islands (Department of Fisheries and Fauna 1950).

In 1969, following its purchase of Stojanovic's processing vessel *East Winds*, *West Coast Traders* obtained an exclusive license for the same area allotted to Stojanovic with the exception of the Muiron Islands and the Montebello Islands (approximately 50 miles of coastline) (Department of Fisheries and Fauna 1969a). The vessel had a capacity of 30 t and took approximately 3 weeks to fill at which point the consignment would be unloaded at Learmonth jetty (McGowan 2005).

Explicit catch returns for Stojanovic/*West Coast Traders* appear in the State records for six seasons and the remainder appeared as cumulative statewide catch records. This documentary evidence suggests that at least 16,468 Green turtles were harvested from the exclusive license area (a live weight of approximately 1,342 t). Oral history testimonies suggest that approximately 3,000 turtles were harvested from the exclusive license area on an annual basis (McGowan 2005).

Unlike *Tropical Traders*, *West Coast Traders* processed its turtles into various export components onboard the vessel. This involved the separation of the shell, flippers and the red and green meat, and the use of a boiler to boil the connecting material between the breast plates (McGowan 2005).

A detailed description of the method used for fishing and processing the turtles was described in the July 1969 edition of the Monthly Staff Bulletin of the Department of Fisheries and Fauna:

> Each licensed freezer boat has several small 16 foot scooter catcher boats, powered by 40 H.P. outboard motors. These scooter boats operate within a one mile radius of the mother freezer boat, in the relatively shallow water inside the offshore reefs, where the turtles graze on the brown and green algae of the rocky sea bed. When a turtle is located it is harpooned from the scooter boats as it races through water from 3 to 8 feet deep. On attaining a full load of about 10 turtles, the scooter boats unload their catch onboard the freezer boat for processing. Turtles are gutted, beheaded, washed, drained and blast frozen. Each carcass weighs about 120 lbs dressed. When the freezer boat attains a full load, usually about 300 turtles taken in about 3 days fishing, it returns to port to unload its catch. From the port of landing the catch is taken by freezer trucks to Robb's Jetty, near Fremantle, for further processing and storage pending export. Most of the best quality cuts of meat come from the muscles of the fore-flippers. Very little wastage of the landed dressed turtle occurs.

During the mid to late 1960s and early 1970s, a number of additional applications for exclusive licenses to take turtle were received by the Fisheries Department. These applications were refused due to Departmental concerns regarding the sustainability of the fishery:

> The Department is concerned that the present rate of exploitation might cause a serious decline in turtle population numbers. It has decided not to grant any additional licenses until data is obtained to give an indication of the turtle populations and to assist in determining maximum safety fishing effort in the future. (Department of Fisheries and Fauna 1969c)

The Department did, however, note that even though an exclusive license couldn't be granted, a person holding a current professional fisherman's license was permitted to engage in the taking of turtle for gain or reward (pursuant to provisions under the *Fisheries Act 1905–1964*). This practice was permitted up until around 1970 when turtles became fully protected under both the *Fauna Conservation Act* and the *Fisheries Act* unless taken under the authority of an exclusive license (Department of Fisheries and Fauna 1950).

From July 1968, Western Australia was the only state in the country still allowing commercial turtle fishing after legislation was introduced to Queensland to fully protect all turtle species. In May 1973, the Minister for Fisheries and Fauna announced that the commercial harvest of turtles in Western Australia would cease from 30 June 1973 (Department of Fisheries and Fauna 1950). *West Coast Traders* was given dispensation to cease its fishing operations by July 25 1973 to allow it to fulfill outstanding orders (Department of Fisheries and Fauna 1969b).

From 1871 until the closure of the Western Australian commercial turtle fishing industry, at least 61 applications for exclusive turtle fishing licenses were received by the state's Government. Exclusive licenses were issued for at least 28 of these applications and ten licensees' commenced operations. The main fishing grounds targeted were the Lacepede Islands, Montebello Islands, the islands of the Dampier Archipelago, islands of the Exmouth Gulf, North West Cape and Yampi Sound.

Catch Records: Hawksbill Turtle Fishery

By the mid nineteenth century, the hawksbill turtle was recognised as a potentially valuable source of tortoise-shell or bekko (Daley et al. 2008). Other turtle shells were considered inferior as they did not possess the thickness and colour pattern on the scute (Mack et al. 1995). The first records of exports of tortoise shell from Western Australia appear in 1869 when 643 pounds of tortoise shell, valued at £ 482.5.0 (15/– per lb) were exported to South Australia (434 lb) and the United Kingdom (200 lb) (Colony of Western Australia 1869).[2]

Between 1869 and 1953, annual exports of tortoise shell appeared regularly in the Western Australian trade records and these figures are reproduced in Appendix A in Halkyard (2009). For certain years, only the value of the exported tortoise shell appears in the trade tables. In these instances, the weight has been inferred from the export value, using an average of the values for the preceding and subsequent years (shillings per pound weight). Export figures were cross-referenced with national and international imports into Western Australia to ensure that all tortoise

---

[2] Total export amount would equate to £ 34,537.40 or £ 53.71 per lb of tortoise shell in modern day (2008) times (using the retail price index). This equates to AUD $ 16,223.12 or $ 2,523 per lb of tortoise shell (conversion rate based on 2008 average of 2.1289 pounds to one Australian dollar). Source: Measuring Worth (2009), URL: http://www.measuringworth.com/ukcompare/index.php, Accessed 16th Feb 2010.

shell exports originated from Western Australia and were not re-exported from other countries or parts of Australia.

Halkyard (2009) shows that tortoise shell production occurred on a relatively small scale until the 1920s when the annual export trade consistently exceeded 1,000 lb of tortoise shell. This increase in production may have been prompted by the increase in the value of tortoise shell in 1917 to around 20 shillings per pound weight. At these rates, one animal could potentially fetch £ 4 on the export market at a time when the average weekly rate of wage payable to adult male workers in Western Australia was 68 shillings and 11 pence (£ 3 8s 11d).[3] Immediately after this peak, values dropped below pre-1920 levels and for the next two decades prices fluctuated between seven shillings and nine shillings per pound weight. For the most part, production of tortoise shell remained above pre-1920 levels with exports occurring on annual basis up until the WWII.

Export data is deficient during WWII when publication of international trade figures in the State's statistical register was suspended. The weight of exported tortoise shell spiked dramatically in 1948 and 1941 when a combined weight of 117cwt of tortoise shell was exported to the USA. This corresponded with the lowest values for tortoise shell in the history of the industry; less than one shilling per pound weight in 1948 and three shillings per pound weight in 1949. The highest value of tortoise shell was obtained between 1950 and 1953, which also represented a marked decrease in tortoise shell production (Halkyard 2009).

From 1953 onwards, tortoise shell was no longer listed under a separate category in the export tables. Instead, there appears a line item for 'Shells, unmanufactured (other)' and it is not possible to ascertain how much of this comprises tortoise shell. It is likely that the amount was minimal given that in the 1950s synthetic materials replaced the use of tortoise shell and the market for the natural product collapsed (Daley et al. 2008).

The evidence suggests that by 1953 at least 45,073 lb (20,445 kg) of tortoise shell had been exported from Western Australia. Assuming that an individual Hawksbill turtle has a carapace weight of between 1 and 2 kg (Mack et al. 1995), between 10,222 and 20,445 hawksbill turtles were harvested from northern Western Australia in 84 years. This is somewhat comparable to the Queensland tortoise shell industry in which at least 86,020 lb of tortoise shell was exported between 1871 and 1938; a harvest which had a significant impact on hawksbill turtle populations in the north of the State (Daley et al. 2008). Anecdotal reports suggest a similar outcome for Western Australian hawksbill turtle populations as a result of this fishing effort:

> The hawksbill turtle was once abundant and heavily exploited commercially for its shell. Today, its appearance is a relatively rare occurrence along the coast. (Department of Fisheries and Fauna 1969c)

---

[3] Weighted average nominal weekly rate of wage payable to adult male workers for a full week's work, 31st December 1917. Source: *Year Book of Australia, No. 11* (1918).

The export trade tables show that the principal export market for Western Australian tortoise shell was the United Kingdom. Other export markets included Singapore, British Malaya, Japan, Ceylon, the Straits Settlement, India, France, Italy, USA, Germany, Austria, and the other Australian States.

The hawksbill fishery was closely associated with the pearling and beche-de-mer industries, and a large proportion of the hawksbill harvest was carried out by pearlers holding professional fisherman's licenses. At the outbreak of war in the Pacific in 1941, the pearling industry ceased to operate. The subsequent influence on the tortoise shell industry is reflected by the absence of an export trade during the war years. Pearling operations did not resume in Western Australia until 1946–1947 (ABS 1951).

It is worth noting that prior to European exploitation, Makassan fishermen travelling out of Indonesia harvested tortoiseshell from the Kimberley coast and possibly as far south as the Pilbara coast. It is thought that this fishery ceased in about 1900 (Limpus 2002). Attempts to quantify this fishery are outside the scope of this chapter.

Catch Records: Green Turtle Fishery

Documentary evidence suggests that between 40,077 and 55,125 Green turtles were harvested from Western Australian waters between 1837 and 1973. The bulk of this fishing effort occurred in the decade prior to the industry being closed down and was concentrated in the waters off Coral Bay, Exmouth, Onslow, and the adjacent offshore islands (i.e. fishing blocks 2313, 2213, 2113 and 2114, see Appendices B and C in Halkyard 2009). Other fishing localities included Derby, Cape Cuvier, Denham, Carnarvon and Mandurah. During the peak of the industry, up to two thirds of the take is likely to have been female turtles (Plug 2005b).

Oral history accounts suggest an even greater quantity of green turtles was harvested from these fishing localities. *Tropical Traders* harvested between 3,500 and 4,000 turtles during each year of its operations, with the exception of its first two seasons when approximately 1,000 turtles were harvested Plug 2005b; Weaver 1998). In total, *Tropical Traders* probably harvested up to 42,000 Green turtles between 1960 and 1972.

The exclusive license first operated by Stan Stojanovich, then *West Coast Traders*, reportedly harvested 3,000 turtles each season, equating to a harvest of 27,000 turtles between 1964 and 1973 (Halkyard 2005). Stojanovic also harvested turtles 2 years prior to applying for an exclusive license for the same turtle grounds, although it is unclear how many turtles were harvested during this time (Fisheries Department 1964).

Based on the catch rates provided by the fishermen themselves, at least 69,000 Green turtles were harvested from Western Australian waters ranging from the southern end of the Ningaloo Reef to the Montebello Islands in the space of 13 years. This evidence demonstrates that each licensed fishing vessel consistently filled its maximum allowable quota each season. Turtle fishing contracts stipulated a minimum weight for turtles of 120 lb (dressed), and thus the harvest would have consisted of adult and large, immature green turtles (Weaver 1998; Limpus 2002).

Public Perceptions and Sustainable Management

The Fisheries Department, in its various incarnations throughout the twentieth century, was the Western Australian government agency responsible for administering the commercial harvest of turtles. Concerns about the sustainability of the turtle fishing industry were acknowledged by industry regulators as early as 1901 (Fisheries Department 1901a; Fig. 11.1).

In recognising the need for a sustainable fishery, the Department proposed several management strategies, but it is unclear to what extent they might have been implemented and enforced. For example, in 1923, size and weight limits for green and hawksbill turtles were recommended by the Chief Inspector of Fisheries. It was suggested that weights of green turtles taken should be fixed at 120 lb and the minimum size of hawksbill turtles taken should be fixed at 24 in. (end to end of carapace) (Fisheries Department 1923). Documentary evidence indicates that these size restrictions were not necessarily enforced because in 1959, turtles as small as 100 lb were harvested by *North West Enterprises* (Fisheries Department 1958). Conversely, contracts for licensees such as *Tropical Traders* specified that all turtles processed had to weigh 120 pounds dressed (Weaver 1998).

Similarly, the take of nesting turtles by commercial fishermen is confirmed by documentary and oral history evidence, despite the practice contravening lease conditions (Department of Fisheries and Fauna 1950). Thus it would appear that even though the management frameworks for a sustainable fishery were in place, the industry operated in the absence of surveillance and enforcement from the agency responsible for administering the fishery. Moreover, it was acknowledged by the Minister for Fisheries and Fauna in 1970:

> There is no direct supervision of the operations of the license holders as it is the licensees' responsibility to patrol their own areas. (20 August 1970, in Department of Fisheries and Fauna 1950)

It is unclear to what degree the industry was able to self-regulate but oral history evidence suggests that the practice of harvesting nesting turtles was not a common occurrence given the difficulties associated with handling animals on the beach, tainting the product with sand and landing vessels on the shore (Weaver 1998; Halkyard 2005; McGowan 2005).

Criticisms of the commercial turtle harvest appear frequently in the documentary evidence from the early 1960s, along with pressure on the government apply catch limits, implement closures to certain areas or disallow the industry entirely. Reports of observable declines in localised turtle populations also appear throughout the documentary evidence, particularly in the late 1960s and early 1970s and were typically attributed to the commercial fishery (Department of Fisheries and Fauna 1950)

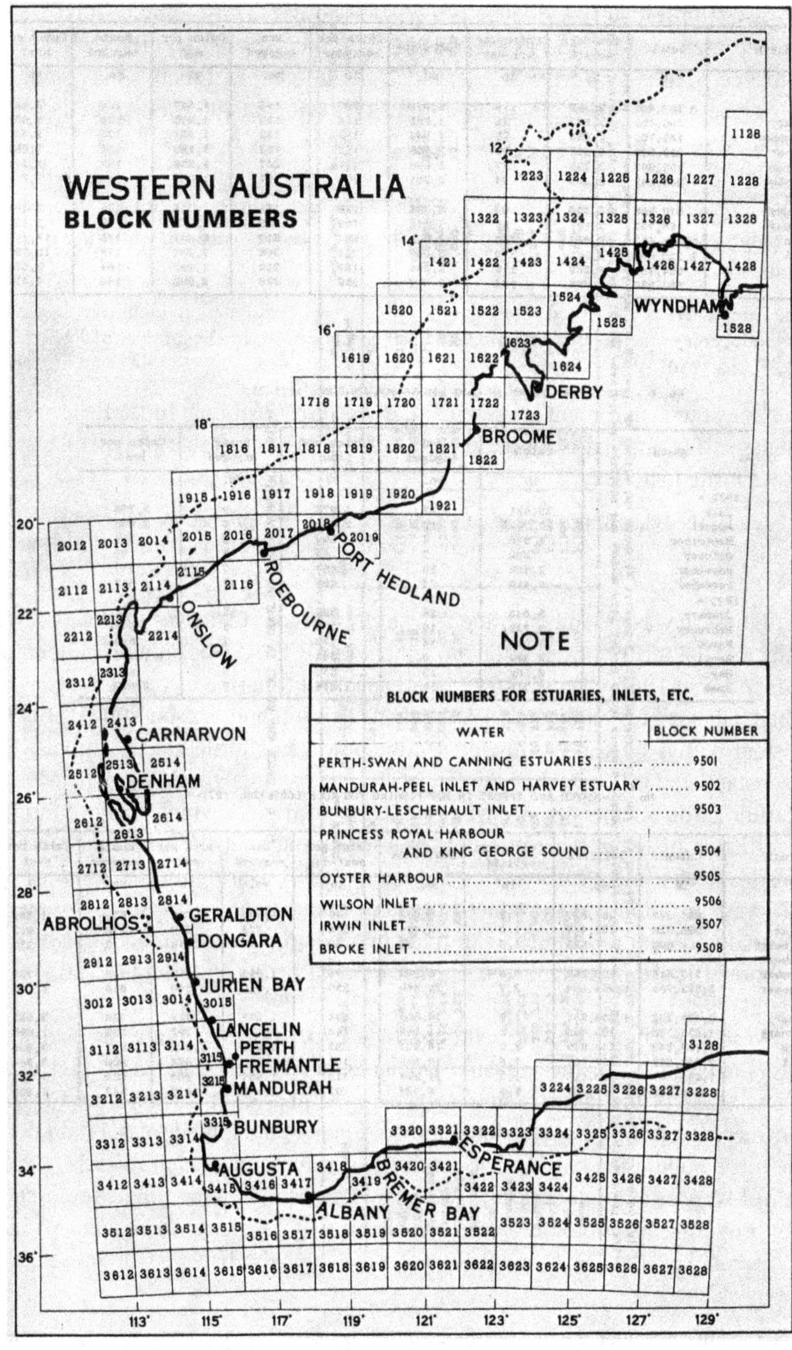

Fig. 11.1 Map of Western Australian fishing blocks, 1968–1969. (Source: Commonwealth Bureau of Census and Statistics, Western Australian Office (1971))

However, oral history evidence refutes this notion. Those directly involved in the industry reported no significant changes in the success of the turtle harvest from year to year: fishing effort was consistent, the quota was reliably harvested and turtle numbers were not observed to diminish (McGowan 2005; Plug 2005a, b).

In the absence of scientific data to confirm one way or the other, the Western Australian Government presented conflicting views regarding the sustainability of the industry. On several occasions throughout the late 1960s and early 1970s it acknowledged its concerns about aspects of the turtle fishing industry and the existing exploitation of turtles:

> The Department is concerned about that the present rate of exploitation might cause a serious reduction in the population numbers. (24 February 1969, in Department of Fisheries and Fauna 1950)

Conversely, the Department argued that the Western Australian turtle fishery could be managed for rational exploitation and was not as vulnerable to collapse as the overseas turtle fisheries:

> It is established that turtle fisheries are vulnerable to over-exploitation, but we have no information to suggest that this is occurring here. (20 August 1970, in Department of Fisheries and Fauna 1950)

Nevertheless, it was around this time that public attitudes to turtle conservation were shifting and the turtle conservation movement in Australia was gaining momentum, as evidenced in 1968 when the take of all species of marine turtles in Queensland was banned except by indigenous Australians (Tisdell and Wilson 2005). This was consistent with a shift in worldwide sentiment towards turtle conservation, which was substantiated in 1969 when the IUCN arranged an inaugural 'Working Meeting of Marine Turtle Specialists' in Switzerland, the purpose of which was to formulate a coordinated plan for sea turtle conservation throughout the tropical regions of the world. (Department of Fisheries and Fauna 1950). During the 1970s and 1980s protective legislation was introduced in many countries and harvesting for international trade gradually declined with trade controls pursuant to the Convention on International Trade in Endangered Species of Wild Fauna and Flora (CITES) (Carrillo et al. 1999).

In Western Australia, public sentiment was a major contributing factor in altering the management of the industry. Documentary evidence indicates that during the last decade of the industry, in response to the pressure placed on the Fisheries Department to prohibit turtle fishing, the government attempted to ascertain details of the turtle population based on its analysis of commercial catch records (Department of Fisheries and Fauna 1950). Vessels were required to complete research log books providing information to the Department of Fisheries and Fauna with the view of enabling the Department to assess whether the fishery was in danger of being overfished:

> Without information on the history of the turtle fisheries, it is possible that pressure on the Government by some members of the community may influence a decision to ban the taking of turtles, as has been done in most parts of the world. However, if data available does show that the resource is being exploited in a rational manner and that the population is not

> trending downwards, then requests for turtle bans can be answered. It is for these reasons
> that I am endeavoring to obtain information about the turtle fishery. (26 September 1968, in
> Department of Fisheries and Fauna 1950)

Public perceptions and consequent political pressure altered the industry in other ways. For example, closures of turtle fishing grounds were implemented in 1968 when the Muiron Islands were removed from the exclusive license of *West Coast Traders*, most likely in response to concerns raised by members of the community considering the feasibility of establishing tourism facilities on the islands. They argued that turtle fishing and tourism were incompatible (Department of Fisheries and Fauna 1950).

In 1969, the Western Australian Government imposed a 50 mile restriction on the exclusive licenses for *Tropical Traders* and *West Coast Traders* (formerly 75 miles of coastline) (Department of Fisheries and Fauna 1969b). At the time their licenses were renewed in 1971, *Tropical Traders* and *West Coast Traders* were advised that it would be unlikely that their licenses would again be renewed in 1973. Additional license conditions were also imposed at this time to more closely control the fishing activities of the boats engaged in the fishery (Department of Fisheries and Fauna 1963, 1969b).

Oral histories challenge the perception that public sentiment was one of opposition to the turtle harvesting. Commercial fishermen note that they maintained public interest in their activities, particularly in the early 1960s, and that they never felt compelled to keep their activities out of the public eye (McGowan 2005; Plug 2005a).

For a number of years prior to the cessation of the industry, the Western Australian Government also attempted to engage Dr. H. R. Bustard, considered 'a world authority on turtles', to undertake a detailed survey of the marine turtle populations in Western Australia and to assess the sustainability of the fishery (Department of Fisheries and Fauna 1950). However, the records suggest that this never happened.

Discussion

The global depletion and local extinction of marine turtle populations as a consequence of large-scale commercial exploitation has been well documented e.g. (Bjorndal and Jackson 2003; King 1995; Dodd 1995; Lutcavage et al. 1997). The life history of marine turtles (i.e. long-lived, slow-maturing, low recruit survival rate and high nesting/feeding site fidelity) makes them especially vulnerable to the effects of exploitation (Heppell and Crowder 1996; Limpus 2002; Daley et al. 2008). Previous studies have shown that even small, long-term increases in annual mortality from anthropogenic sources above natural mortality levels will cause population declines (Limpus 2002).

This chapter demonstrates that significant commercial fishing of marine turtles occurred in Western Australian waters and that the harvest was most intense during the 1970s prior to the cessation of the industry, coinciding with reports of localised depletions. While there was some regulation by the Western Australian Government, there was little scientific monitoring. In the absence of monitoring data, it could be argued that a harvest of approximately 69,000 green turtles and 20,445 hawksbill turtles exceeded the sustainable harvest and this is likely to have had a substantial impact on Western Australian marine turtle populations.

Given the requirement to harvest large immature or adult turtles during the peak of the industry (as a consequence of minimum size restrictions), it could also be argued that anthropogenic pressures are likely to have further exacerbated the loss of adult turtles from the population. Adult survival rates have been demonstrated to be the most sensitive life history stage and high survivorship of adult marine turtles is required to maintain population stability (Heppell and Crowder 1996; Limpus 2002; Daley et al. 2008).

For fisheries where harvesting is biased towards females, sustainable harvest rates are further reduced (Choquenot 1996). While other marine turtle fisheries have impacted disproportionately on female turtles at the critical life stage when they come ashore to nest, this did not appear to be an overly common practice in the Western Australian commercial turtle fishery. Nevertheless, oral history evidence indicates that a large proportion of the overall catch (up to two-thirds) may have consisted of female turtles (Plug 2005b).

During the peak of the industry in the 1960s and early 1970s, the bulk of the fishing effort took place between March and August, outside of the summer breeding season. Given that green turtles exhibit high levels of fidelity to foraging areas over extended time periods, foraging populations would have been most vulnerable to the large-scale, concentrated fishing effort in the two exclusive license areas (Broderick et al. 2007; Daley et al. 2008). Consequently, there would have been a serious reduction in the localised green turtle foraging populations between Point Maud and the Montebello Islands. This is supported by oral history evidence that turtles were always harvested from the back of the reef where they were 'feeding off the weed' and that none were caught in close to shore (Weaver 1998). Limpus (2002) also suggested that the harvest of breeding migrants aggregated for courtship as well as internesting females would have resulted in some reduction of the nesting population.

It is likely that the fishing effort for the tortoise shell industry had a significant impact on hawksbill turtle populations in the State's north; however, it is difficult to pinpoint where the fishing effort was concentrated and which foraging and/or breeding aggregations were most impacted. It should also be noted that oral history evidence from those directly involved with the fishery contradicts the notion that the commercial turtle harvest caused a decline in green turtle numbers. Furthermore, the commercial turtle fishermen reported no significant changes in the success of the turtle harvest from year to year: fishing effort was consistent, the quota was reliably harvested and turtle numbers were not observed to diminish (McGowan 2005; Plug 2005a, b). Even if a decline had been observed by the fishermen, this would not necessarily mean that the harvests were unsustainable (Carrillo et al. 1999).

Anecdotal reports of declines in the numbers of nesting green turtles on the north west coast may also be attributed to factors other than over-exploitation such as large inter-annual differences in the proportions of females making breeding migrations and nesting (Daley et al. 2008). On the other hand, it has been documented that the effects of a commercial turtle harvest may not be immediately apparent and that the actual magnitude of the effect on marine turtles may not appear until decades later (Lutcavage et al. 1997; Daley et al. 2008). This may give the false impression that continued exploitation will not be detrimental to the population being harvested.

Despite the past commercial turtle harvest, Western Australian green and hawksbill turtle populations are at present the largest in the Indo-Pacific region and some of the largest remaining in the world (DEC 2008). Exploitation of the Western Australian turtle fishing industry never reached the magnitude of turtle fisheries elsewhere in the world where animals were harvested in their hundreds of thousands (Lutcavage et al. 1997; McClenachan et al. 2006). Furthermore, the turtle fishing industry reached its peak in the 1960s and the early 1970s, which was a belated establishment relative to turtle fisheries conducted elsewhere. In Queensland, for example, a substantial turtle meat and soup industry had been established by 1896 (Daley et al. 2008). At one of the earliest recorded commercial turtle fisheries in the Cayman Islands, exploitation of green turtles commenced in 1655 and the fishery had collapsed by 1790 (Bjorndal and Jackson 2003).

There a number of reasons why the Western Australian turtle fishing industry never achieved the scale commercial exploitation documented elsewhere, including: the lack of suitable refrigeration points along the north-west coast; improper treatment of raw material; insufficient capital; insufficient export market; shortages of suitable labour; a lack of knowledge of local conditions; and the large distances to the export markets. Competition with established markets in Queensland and overseas may also have been prohibitive. The extent and nature of the harvest was also controlled to a large degree by the State's management agencies.

The estimates of take and fishing effort presented in this chapter may underestimate the actual historical exploitation that took place. Export trade figures for tortoiseshell include only interstate and international trade figures and there is no data that indicates the amount of shell that was collected for local Western Australian markets (Halkyard 2009). Those possessing exclusive licenses and professional fishermens' licenses were required to submit catch returns, including turtles, to the Fisheries Department, however many of these records appear to be missing from the archives.

Although commercial fishing pressure on Western Australia's marine turtle populations ceased nearly 40 years ago, it is likely that the historical harvest has increased the vulnerability of green and hawksbill turtles to modern-day pressures. The ongoing management of human impacts on marine turtles is essential to ensure the long-term conservation and viability of marine turtles in the State.

Acknowledgments I would like to extend my deepest gratitude to the individuals involved in the commercial turtle fishing industry who shared their knowledge and experience in such an open and generous manner. Their contribution has been fundamental to the preparation of this chapter and I am sincerely thankful to have been given the opportunity to document their stories and insights. Thank you to everyone who took the time to review my draft manuscripts and provide me with much appreciated feedback and suggestions. Thanks also to staff at Department of Fisheries,

Department of Environment and Conservation, State Library of Western Australia and the State Records Office of Western Australia for their invaluable assistance with my archival research. Full bibliographic references to archival sources and appendices listing export trade figures and catch records can be found in Halkyard (2009). This work has been funded courtesy of the History of Marine Animal Populations project and supported by the Western Australian Department of Environment and Conservation.

References

ABS (1951) Year book Australia, 1951. Australian Bureau of Statistics, Canberra

Bjorndal KA, Jackson JBC (2003) Roles of seaturtles in marine ecosystems: reconstructing the past. In: Lutz P, Musick JA, Wyneken J (eds) The biology of sea turtles, vol II. CRC Press, Boca Raton

Broderick AC, Coyne MS, Fuller WJ, Glen F, Godley BJ (2007) Fidelity and over-wintering of sea turtles. Proc R Soc Biol Sci 274(1617):1533–1539

Carrillo E, Webb GJH, Manolis SC (1999) Hawksbill turtles (Eretmochelys imbricata) in Cuba: an assessment of the historical harvest and its impacts. Chelonian Conserv Biol 3(2):264–280

Carse WE (1925) Notebook [manuscript]: his working notes when employed at the Roebourne Produce Company. State Library of Western Australia, Perth

Choquenot D (1996) The ecological basis of wildlife harvesting. In: Bomford M, Caughley J (eds) Sustainable use of wildlife by Aboriginal peoples and Torres Strait Islanders. Australian Government Publishing Service, Canberra

Cox JM (1977) Barrow island: a historical documentation. State Library of Western Australia, Perth

Daley B, Griggs P, Marsh H (2008) Exploiting marine wildlife in Queensland: the commercial dugong and marine turtle fisheries, 1847–1969. Australian Econ Hist Rev 48(3):227–265

Department of Aborigines and Fisheries (1909) Turtles dried: calipash and calipee. File 1911/1008, Cons 652, State Records Office of Western Australia.

Department of Aborigines and Fisheries (1911) Turtles dried: calipash and calipee. 1909–1920. File 1911/1008, Cons 652, State Records Office of Western Australia.

Department of Aborigines and Fisheries (1919) H. Baron-Rodway-Exclusive licence to farm turtles-Cape Preston to Cape Lambert. 1913–1921. File 1919/0025, vol. 1, Cons 477, State Records Office of Western Australia

Department of Conservation and Land Management [CALM] (2005) Management plan for the Ningaloo marine park and Muiron islands marine management area 2005–2015: Management Plan No. 52. Perth, Western Australia

Department of Environment and Conservation [DEC] (2008) Marine turtle recovery plan for Western Australia—draft. Perth, Western Australia

Department of Fisheries and Fauna (1950) W. A. Turtles. 1950–1974. File 1950/0248 v1, fol. 210, Cons 1598, State Records Office of Western Australia

Department of Fisheries and Fauna (1963) Fisheries-Turtles-tropical traders ltd application excl. Licence to take Turtles-In the area from Pt Maud. File 1963/0195, Cons 1759, State Records Office of Western Australia

Department of Fisheries and Fauna (1966a) Fisheries-Western Australian fisheries statistics (Monthly Tabulations-1965). 1965–1966. File 1966/0210, v.1, Cons 1552, State Records Office of Western Australia

Department of Fisheries and Fauna (1966b) Statistics-general. 1960–1979. File 1966/493 v1, Cons 4169, State Records Office of Western Australia

Department of Fisheries and Fauna (1969a) Fisheries-Application for exclusive licence to take turtles. 1969–1973. File 1969/0209, Cons 1553, State Records Office of Western Australia

Department of Fisheries and Fauna (1969b) Fisheries-West Coast Traders Pty Ltd-exclusive license to take turtles. 1969–1973. File 1969/0239, Cons 1565, State Records Office of Western Australia

Department of Fisheries and Fauna (1969c) The turtle fishery of Western Australia. Mon Staff Bull 18–19:7

Department of Industrial Development (1938) Westella Canning Co. re Manufacture of turtle soup, canning of crayfish, sheeps tongues etc. File 1938/0001, Cons 961, State Records Office of Western Australia

Department of Industrial Development (1959) North-West turtle industry-treatment of turtles for export-general. 1959–1961. File 1959/0015, Cons 961, State Records Office of Western Australia

Dodd JK (1995) Does sea turtle aquaculture benefit conservation? In: Bjorndal KA (ed) Biology and conservation of sea turtles. Smithsonian Institution Press, USA

Duckett B (1990) From Guano to gas: the story of Australia's remote North West. Woodside Offshore Petroleum, Perth

Durant C (2004) Rockingham's turtle factory. The Rockingham Historian, August 2004

Fisheries Department (1901a) Pearling and turtling industry in North West. Report on. File 1901/0344, Cons 477, State Records Office of Western Australia

Fisheries Department (1901b) Turtling industry in WA. File 1901/0888, Cons 477, State Records Office of Western Australia

Fisheries Department (1919) Fisheries-Turtles-Exclusive Licence to Farm H. Barron-Rodway. 1921–1937. File 1919/0025 v2, Cons 477, State Records Office of Western Australia

Fisheries Department (1920) Fisheries-North West Australia Development and Sea Products Limited-Excl. Lic. take turtles, dugong, etc-Fish-Exclusive Licence. 1920–1930. File 1920/0016, Fol. 20, Cons 477, State Records Office of Western Australia

Fisheries Department (1921a) Pearling & Fisheries-Broome-Annual Report 1921. 1921–1927. File 1921/0157, Cons 477, State Records Office of Western Australia

Fisheries Department (1921b) Fisheries-information re possibilities of WA fishing industries (Whaling, Turtle, Dugong). 1938–1956. File 1921/0170, Cons 477, State Records Office of Western Australia

Fisheries Department (1922) Fisheries-Broome-Preserving Company-Excl. Lic. To take turtle from Lacepede and adjoining Islands. 1922–1937. File 1922/0058, Cons 477, State Records Office of Western Australia

Fisheries Department (1923) Fisheries-Turtles-Fixing Size and Weight Limit. 1923–1928. File 1923/0164, Cons 477, State Records Office of Western Australia

Fisheries Department (1924) Fisheries-W. Benstead-Excl. Lic.-Turtles-DeLambre, LeGendre, Forrestiers, Bedout, DePuch and Turtle Islands. 1924–1931. File 1924/0015, Cons 477, State Records Office of Western Australia

Fisheries Department (1934) Fisheries-Australian Canning Company-Excl. Lic turtles and dugongs at Monte Bello Islands. 1931–1959. File 1934/0012, Cons 477, State Records Office of Western Australia

Fisheries Department (1939) Fisheries-V. Gardiner & W.J. Rowden-application for an Exclusive license to fish turtles of the Montebello Group. 1939–1940. File 1939/0094, Cons 477, State Records Office of Western Australia

Fisheries Department (1958) Fisheries-James Mazza & Alfred R.E. Russell-Exclusive License to take turtles from North West Cape-Wyndham. 1958–1859. File 1958/0148, Cons 477, State Records Office of Western Australia

Fisheries Department (1964) Stojanovic-exclusive licence to take turtles. 1964–1969. File 1964/0235, Cons 1598, State Records Office of Western Australia

Fraser MAC (1896) Fish and fisheries of Western Australia. Collected papers and reports (Western Australia) 1876–1929. Fisheries Western Australia Publications, Perth

Gale CF (1901) Report on the fishing industry for the year 1900 by the chief inspector of fisheries. Fisheries Department, Western Australia

Gale CF (1904) Report on the fishing industry for the year 1903 by the chief inspector of fisheries. Fisheries Department, Western Australia

Halkyard BR (2005) Historical exploitation of turtles and lobsters at Ningaloo Reef. Department of Conservation and Land Management and Murdoch University

Halkyard BR (2009) Exploiting green and hawksbill turtles in Western Australia: a case study of the commercial marine turtle fishery, 1869–1973. A HMAP Asia Project Paper. Asia Research Centre Working Paper no. 160, Murdoch University

Heppell SS, Crowder LB (1996) Analysis of a fisheries model for harvest of hawksbill sea turtles (Eretmochelys imbricata). Conserv Biol 10(3):874–880

IUCN (2009) IUCN list of threatened species. Version 2009.2

King PP (1827) Narrative of a survey of the intertropcial and Western Coasts of Australia performed between the years 1818 and 1822. John Murray, London

King W (1995) Historical review of the decline of the green turtle and the hawksbill. In: Bjorndal KA (ed) Biology and conservation of sea turtles. Smithsonian Institution Press, United States of America

Lands and Surveys Department, Western Australia (1881) William MacVean—enquiring about leasing all the islands between Cossack and the N.W. Cape for turtle and other fishing purposes, also pastoral purposes, also Melville Island. A0353, Cons 541, State Records Office of Western Australia

Limpus CJ (2002) Western Australian marine turtle review. Department of Conservation and Land Management Western Australia, Perth

Lutcavage ME, Plotkin P, Witherington B, Lutz P (1997) Human impacts on sea turtle survival. In: Lutz P, Musick JA, Wyneken J (eds) The biology of sea turtles. CRC Press, Boca Raton

Mack D, Duplaix N, Wells S (1995) Sea turtles, animals of invisible parts: international trade in sea turtle products. In: Bjorndal KA (ed) Biology and conservation of sea turtles. Smithsonian Institution Press, United States of America

McClenachan L, Jackson JBC, Newman MJH (2006) Conservation implications of historic sea turtle nesting beach loss. Front Ecol Environ 4(6):290–296

McGowan D (2005) Interview with the author, 20 November 2005

Plug D (2005a) Interview with the author, 15 November 2005

Plug M (2005b) Interview with the author, 15 November 2005

Stokes JL (1846) Discoveries in Australia, with and account of the coasts and rivers explored and surveyed during the Great Voyage of HMS Beagle in the years 1837–43. T and W Boone, London

Taggart N (1984) Rockingham looks back: a history of the Rockingham district 1829–1982. Rockingham District Historical Society (Inc.), Rockingham

Tisdell CA, Wilson C (2005) Does tourism contribute to sea turtle conservation? Is the flagship status of turtles advantageous? MAST 2005 3(2):145–167

Weaver P (1998) An oral history of Ningaloo reef. Edith Cowan University, Perth

Chapter 12
Shifting Baselines or Shifting Currents? An Environmental History of Fish and Fishing in the South-West Capes Region of Western Australia

Andrea Gaynor

Abstract The South-west Capes region of Western Australia is one of high marine biodiversity but relatively low productivity. Still the region's waters have long provided food for the local Noongar people and sustained commercial and recreational fishing since the nineteenth century, when activities were loosely regulated, if at all. But from the mid- to late-twentieth century, as catch rates apparently declined while the popularity and reach of recreational fishing increased, policies governing fishing in the region became increasingly restrictive and fiercely contested. This chapter therefore endeavours to disentangle the strands of policy, perception, and fish populations in the Capes region, evaluating evidence of change in the region's fish populations over the long run, and accounting for it with reference to social contexts, fishing intensity and practices, and change in the regional environment. It ultimately suggests that the movements and abundance of fish have varied considerably over time due to biophysical and ecological influences, and claims of depletion have sometimes reflected cultural anxieties rather than environmental change. However, there is also a long history of human interventions in the region's marine ecosystems. Such interventions, shaped by complex cultural and economic factors, have left short- and longer-term imprints on the region's ecosystems.

Keywords Marine environmental history · Ngari Capes Marine Park · South-west capes region · Fishing history Western Australia · Leeuwin current

Environmental issues featured prominently in the lead up to the Western Australian state election held in February 2001, with then opposition leader Geoff Gallop making a range of election promises relating to old growth forest, ecotourism and conservation, coastal development and dryland salinity. Among the promises was a proposal to create a marine park in the state's south-west, which would encompass Geographe Bay, the coast between Capes Leeuwin and Naturaliste, and Hardy Inlet (Fig. 12.1). The proposed park included several sanctuary zones in which

A. Gaynor (✉)
History, The University of Western Australia, 35 Stirling Highway,
6009, Crawley, WA, Australia
e-mail: andrea.gaynor@uwa.edu.au

J. Christensen, M. Tull (eds.), *Historical Perspectives of Fisheries Exploitation in the Indo-Pacific*, MARE Publication Series 12, DOI 10.1007/978-94-017-8727-7_12,
© Springer Science+Business Media Dordrecht 2014

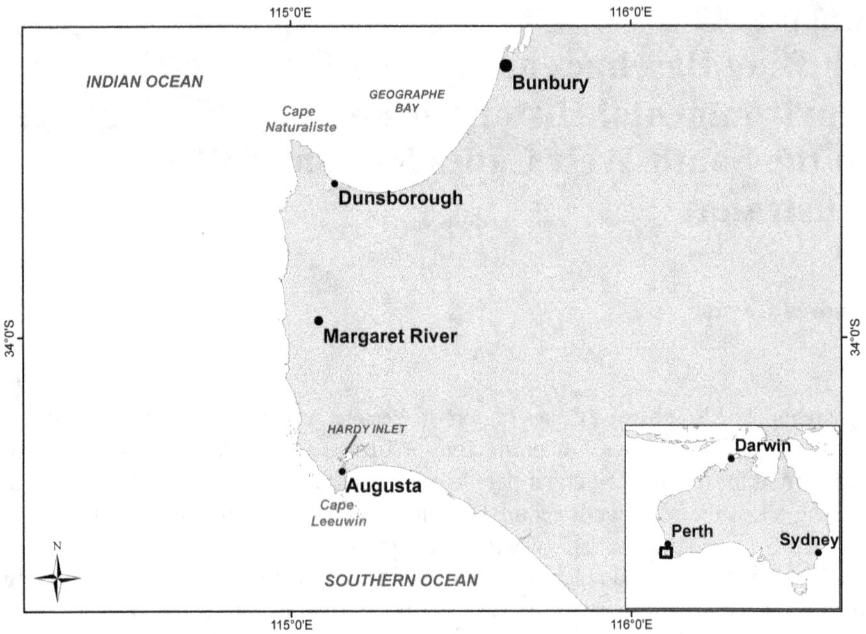

Fig. 12.1 Map of the South-west Capes region and surrounds

fishing would be prohibited. Gallop won the election and the planning process for the park was commenced but later stalled after the period of public consultation, as the increasingly beleaguered Labor government pursued other priorities. The Ngari Capes Marine Park was finally declared in 2012, including 15 sanctuary zones in which fishing will be prohibited. At this time the region was also subject to a marine bioregional planning process being undertaken by the Commonwealth Government (DEWHA 2009). The planning and proclamation of marine parks in the region has generated considerable public debate, with both proponents and opponents increasingly turning to historical records to support claims of degradation and depletion over time on the one hand, or sustained and effective fisheries management for ecosystem health on the other. In this context, there is an obvious need to better understand the history of fish and fishing in the region.

Although the region's fisheries have been the subject of considerable scientific research over the past four decades, there is still much to learn about the biology and ecology of the region's marine life. Less still is known of the way in which the region's marine resources have been exploited in the past. Some archaeological and anthropological studies have shed a little scholarly light on the long history of Aboriginal fishing (Meagher 1974–1975; Dortch 1997); a few researchers have considered aspects of commercial and recreational fishing in the region as part of broader studies (see for example Lenanton 1984; Tull 1992; Christensen 2009); and research reports have employed fisheries data stretching back to the 1970s in order to evaluate the condition of fish stocks (see for example Wise et al. 2007; Hall and Wise 2011). However, to date no historical study has focused on fishing in the region from Aboriginal times to the 1970s and placed recent trends in the context of this longer history.

This chapter therefore aims to evaluate evidence of change in the region's fish populations over the long run, and account for it with reference to social contexts, fishing intensity and practices, and change in the regional environment. That is, it endeavours to explore, as far as the evidence will allow, the relationship between fish, fishing and environmental variation in the region. It ultimately suggests that the movements and abundance of fish have varied considerably over time due to biophysical and ecological influences, and claims of depletion have sometimes reflected cultural anxieties rather than environmental change. However, there is also a long history of human interventions in the region's marine ecosystems. Such interventions, shaped by complex cultural and economic factors, have left short- and longer-term imprints on the region's ecosystems.

The Capes Region

The marine life of the Capes region is characterised by high biodiversity. This is partly due to the region's diverse habitats, which range from the protected seagrass meadows of Geographe Bay to the high energy coastline and limestone reefs between the Capes and the estuarine environment of Hardy Inlet. It is also due to the pervasive influence of the Leeuwin Current, which transports warm water of low salinity southward from the tropics. This enables tropical species to survive—temporarily or permanently—further south than they would otherwise, placing the Capes region within the overlap between the range of northern tropical and southern temperate species (Thomson-Dans et al. 2002–2003). However, the marine productivity of the Capes region, especially of finfish, overall is low. As Western Australian Superintendent of Fisheries A. J. Fraser remarked in 1953:

> I do not for a moment suggest that Western Australia's fishery resources are unlimited… in contradistinction to other parts of the world, Nature was somewhat niggardly when she endowed our fisheries. We certainly have many species of fish, but we have a smallish number of individuals of each species and they could possibly, without proper management, in the long run become depleted. It is essential therefore that we take very good care of what we have. (Fraser 1953)

Fish stocks are limited by a range of factors, one being the relatively narrow continental shelves (Tull 1992). Another is that the Leeuwin Current waters are low in nutrients, so do not support the abundance of pelagic planktivorous fish found in other eastern boundary current systems such as the Benguela and Humboldt (Lenanton et al. 1991). The main sources of nutrients in the region are surface water runoff, groundwater, and inshore biological processes, so finfish are mainly found shoreward of the current. There is a winter nutrient peak in the region's inshore waters, presumably from the runoff of surface water which occurs mainly in the winter months. However, primary production during this period is limited by light, reduced during winter by the shorter day length and turbid water (Hanson et al. 2005).

In addition to the Leeuwin Current, the region is influenced by the Capes Current, a summer inshore countercurrent flowing equatorward past Cape Naturaliste (Pearce and Pattiaratchi 1999). This current, which originates between Capes Leeuwin and Naturaliste, augmented by waters from the south east of Cape Leeuwin (Gersbach et al. 1999), is probably a significant influence on economic finfish species, facilitating or impeding the movement and feeding of pelagic species such as Western Australian salmon (*Arripis truttaceus*), pilchard (*Sardinops sagax*), and Australian herring (*Arripis georgianus*), all of which spawn off the south-west coast (Pattiaratchi 2007).

The Capes marine environment is subject to considerable variation, both natural and anthropogenic. The Leeuwin Current varies from month to month, flowing more strongly in winter than summer, and from year to year, with variations associated with the El Niño Southern Oscillation (ENSO). The Capes Current depends on wind velocities which vary seasonally and annually, as well as interaction with the Leeuwin Current. Within the area affected by the Capes Current, the most intense upwelling occurs mid-way between Capes Leeuwin and Naturaliste under summer conditions of strong southerly winds (Pattiaratchi 2007). Although this water is still nutrient-poor in world terms, it supports primary production which is relatively high for this region, and which forms the basis of extended food chains involving aggregations of small pelagic fish and predators including larger fish, sharks, dolphins and seabirds (Pattiaratchi 2007). Climate change may be affecting the key marine ecological processes around the Capes, as the Leeuwin Current is predicted to weaken under conditions of global warming (Pattiaratchi and Buchan 1991; Feng et al. 2009). There is also less runoff of nutrient-rich surface water due to declining winter rainfall since the mid-1970s, which has reduced streamflows in the region by up to 50% (Pearcey and Terry 2005), while more and more of the region's surface water is impounded for agricultural use.

The historical sources relating to fish and fishing in the region are patchy at best, reflecting its distance from the administrative centre, under-resourcing of management agencies, and low priority given to preserving 'mundane' records (such as district monthly reports), particularly prior to the 1960s. The main primary sources relied upon here are the reports of district inspectors (1901 on); departmental and ABS statistics (1941 on); archived correspondence, forms and reports; newspaper reports; and a series of oral histories with local fishers conducted in 2005–2006. Of the early inspectors' reports, relatively few survive; the statistical data prior to 1975 is also frustratingly incomplete. Furthermore, the surviving numerical data is often confounded by the nature of Capes fisheries: multi-species, multi-method, and, particularly from the 1970s, with a significant but largely unquantified recreational component. Still, taken together, the available sources illuminate something of the fisheries' past, suggesting a long-term picture of decline in some species, against a background of considerable natural variability.

Capes Fisheries Past

Prior to establishment of the first British permanent settlements in the 1830s, there were clearly sufficient fish for the local Noongar people to catch significant quantities by constructing fish traps in the mouths of estuaries, spearing the fish or catching them by hand. The Capes region has been occupied by Aboriginal people for at least 48,000 years (Turney et al. 2001), and it is likely that they engaged in fishing activities of various kinds for much or all of that time. Having no means of water transport, Noongar fishing was confined to shallow waters, but by the nineteenth century fishing had nevertheless become a key subsistence activity and prime fishing sites had high economic, social and ceremonial values (Dortch 1997). One of the mineralogists of the Baudin expedition, arriving in Geographe Bay in May of 1801, reported that he had found a small lagoon with its mouth 'barred across with rough wooden stakes that the natives plant there to catch fish brought in by the rising tide' (Baudin 1974). Large quantities of fish might be caught by men and women gathered to drive them into such traps set in rivers or estuaries, with excess fish left to die, buried for later use, or cooked and wrapped in soft bark. There are no accounts of Aboriginal people using nets, lines, or narcotic or poisonous plants, although some used bait such as ground shellfish to attract fish for spearing (Meagher 1974–1975; Gaynor et al. 2008). Captain Baudin himself reported an encounter with 'a native' who was 'up to his waist in water, busy spearing fish' (Baudin 1974; also reported by Peron 1975). It is highly unlikely that this latter method of fishing would yield much of a meal today; this in itself suggests a change in abundance or behaviour of fish, or both.

On the issue of fish abundance, the extant records of exploration and settlement are equivocal. For example, in May 1801 Nicholas Baudin wrote that 'fish are very rare' in Geographe Bay (Baudin 1974). By contrast, in March 1827 James Stirling described the fishing prospects in much the same glowing terms as his evaluation of the land, which proved incorrect. He did, however, claim that in the broad lower west coast region 'a Boat with one or two Men in her might be filled [with fish] in a few hours' (Stirling 2005), though whether this was conjecture or observation is unclear. Such conflicting reports support an image of historically variable fish abundance, spatially and temporally. Indigenous people had learnt how to successfully exploit this variable abundance, but visitors to the region had no such experience, so their encounters with fish were largely a matter of chance.

The first British settlement established in the region was at Augusta, where a small number of families landed in May 1830. In 1834 members of one of these families, the Bussells, left Augusta and established themselves on the Vasse River at the site of what would become Busselton; Bunbury was established in 1838. From the 1840s, Busselton became the centre of an emerging dairy industry, and by the mid-nineteenth century it had become a port for the export of timber, though in 1850 the non-Indigenous population of the Sussex district, encompassing the whole of the Capes area, was only 209 people (Jennings 1983). As the region's population grew, there was doubtless a movement from subsistence fishing to a division of labour in which some families began to catch and sell fish. However, the detail of such a transition is now lost to us, as the earliest surviving detailed accounts of commercial fishing in the region were created in the late nineteenth century by a fledgling fisheries bureaucracy.

Fisheries in the region were effectively unregulated before the advent of the *Fishery Act 1889*, 'An Act for the Protection of Fish', which established a framework for legal fishing. This involved a schedule of minimum weights for various commercial species and empowered the Governor to appoint Inspectors, as well as specify permitted net mesh dimensions, closed seasons and closed waters, and prohibited methods. In 1893, a Fisheries sub-Department was created within the Department of the Commissioner for Crown Lands and its officers were given the responsibility of enforcing fisheries regulations (Anon 1984). With meagre resources, however, effective oversight of fisheries was impossible: in the south-west one Inspector, based first in Busselton and then in Bunbury, was expected to supervise coastal fishing from Bunbury to beyond Augusta, as well as two major estuaries, more minor inlets, and inland waters, equipped with only a bicycle and rowboat.

Under such circumstances, one might expect extensive abuse of the resource. However, populations in the region were small, and the only effective way to transport a catch to the metropolitan centre of Perth, 175 km to the north, was via rail, a line between Perth and Bunbury having been completed in 1893. Perth at that time was also a relatively small market, with a non-Indigenous population of only 9,617 in 1891 (Fraser 1895). Furthermore, ice was scarce and freight expensive, so sending fish to Perth could be a risky financial proposition. By 1897, the Perth Ice and Refrigeration Company had begun to run a railway car with ice to Busselton daily, buying fish from the local fishermen to send to Perth, but in October the Manager declared that:

> we very much regret that through the many times it has had to leave Busselton empty and the many other times when it contained only three or four schnapper, we have been compelled to request the Railway Department to discontinue running the car on our account... we have lost heavily in waiting for the fishermen to get... wonderful catches, but up to the present we have not received them and if we wait until the Busselton fishermen endeavour to supply them, I am afraid that we shall have to wait a very long time. (Hancock 1897)

Why this scarcity of fish? If local prices were often better than those offered by the Company, fishers may well have been content with selling small catches to the small local market. Perhaps it was also the case, as Malcolm Tull suggests, that 'the lack of effort by many fishermen suggests that for them fishing is a way of life rather than an occupation' (Tull 1992). In any event, it appears likely at this time that market factors played a part in limiting the fishing effort.

By 1901, in his first annual report from the district, the Busselton Inspector of Fisheries reported that there were 20 fishermen and 10 boats, and that the fishing was good; a diverse range of species was caught, including mullet, pilchard, whiting, herring, snapper, dhufish, salmon, tailor, salmon trout and silver bream. By this time, most of the fish was sent by rail to Perth and regional towns, while around 180 kg was sold locally per week. Other than inclement weather, the only problems reported at this time were with fish-eating cormorants, which attracted a bounty of 6d per head; the inspector wanted that extended to pelicans, of which there were 'a great number' in the region. Also the entrance to the Wonnerup estuary was 'choked with sand and seaweed' leading to fish kills (Chief Inspector of Fisheries 1901). This was likely a result of agricultural development in the catchment, which was quite extensive by that time. With regard to effort, in 1912 the Inspector remarked that Bunbury and Busselton fishermen would only fish in the shallows or haul from the beach; they didn't like fishing in deeper water (Department of Aborigines and Fisheries 1912). Early reports indicate that in any event weather prevented much offshore fishing, especially in the winter months. Gear was rudimentary: basic nets were set or hauled; snapper and dhufish were caught with manual hand lines. In 1910 there was only one 'big boat' (Department of Aborigines and Fisheries 1910). Even in 1952, of the 68 boats licensed in the Bunbury district, 30 were propelled by oars or sail; 26 of the motorised boats were 25 ft or under; and only 2 motorised boats were over 35 ft. The total value of boats and gear was £ 43,020 (Fisheries Department 1960).

In the 1940s, as the war disrupted food imports, and some locally-produced food was diverted to feed servicemen and women in Australasia, the price of fish increased, and as Chief Inspector A. J. Fraser put it in 1953, 'Whereas formerly he [the Western Australian fisherman] was on the bread-line... and regarded by many as a nuisance who was spoiling their angling, the fisherman all at once found himself earning good money, suddenly needed' (Fraser 1953). This provided the impetus for a gradual transformation of the industry from one characterised by low investment and low returns, to greater investment for larger returns. From the 1950s, we therefore see the Capes region's fishing fleet being upgraded. By 1968, of the 65 licensed boats only 16 were not motorised, and 19 were over 25 ft. The value of boats and gear had risen to $ 310,226 (Fisheries Department 1960).[1] The introduction of motors, and larger boats, helped to overcome the obstacle of weather and extend the fishing season. From 1955, black and white sounders were available, but it is unclear how widely they were adopted in the south-west (Wise et al. 2007). More technological innovations with the potential to considerably increase efficiency followed: hydraulic reels, useful for handline fishers targeting dhufish, were (gradually) adopted from the mid-1970s; colour sounders were becoming more prevalent in the mid-1980s; and the technology responsible for the greatest increase in fishing efficiency, GPS, was available from the mid-1980s and widely adopted in the early 1990s (Wise et al. 2007).

Participation in any of the state's fisheries was unrestricted until 1963 when, amid concerns about declining productivity, entry to the Western Rock lobster fishery was limited in order to constrain growth in commercial fishing effort and enable more sustainable exploitation of the resource (Rogers 2000; Morgan 2001). The Shark Bay and Exmouth Gulf Prawn fisheries were declared limited entry soon after. Access to some south-west fisheries was limited from 1969, when there was a deliberate reduction in the number of licenses for estuarine fishing in the region (Lenanton 1984; Fisheries Department 1999). The South-west Coast Salmon Limited Entry Fishery was created in 1976. However, access to the remainder of the

---

[1] When the Australian dollar was introduced in February 1966, the exchange rate was £ 1 = $ 2.

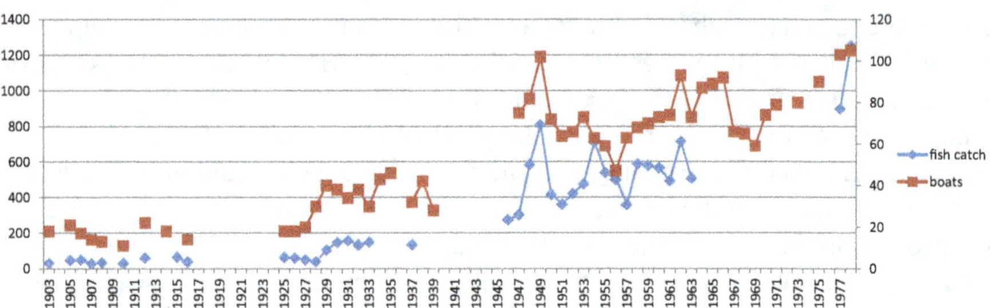

Fig. 12.2 Catch of finfish (tons) and licensed boats, Bunbury district, 1903–1978

state's fisheries remained unrestricted and the fishing fleet continued to expand until 1983, when in light of concerns over excess fishing capacity, a moratorium was placed on the issue of new fishing boat licenses. Subsequently, more Managed Fisheries were created,[2] and legislation was introduced enabling the creation of voluntary fishing license buy-back schemes. Together these measures saw the state's fishing fleet shrink from its peak of 1,615 units in 1985, to 1,361 units in 1998 (Rogers 2000). However, access to one of the region's major fisheries was only limited in January 2008, in anticipation of the creation of the West Coast Demersal Scalefish Interim Managed Fishery in January of the following year (IFAAC 2010).

Turning, then, to the available catch and effort data for the region, the only extended run of data comprises the number of fishing boats registered and reported catch for the Bunbury district, which encompassed all fisheries from around Myalup on the west coast to Broke Inlet on the south coast, including the inshore and offshore fisheries around the Capes, and the Leschenault and Hardy Inlets (Fig. 12.2). The data is highly confounded, as it includes multiple fisheries, and does not include the catch or effort of boats from the Fremantle-based fleet, which often visited the region. Nor does it account for changing technology or the changing nature of the fisheries, for example in terms of expansion within the region; shifts in methods and species targeted (and in particular the expansion of deep-sea shark fishing from the 1940s); the increasing recreational catch and effort; or changes in technology, over time. It also relies heavily on fishers' reports of their catch, without independent verification.

Nevertheless, it does point usefully, if vaguely, to some changes in the regional fisheries over time. Most generally, it shows a trend toward increasing catch per unit effort (CPUE) over the period, reflecting increasing efficiency and the absence of any sustained general collapse in fish stocks over the region as a whole (Fig. 12.3). Looking at the data in more detail, it indicates a fairly stable number of boats, total catch, and CPUE from 1903 to 1927 then, after a dip in the late 1920s, an increase in boats, catch and CPUE from 1930. As the measure of CPUE here, namely licensed boats, does not account for how often the boats were used, it is quite likely that the

---

[2] Managed Fisheries were the successor to Limited Entry Fisheries. By 1998 there were 31 Managed Fisheries and one Interim Managed Fishery (Crowe et al. 1999).

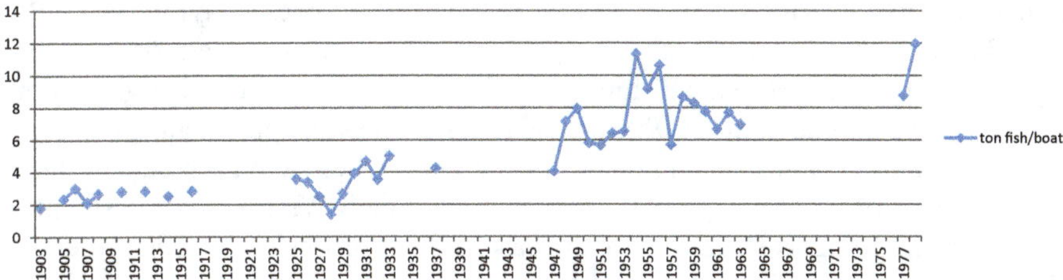

Fig. 12.3 Catch per unit effort indicator: finfish catch (tons) per licensed boat, Bunbury district, 1903–1978

increase in CPUE at this time is the result of an increase in the number of days spent fishing, as fishers struggled to make a living during the Great Depression. After the Second World War (WWII), there appears to have been a rapid increase in boats, catch and CPUE, perhaps associated with the expansion of shark fishing in the region (McAuley and Simpfendorfer 2003). The number of locally-registered boats then fell from 1949 to 1956, before increasing again. However, the fish stocks appear to have come under pressure from 1955 to 1963, a period of declining CPUE.

Given the problems associated with using such aggregated data, it is necessary to look in more detail at individual fisheries in order to arrive at a better understanding of the relationship between fish, fishing and natural variation in the region. To this end, two species of fish of commercial and recreational significance, both relatively well-represented in the historical sources but of diverse biology, were chosen for further investigation. One, snapper, is a demersal fish; the other, Australian herring, is pelagic.

Snapper

Snapper (*Pagrus auratus*) is a long-lived, slow-growing carnivorous demersal scalefish usually known locally as 'pink snapper', and historically as 'schnapper'. Distributed throughout temperate coastal Australasian waters, the adult fish form spawning aggregations in particular sheltered marine embayments, which are also important nursery areas. On the lower west coast of Western Australia, the maturing fish move further offshore after 2–4 years (Wakefield et al. 2011). Adult fish then return to inshore waters to spawn, and irregularly at other times of year to feed.

Snapper have long been a prized table fish, commanding good market prices. They were therefore extensively targeted by line fishers in the early twentieth century, though relatively few were operating at the Capes, and with unsophisticated boats and technology. In spite of that, Inspectors' reports suggest localised depletion at a very early stage of the fishery's development. For example, in 1902, Inspector Locke reported from Busselton that snapper were among the fish caught, and 'the supply of most species has been plentiful, with the exception of whiting' (Locke 1902). Likewise, in his 1903 annual report, F. W. Taylor, Bunbury Inspector of

Fisheries, noted that the snapper 'was fairly plentiful at times' (Taylor 1903). By 1905, however, he observed that 'Schnapper Fishing has been very erratic, and owing to the long distances to the Fishing grounds now, it does not return large profits' (Taylor 1905). Similarly, in his 1906 annual report, 'The schnapper fishing has been erratic, at times very scarce, and other times good catches have been taken. There has been a lot of prospecting for schnapper in the South-West during the summer months, boats going as far as the South coast in quest of fresh schnapper grounds' (Taylor 1906). These reports reflect the natural variability of the region and movement of the fish, but the references to the need for fishing boats to move further afield to find snapper may also be taken to suggest an early and rapid localised depletion of 'virgin' abundance, which the Inspectors evidently accepted as a normal, inevitable event. This is perhaps similar to the experience of northern New Zealand where, although the initial abundance was likely greater than at the Capes, localised depletion of snapper was first observed in the early twentieth century, but the species did not subsequently become rare (Parsons et al. 2009).

Other reports also suggest some early concern about the snapper stocks in the state's waters. In 1931, the Chief Inspector of Fisheries wrote that 'The question of increasing the minimum length at which these fish may be taken is one continually in my mind, but the present moment, is, I fear, inopportune' (probably because of the circumstances of the Great Depression, in which restricting access to a wild food item would have been politically difficult) (Chief Inspector of Fisheries 1931).[3] Evidently much offshore snapper fishing before WWII was undertaken by Italian migrants (Chief Inspector of Fisheries 1942). When many of these people were interned as 'enemy aliens' during the war, the species had some respite, and at least one report from 1947 suggested some increase in numbers along the lower west coast (Bramley 1947).[4] The sparse available monthly and annual returns of snapper catches (line only) shows variable but declining returns (Fig. 12.4). The downward trend is particularly evident in the 1970s and 1980s, especially considering the improvements in technology referred to above: bigger and better boats, hydraulic winches, and sounders.

The available catch records are interesting not least because they point to a shift in effort in the region from snapper to dhufish: in 1916, only 43 kg of dhufish were landed in the year, compared with 1,145 kg of snapper (Fisheries Department 1917). By 1941 the tables had turned, as 8,352 kg of dhufish were landed compared with 1,677 kg of snapper (Fisheries Department 1945). This shift is worth investigating in more detail. West Australian dhufish (*Glaucosoma hebriacum*) is a reef-associated demersal scalefish that grows to over 1 m in length. Although dhufish is endemic to the waters off Western Australia, both it and the more cosmopolitan snapper were recognised early on as highly desirable table fish and appeared, for example, on late nineteenth-century restaurant menus (*West Australian* 20 April

---

[3] The minimum size was at that time 11 inches (279 mm); by 2010 it had been increased to 500 mm, or 410 mm north of 31°S.

[4] For 2 weeks in late April 1947, anglers caught up to 5 snapper a night, weighing between 19 and 24 lb (8.6–10.9 kg) at the Bar, Mandurah, for the first time in living memory.

Month	catch kg	men	Boats	days	hooks	catch/boat	catch/man	catch/hookday
Sep 1911	356*	2	1			356*	178*	
Oct 1911	375*	4	3			125*	94*	
Jan 1964	2	1	1	15	4	2	2	0.4
Jan 1965	91	1	1	4	2	91	91	11
July 1965	345	1	1	9	20	345	345	2
Feb 1966	617	2	1	7	10	617	308	8.6
Year	catch kg	men	Boats	days	hooks	catch/boat	catch/man	catch/hookday
1916	1145							
1941	1677							
1943	1223							
1975-6	813	26	10	119	37	81	31	1.58
1976-7	628	8	6	76	17	105	78	2.84
1977-8	603	22	11	120	35	46	22.9	1.42
1978-9	1438	20	10	178	36	72	144	2.95
1979-80	388	12	6	61	17	65	32	2.23
1980-81	374	15	7	72	30	53	25	1.15
1981-82	213	9	5	41	24	43	24	1.3
1982-83	366	26	10	63	55	37	14	1.15

* Catch includes snapper, dhufish and whiting; however, quantities of the latter two species are likely to have been negligible.

Fig. 12.4 Commercial handline snapper catches 1911–1983

1895, p. 5. See 'Classified Advertising', Menu for the Duke of York Restaurant). It seems likely that snapper was initially valued more highly than dhufish, given its treatment on menus and in discussion of fisheries (*Western Mail* 13 February 1892, p. 26; *West Australian* 20 April 1895, p. 5). In 1900 snapper (of average weight 12 lb/5.4 kg) fetched 4s each at the Perth wholesale market, whereas dhufish (of unknown weight, though likely larger) sold for 6s each (*West Australian* 3 November 1900, p. 3). By 1911 the two species attracted the same price per pound (*West Australian* 26 January 1911, p. 5) and by the 1930s, dhufish routinely fetched around 3 pence more per pound than snapper (*West Australian* 28 September 1937, p. 14; 13 August 1938, p. 14). In March 1941, the price of dhufish reached a record £ 2/01/2 per pound (*West Australian* 8 March 1941, p. 4). This change in the market doubtless played a role in the shift to targeting dhufish in the region, though given that snapper continued to command good prices, it is perhaps still surprising that they were not more often caught if they were abundant.

In 1941, the Chief Inspector of Fisheries reported that although Cockburn Sound (approximately 150 km to the north of Geographe Bay) was one of the most important spawning grounds for snapper in the state, '"school" snapper, i.e. snapper which have schooled for the purpose of spawning, have been marketed from various localities', including the Bunbury district (Chief Inspector of Fisheries 1941). In the context of declining abundance, these 'distributed' schooling events might have become less common, at least in shallower waters, making snapper harder to target in the region than the dhufish that routinely school in the vicinity of Cape Naturaliste to spawn (John Nelson, personal communication with the author, 31 October 2010). By the 1940s, most of the state's snapper was supplied from the abundant resources of the mid-west coast and Shark Bay, hundreds of kilometers to the north. In the Capes region snapper became by-catch, although still a valuable one, that was usually retained and sold. Snapper were also highly sought-after by the increasing number of recreational fishers in the region, many of whom had access to boats.

In the late 1970s, a study of the marine resources of the waters of Bunbury and Geographe Bay caught more dhufish than snapper on set lines, though neither species was prolific (Walker 1979).[5] When changes to recreational bag limits were being canvassed in 1984, the Busselton fisheries inspector considered both snapper and dhufish to have declined, and noted that there was considerable community concern over the demersal scalefish. Oral histories recorded in 2005/2006 yielded multiple observations of declining abundance of snapper and dhufish (Gaynor et al. 2008). By 2007, snapper and dhufish were classified as 'being overfished' over their range on the basis of age-length modelling; fishing effort was accordingly to be reduced by at least 50% (Wise et al. 2007). This led to a prohibition on the commercial take of demersal scalefish in the Metropolitan Zone in November 2007; creation of the West Coast Demersal Scalefish Interim Managed Fishery in January 2009; closure of the recreational fishery for snapper and dhufish (along with other key demersal species) for 2 months per year; a decrease in recreational bag limits (2 snapper, 1 dhufish per fisher), and an increase in the minimum size.

In 2011, a scientific study of commercial fishery data collected between 1976 and 2005 found that in the West Coast Bioregion (which includes the Capes region), there was, overall, no evidence to suggest the occurrence of 'fishing down the food web' (that is, depleting higher-order predators then moving on to exploit species occupying progressively lower trophic levels). Rather, mean trophic levels (MTLs) and mean maximum length of fish caught increased then stabilised over the 30 year period (Hall and Wise 2011). However, it would be unwise to draw too much comfort from this conclusion in relation to the status of specific species in the Capes region, for several reasons. Firstly, as the authors acknowledge, the data was aggregated over large regions, and so cannot detect whether the indices were maintained by expanding the fishery into previously less exploited areas (or, for that matter, by other changes in fishing practice, such as technologically-driven increases in gear efficiency or economically-driven shifts in species targeted). Similarly, it assumes that within each region each species represents a single breeding stock, when the existence of sub-populations and separate breeding stocks are increasingly being recognised. The authors also acknowledge that quantitative data on fish diet is lacking even for key species (including snapper), which somewhat reduces the certainty of the modelling, as does uncertainty around the exact trophic levels of different species. Finally, the consensus that change in MTL is the best measure of fishery sustainability in all contexts has recently been challenged by a research finding that catch and ecosystem MTLs often diverge, and fisheries can collapse even when MTL is increasing, depending on patterns of development (Branch et al. 2010). It would therefore be unwise on the basis of this report to ignore other sources of

---

[5] For 6,300 hooks set, 12 dhufish and 9 snapper were caught.

historical information that suggest declining abundance of particular species, or ecosystems in trouble.

Overall, therefore, the picture for snapper is one of historical variability: the area was probably never brimming with the species as Shark Bay and the north of New Zealand were, but there was possibly a quite rapid initial localised decline in abundance, followed by a further uneven but sustained decline, associated with commercial and recreational fishing effort.

Australian Herring

The Australian herring or tommy ruff (*Arripis georgianus*) is a small pelagic fish found in southern Australian waters. Most juveniles of the adult population found in Western Australian waters mature in South Australia and on the south coast of Western Australia, then migrate west to spawn in a run that commences in summer and usually reaches the west coast in autumn, with spawning reaching a restricted peak in May. This spawning run heralds the beginning of the main fishing season for herring, and most of the commercial catch is comprised of fish migrating to spawn, although anglers target the species on the lower west and south coasts all year round. An area north of Cape Leeuwin is the main spawning ground for the species, although spawning probably also occurs across its range in Western Australia (White 1980). Once spawning is complete, the adult fish tend to remain in the south west region. Some larvae hatching inshore also remain in the area, forming part of a resident population, but most are carried eastward by the Leeuwin Current to South Australian or even Victorian waters, from where they will begin their westward journey at around 2 years of age (Ayvazian et al. 2000). Following a study of recruitment indices in the years 1996–1998, and a longer-term South Australian study, it has been proposed that recruitment to the Western Australian fishery is influenced by the Leeuwin Current, with more adults recruiting to the west coast following weak flows, and the south coast following strong flows, although factors such as the strength and direction of prevailing winds are also implicated (Ayvazian et al. 2000). Certainly this relationship is reflected in historical years of high abundance in the South-west, with 1953, 1960, 1974, 1981, 1982 and 1983 each following 2 years of relatively weak flows.

Herring have long been valued as a 'bread and butter' fish, and remain popular among anglers looking for both some sport and a meal. In Western Australia most of the commercial catch has historically been taken from the south coast, with a small percentage coming from Geographe Bay and Bunbury (Walker 1987). The commercial herring catch has historically varied according to a range of factors, including the size of the catch from the preceding run of Australian salmon (*Arripis trutta*), with more effort in years of a poor salmon catch, and less in good salmon years (Walker 1987). Demand has also played an important role. Between 1970/1971 and 1976/1977 a market for herring as rock lobster bait was established, and as demand for the fish rose, prices increased from around $ 60 per tonne in 1970, to $ 300 per tonne in 1979 (White 1980). This market was opened up to Geographe Bay fishers with the arrival of West Ocean Canning at Busselton in 1975/1976. Within 2 years, they were buying most of the herring caught within the region and processing it for bait (White 1980). This appears to have led to an increase in effort in the region, and in the 1980s record catches were taken (ABS 1966/1967 to 1989/1990). With a few minor exceptions until the 1983 season the Western Australian herring fishery was unregulated. However, as the value of herring increased so did conflict among fishing teams, and regulations relating to entry to the fishery and use of herring traps followed soon after (Walker 1987).

Rock lobster fishers had been catching their own herring to use as bait, mainly around Rottnest Island (approx. 175 km to the north of the Capes region) since the 1950s, but by 1964 conflict with recreational fishers had led to a seasonal restriction on netting activities in that area. When conflict continued, a ban on netting in the area was implemented in 1973 (Walker 1987). The earliest preserved written claim of a decline in herring comes from 1960, when one G. Reid complained that the abundance of herring at metropolitan beaches was much less than in the 1920s and 1930s (Reid 1960). It is difficult to know how to interpret this complaint, as commercial catches were rising in this period, and the late 1950s saw a weak Leeuwin Current which should have enabled more herring to recruit to the west coast. Furthermore, contemporary newspaper reports suggest that 1960 was in fact a bumper year for migratory herring, with large hauls reported. Once the main run of fish arrived, one party of anglers caught 15 dozen herring in two trips, and another caught 38 dozen herring and garfish in 7 h off northern beaches (Davidson 1960a, b). In the Capes region, oral histories recorded in 2005–2006 also suggest that herring were abundant in the late 1950s and early 1960s (Gaynor et al. 2008). Made in the context of rising concern over the effects of the netting around Rottnest, it is likely that the claim that herring had declined was driven by a deeply-held belief that commercial netting was causing long-term damage to the fishery and depriving anglers of their fair share of the resource,[6] though it may also have been the case that the size of the commercial and recreational catch combined reduced the resident herring population in popular fishing areas.

The next claim that herring were declining appeared in the mid-1970s. This was after the ban on netting at Rottnest had been declared, although the author seemed unaware of this development, as the complaint was mainly directed at the Rottnest netting (Mumme 1976). Writing in December 1976, the complainant (for a trade union) described a decline over 'the last two seasons', but in 1977 the Fisheries and Wildlife Department's chief scalefish research officer reported that 'Herring were abnormally abundant in the metro area during 1974/1975 season & anglers made exceptionally good catches up until about March 1976' (correspondence in Fisheries Department 1947). However, the 1975–1976 herring season was a poor one for both amateurs and professionals, as the main run of fish did not make it as far north

---

[6] This is particularly so because the complaint contains some obvious falsehoods: for example, given the scale of herring catches by anglers in 1960, it simply could not be true that 'In 1960, after 10 years of systematic netting, the total herring catch by all the anglers during the whole season would not represent one good day's catch even to the inexperienced angler of 1920' (Reid 1960).

as usual. This was likely due to strong Leeuwin Current flows in 1974 and 1975. Interestingly, the commercial catch of herring reached 1,200 t in 1972–1973, the highest to that point, and remained significant at 800 t in 1974–1975. Given the reports of herring abundance in that year, it appears that the commercial catch (mostly on the south coast) was having little, if any, effect on the apparent abundance of fish reaching the west coast.

A subsequent controversy over herring decline, in November 1979, did not follow strong Leeuwin Current flows, and commercial catches in the preceding years had been only average, ranging between 500 and 800 t. As anglers' claims of dramatic depletion at this time are not borne out by the available evidence, it seems highly likely that their perceptions arose out of their moral indignation over the use of a sought-after angling species as lobster bait, rather than any significant decline in herring numbers. Indeed, in a newspaper article published at the height of the controversy, Australian Anglers Association publicity officer Laurie Birchall stated that:

> We don't mind amateur anglers catching 50 or 100 to take home for family and friends… And we don't mind them being caught by the tonne if it's for food. However we do object to unrestricted slaughter for fertiliser, petfood and craybait when alternatives are available (*Sunday Times* 15 April 1979, p. 70).[7]

It therefore seems that past perceptions of herring abundance may reflect the effect of the Leeuwin Current on the abundance of the fish in the region, but even more so the immediate political context of resource conflict between professional and recreational fishers.

A 2008 study of angling club herring catch records for the years 1977–2008 raised the possibility of declining abundance in more recent years, showing a dramatic reduction in the mean number of herring caught per angler per trip (from 22.9 in 1984 to 1.7 in 2008), and a considerable reduction in the percentage of trips resulting in a bag limit of the species, from 26.6 % in 1984 (bag limit of 50), to 12.1 % in 2008 (bag limit of 8) (Pember 2008). However, caution must be used in interpreting angling club data over the long run, as it is very likely to reflect changes in a range of factors other than abundance of fish.

The club in question has run beach and rock field day competitions since 1974, including autumn and winter field days in the Capes region. The data available for analysis included the number and weight of each species caught by each angler on each field day between 1976 and 2008, bar 13 field days for which the records were either not created or not kept. In that time the club, either on its own initiative or in line with state legislation, implemented bag limits for herring, starting with 50 fish per angler in mid-1980. This was then reduced to 40 in 1992, 20 in 2000 and 8 in mid-2007. The autumn field trip records reveal that for the 3 years prior to imposition of a bag limit in 1981, the average catches for the top 5 anglers fluctuated around 40–50 herring; for the top 10 around 30; and for all anglers around 20. On imposition of the bag limit of 50, the top 5 and top 10 anglers caught the bag limit, or very close to it, in each year bar one (Pember 2008). This strongly suggests a cultural phenomenon at work, in which the bag limit becomes a target. From 1994,

---

[7] Quoted in Sergeant Baker (pseud.). Later in the same article, it was acknowledged that herring provided most of the catch at an angling club field day the previous weekend.

however, the bag limit was not reached with such regularity. The winter field trip data yields similar findings, although as this is not peak season for herring, the bag limits were never reached with quite such regularity.

How can we explain the reduction in the percentage of anglers catching a bag limit of herring on field days? Were there fewer fish to be caught, or was it more the result of changes in angling practices? The commercial take of herring in WA reached an all-time high of 1,537 t in 1991; as the fourth year with a catch over 1,200 t, this may, in conjunction with the recreational take, have reduced the breeding stock to an unsustainably low level, and led to a reduction in the abundance of herring available for anglers to catch in subsequent years. However, from 1995 the price of Australian herring fell, most likely due to higher imports of cheaper North Sea herring for lobster bait. By 1999 demand for Australian herring was very low, with reports of frozen fish remaining unsold for over a year, and herring being discarded from trap nets (Ayvazian et al. 2000). The commercial catch therefore fell rapidly, to around 800 t by 1993. By 2008, it had fallen to only 217 t. Although the commercial take in the early 1990s may have reduced the availability of fish for anglers in the short term, given its reduction over subsequent years it is unlikely to have been responsible for a sustained decline.

Cultural factors doubtless also played a role in the failure of even the best anglers to routinely catch the bag limit even after it was reduced to 20. By the late 1980s, the club had changed the scoring system such that instead of receiving one point per fish and three points per kg of fish weighed in on a field trip, the fish caught were differentiated by species, and more points awarded for 'premium' fish. This would presumably have enticed anglers to spend more time targeting higher-scoring fish than herring. Any such response would have been reinforced by a broader shift in angling culture in this period. Rather than 'bread and butter' species such as herring, more anglers sought larger fish, which were seen as presenting a greater personal challenge than the shrinking legal bags of small fish (Brenton Pember, personal communication with the author, 20 January 2011). This shift is reflected in newspaper fishing columns, which even in the 1970s included stories of large herring catches, but by the 2000s typically discussed the conquest of larger table species, and sports fishing feats. The rise of sports fishing, centred around the skill of catching the fish rather than filling the freezer, was accompanied by the growing prevalence of 'catch and release' over this period. This was briefly accommodated by the club's rules via an honour system, but only from 2005–2007; outside of this period club anglers releasing fish would not have had them counted. Finally, there was also a broader shift within the culture of the angling club in question: a past president, Brenton Pember, recalled that in the 1980s the club was very competitive and anglers would fish all night during field trips, but over time it became much more a social club, with less emphasis on the competition (Pember, personal communication). This, too, may well have influenced the catch rates from the mid-1990s.

The mid-1990s were years of average to low Leeuwin Current flow, so any variation at that time was unlikely to have been a result of the Current's influence on recruitment. However, it may be that larger-scale environmental change in the region has begun to affect the herring's fortunes. Although much uncertainty remains, the

Leeuwin Current is predicted to weaken due to climate change (Feng et al. 2009). As we have seen, this has the potential to increase recruitment of herring to the western fishery, although that potential could be undermined by other factors. One such factor is the changing climate of the region, which has seen a step decline of around 15 % in the rainfall over south-west WA since the 1970s, largely due to fewer winter storms in the region (Hope and Foster 2005; Rupercht et al. 2005). It has been hypothesised that the reduction in both Leeuwin Current flow and winter storms will reduce nutrient inputs into the upper ocean, reducing productivity in the region (Feng et al. 2009); perhaps we are seeing these effects already. In 2008, the Department of Fisheries reported that although the usual index of herring abundance, the commercial catch rate, had become increasingly unreliable due to the reduction in fishing effort, other data sources including voluntary logbooks for anglers and fishery-independent sampling suggested a decline in abundance from the late 1990s (Smith and Brown 2009). The lack of reliable data, however, has seen the stock status deemed 'uncertain'.

Conclusion

In the relatively low-productivity fisheries of the Capes region, where fishing effort has varied considerably over time, it is impossible to precisely trace changes in the abundance of fish. However, it is clear that apparent shifts in the abundance of fish have taken place against the backdrop of environmental variation, which is quite pronounced as a result of the variability of the Capes and Leeuwin Currents, as well as changes in the regional climate. In this context, some claims of declining abundance appear to have been misconceptions or tendentious; some reflected likely short-term variation; others suggest longer-term localised depletion. The three cannot be definitively disentangled without supporting information, which is patchy at best in this region for the period prior to 1976. However, an historical perspective, even if incomplete, provides a potentially useful resource for current management, encouraging a cautious approach in the light of (necessarily) incomplete scientific understanding. It also provides a useful context for future Ngari Capes Marine Park monitoring, especially as the interim management plan claims that 'most finfish populations are considered to be in a largely undisturbed state' (DEC 2006). Furthermore the proposed conservation objectives for any reserves established in the area under the Commonwealth marine planning process include preserving transects of land and water 'in its natural and/or unmodified conditions' from the coast, to inshore waters and into the deep ocean (DEWHA 2009). As this research suggests, commercially-significant marine species in the region are far from 'undisturbed', and nature itself can be a source of considerable 'modification'.

References

ABS (1966/1967 to 1989/1990) Fisheries-Western Australia. Australian Bureau of Statistics, Perth

Anon (1984) The history of the department of fisheries and wildlife. FINS 17:7–8

Ayvazian SG, Jones GJ, Fairclough D, Potter IC, Wise BS, Dimmlich WF (2000) Stock assessment of Australian herring. Fisheries Western Australia, Perth

Baudin N (1974) The journal of post captain Nicolas Baudin, commander-in-chief of the Corvettes Geographe and Naturaliste (trans: Christine Cornell). Libraries Board of South Australia, Adelaide

Bramley J (1947) Letter from J. Bramley, inspector of fisheries Mandurah, to the chief inspector of fisheries, 5 May 1947. In: Fish species snapper—general file, Fisheries Department file 1932/30 v1, Cons 5368, State Records Office of Western Australia, Perth

Branch TA, Watson R, Fulton EA, Jennings S, McGilliard CR, Pablico GT, Ricard D, Tracey SR (2010) The trophic fingerprint of marine fisheries. Nature 468:431–435

Chief Inspector of Fisheries (1901) Report of the chief inspector of fisheries for the year 1901. In: Votes & Proceedings of the Parliament of Western Australia, 1902, vol.1, part 2, paper 4. Government of Western Australia, Perth

Chief Inspector of Fisheries (1931) Letter from chief inspector to inspector Linton, 27 April 1931. In: Fish species snapper—general file, Fisheries Department, file 1932/30 v1, Cons 5368, State Records Office of Western Australia, Perth

Chief Inspector of Fisheries (1941) 12 November 1941. In: Fish species snapper—general file, Fisheries Department, file 1932/30 v1, Cons 5368, State Records Office of Western Australia, Perth

Chief Inspector of Fisheries (1942) Letter from chief inspector to chief of division of fisheries CSIR 6 July 1942. In: Fish species snapper—general file, Fisheries Department, file 1932/30 v1, Cons 5368, State Records Office of Western Australia, Perth

Christensen J (2009) Recreational fishing and fisheries management: a HMAP Asia project paper. Perth: Working Paper no. 157, Asia Research Centre, Murdoch University, Perth

Crowe F, Lehre W, Lenanton RCJ (1999) A study into Western Australia's open access and wetline fisheries. Fisheries Research Report 118. Fisheries Western Australia, Perth

Davidson R (1960a) Rod and Line. In: Sunday Times (Perth, WA) 24 April 1960:54

Davidson R (1960b) Rod and Line. Sunday Times (Perth, WA) 1 May 1960:62

DEC WA (2006) Indicative management plan for the proposed geographe Bay/Leeuwin-Naturaliste/Hardy Inlet Marine park. Western Australian Department of Environment and Conservation, Perth

Department of Aborigines and Fisheries (1910) Inspector of fisheries, Bunbury: Bunbury district, monthly reports, 1910. In: Department of Aborigines and Fisheries file 1910/0219, Cons 652, State Records Office of Western Australia, Perth

Department of Aborigines and Fisheries (1912) Monthly reports, Bunbury District. In: Department of Aborigines and Fisheries, file 1912/0333, Cons 652, State Records Office of Western Australia, Perth

DEWHA (2009) Marine bioregional planning in the South-west marine region—areas for further assessment. Australian Government Department of the Environment, Water, Heritage and the Arts, Canberra

Dortch CE (1997) New perceptions of the chronology and development of aboriginal fishing in South-Western Australia. World Archaeol 29:15–35

Feng M, Waite AM, Thompson PA (2009) Climate variability and ocean production in the Leeuwin current system off the west coast of Western Australia. J Roy Soc West Aust 92:67–81

Fisheries Department (1917) Inspector Eaton, Bunbury- report for 12 months ending 31 Dec. 1916. In: Fisheries Department file 1917/0658, Cons. 477, State Records Office of Western Australia, Perth

Fisheries Department (1945) Fisheries—chief inspectors annual report 1939 to 1943 incl. In: Fisheries Department file 1945/0003, Cons. 477, State Records Office of Western Australia, Perth

Fisheries Department (1947) Species-herring. In: Fisheries Department file 1947/0021/V01, Cons. 6397, State Records Office of Western Australia, Perth

Fisheries Department (1960) Statistics—issue of fishing licences in WA—fishing boat statistics. In Fisheries Department file 1960/0210 V1, Cons 4169, State Records Office of Western Australia, Perth

Fisheries Department (1999) Management directions for Western Australia's estuarine and marine embayment fisheries: a strategic approach to management. Fisheries management paper no. 131, Fisheries Western Australia, Perth

Fraser MAC (1895) Western Australian year book 1893–94. Government Printer, Perth

Fraser AJ (1953) The fisheries of Western Australia, Fisheries Department. Government of Western Australia, Perth

Gaynor A, Kendrik A, Westera M (2008) An oral history of fishing and diving in the capes region of South-west Western Australia project CM01a July 2008. University of Western Australia, School of Humanities and School of Plant Biology, Crawley

Gersbach GH, Pattiaratchi CB, Ivey GN, Creswell GR (1999) Upwelling on the South-west coast of Australia-source of the Capes Current? Cont Shelf Res 19:363–400

Hall NG, Wise BS (2011) Development of an ecosystem approach to the monitoring and management of Western Australian fisheries. FRDC report—project 2005/063. Fisheries research report no. 215, Department of Fisheries, Perth

Hancock WM (1897) The Busselton fisheries. West Australian 27 September 1897:7

Hanson CE, Pattiaratchi CB, Waite AM (2005) Seasonal production regimes off South-western Australia: influence of the Capes and Leeuwin Currents on phytoplankton dynamics. Marine Freshw Res 56:1011–1026

Hope P, Foster I (2005) How our rainfall has changed—the South-west climate note 5/05. Indian Ocean Climate Initiative, Perth

IFAAC (2010) Integrated fisheries management draft allocation report—west coast demersal scalefish. Integrated Fisheries Allocation Advisory Committee, Department of Fisheries, Perth

Jennings R (1983) Busselton: 'outstation on the vasse', 1830–1850. Shire of Busselton, Busselton

Lenanton RCJ (1984) The commercial fisheries of temperate Western Australian Estuaries: early settlement to 1975. Department of Fisheries and Wildlife, North Beach

Lenanton RC, Joll L, Penn J, Jones K (1991) The influence of the Leeuwin Current on coastal fisheries of Western Australia. J Roy Soc West Aust 74:101–114

Locke FW (1902) Report by FW Locke. In: Report on the fishing industry for the year 1902. Votes & Proceedings of the Parliament of Western Australia, 1903–04, vol. 2, part 1, paper 19. Government of Western Australia, Perth

McAuley R, Simpfendorfer C (2003) Catch composition of the Western Australian temperate demersal gillnet and demersal longline fisheries, 1994 to 1999. Department of Fisheries, Perth

Meagher S (1974–1975) The food resources of the Aborigines of the South-west of Western Australia. Rec WA Mus 3:14–65

Morgan GR (2001) Initial allocation of harvesting rights in the rock lobster fishery of Western Australia. In: Shotton R (ed) Case studies on the allocation of transferable quota rights in fisheries. Food and Agriculture Organisation of the United Nations, Rome

Mumme CE (1976) Letter from C. E. Mumme, organiser for the Federated Engine Drivers' & Firemen's Union of Workers, to Mr J. J. Harman M. L. A, 22 December 1976. In: Species- herring, Fisheries Department, file 1947/0021/V01, Cons 6397, State Records Office of Western Australia, Perth

Parsons DM, Morrison MA, MacDiarmid AB, Stirling B, Cleaver P, Smith IWG, Butcher M (2009) Risks of shifting baselines highlighted by anecdotal accounts of New Zealand's snapper (Pagrus auratus) fishery. N Z J Mar Freshw Res 43:965–983

Pattiaratchi C (2007) Understanding areas of high productivity within the South-west marine region. Department of the Environment, Water, Heritage and the Arts, Canberra

Pattiaratchi CB, Buchan SJ (1991) Implications of long-term climate change for the Leeuwin Current. J Roy Soc West Aust 74:133–140

Pearce A, Pattiaratchi C (1999) The Capes Current: a summer countercurrent flowing past Cape Leeuwin and Cape Naturaliste, Western Australia. Cont Shelf Res 19:401–420

Pearcey M, Terry C (2005) How our river flows have changed climate note 7/05. Indian Ocean Climate Initiative, Perth

Pember B (2008) Modelling environmental fish abundance from angling club data, ISC assignment 2. Murdoch University, Perth

Peron F (1975) A voyage of discovery to the southern hemisphere, performed by order of the Emperor Napoleon during the years 1801, 1802, 1803 and 1804. Marsh Walsh, North Melbourne

Reid G (1960) Letter from G. Reid to the Minister for Fisheries 20 August 1960. In: Species-herring, Fisheries Department, file 1947/0021/V01, Cons 6397, State Records Office of Western Australia, Perth

Rogers PP (2000) Towards a better future in fisheries management: rights-based fisheries management in Western Australia. In: Shotton R (ed) Use of property rights in fisheries management. Food and Agriculture Organisation of the United Nations, Rome

Rupercht J, Li Y, Campbell E, Hope P (2005) How extreme South-west rainfalls have changed climate note 6/05. Indian Ocean Climate Initiative, Perth

Smith K, Brown J (2009) Australian herring fishery status report. In: Fletcher WJ, Santoro K (eds) State of the fisheries report 2008/09. Department of Fisheries, Perth

Stirling J (2005) Narrative of operations—expedition to Swan river, 1827. In: Shoobert J (ed) Western Australian exploration, vol. 1. Hesperian Press, Victoria Park

Taylor FW (1903) Annual report 1903. In: Chief Inspector of Fisheries, Report on the fishing industry for the year 1903, Annual reports box, Fisheries Western Australia library, Western Australian Fisheries and Marine Research Laboratories, Hillarys

Taylor FW (1905) Annual report 1905. In: Chief inspector of fisheries, Report on the fishing industry for the year 1905, Annual reports box, Fisheries Western Australia library, Western Australian Fisheries and Marine Research Laboratories, Hillarys

Taylor FW (1906) Annual report 1906. In: Chief inspector of fisheries, Report on the fishing industry for the year 1906, Annual reports box, Fisheries Western Australia library, Western Australian Fisheries and Marine Research Laboratories, Hillarys

Thomson-Dans C, Ryan K, Hill A (2002–2003) The Capes coast: a diverse coastal and marine environment. Landscope Summer 18(2):1–8

Tull M (1992) Profits and lifestyle: Western Australia's fishers. Private Enterp Gov Soc: Stud West Aust Hist 13:92–111

Turney CSM, Bird MI, Fifield LK, Roberts RG, Smith M, Dortch CE, Grün R, Lawson E, Ayliffe LK, Miller GH, Dortch J, Cresswell RG (2001) Early human occupation at devil's lair, South-western Australia 50,000 years ago. Quatern Res 55:3–13

Wakefield CB, Fairclough CV, Lenanton RCJ, Potter IC (2011) Spawning and nursery habitat partitioning and movement patterns of Pagrus auratus (Sparidae) on the lower west coast of Australia. Fish Res 109:243–251

Walker MH (1979) An inventory of the marine resources of the Bunbury marine area and Geographe Bay. Department of Fisheries and Wildlife, Perth

Walker MH (1987) The Australian herring fishery in Western Australia, 1973–1985. Fisheries Department, Perth

White TF (1980) Herring and Tailor. Their exploitation and utilization in Western Australia. Report by TF White, Fishing and Allied Industries Committee, March 1980. In: Species-herring, Fisheries Department, file 1947/0021/V01, Cons 6397, State Records Office of Western Australia, Perth

Wise BS, St John J, Lenanton RC (eds) (2007) Spatial scales of exploitation among populations of demersal scalefish: implications for management. Part 1: Stock status of the key indicator species for the demersal scalefish fishery in the West Coast Bioregion. Final FRDC report—project 2003/052. Fisheries research report no. 163, Department of Fisheries, Perth

Chapter 13
Shark Bay Snapper: Science, Policy, and the Decline and Recovery of a Marine Recreational Fishery

Joseph Christensen and Gary Jackson

Abstract Since the mid-1990s Shark Bay's inner gulf snapper fishery has become one of the most intensively-studied and better-understood marine recreational fisheries in Australia. It provides an important case-study of the impact that recreational fishing can have on highly targeted stocks, showing that recreational fishers, by virtue of their greater numbers and their uptake of technologies developed for the commercial sector, can have an equal if not greater total catch than professional fishers in the same or similar fisheries. The Shark Bay case-study also demonstrates the complexity of the challenges associated with sustainably managing marine recreational fisheries. The effectiveness of traditional recreational management measures is increasingly being questioned. As more and more jurisdictions move towards implementing ecosystem-based management approaches, strategies to ensure sustainable harvests will be required for all sectors—commercial, recreational and artisanal alike. This chapter highlights the role that effective biological research and robust management intervention can play in assisting the recovery of a stock fished to the brink of collapse. Whilst the recovery of the inner gulf snapper stocks is continuing, it constitutes one of the few documented examples worldwide of the successful recovery of a marine recreational fishery through the promotion of sustainable, scientifically-based recreational harvest levels. The focus on ecological outcomes and other factors that contributed to the successful restoration of inner Shark Bay's recreational snapper fishery are being recognised as essential elements in the reform of recreational fisheries management elsewhere.

Keywords Amateur fishing history · Recreational fishing history · Recreational fisheries management · Shark Bay · Pink snapper

J. Christensen (✉)
Asia Research Centre, Murdoch University, 90 South Street,
Murdoch, WA 6150, Australia
e-mail: J.Christensen@murdoch.edu.au

G. Jackson
Western Australian Fisheries and Marine Research Laboratories, Department of Fisheries,
Western Australia, PO Box 20, North Beach, WA 6920, Australia
e-mail: Gary.Jackson@fish.wa.gov.au

J. Christensen, M. Tull (eds.), *Historical Perspectives of Fisheries Exploitation in the Indo-Pacific,* MARE Publication Series 12, DOI 10.1007/978-94-017-8727-7_13,

"""

Amateur or recreational fishing has joined commercial and artisanal fishing as the third sector in fisheries exploitation globally, and the evidence of its ecological impacts is challenging traditional notions of a benign and tranquil activity. This is true for both freshwater and marine fisheries, where increasing leisure time and rising affluence in developed nations have seen recreational fishing develop into a mass pursuit that attracts hundreds of thousands of participants annually (Kearney 2002; Aas 2008). Fish-finding technologies that were originally employed in commercial operations have become widely used by the recreational sector, and in some fisheries, recreational fishers now surpass commercial fishers in terms of harvest levels and overall economic impact. Documented cases of stock declines and changes in stock structure driven by increasing effort and efficiency gains of recreational fishers proliferated during the 1990s and 2000s, particularly in marine fisheries located in temperate regions of Europe, North America, Australia, New Zealand and South Africa (Lewin et al. 2004; Cooke and Cowx 2006). One landmark study in the United States estimated that recreational fishers accounted for 4% of the nation's marine fish landings between 1981 and 2002, with recreational fishers responsible for 23% of the harvest of 'populations of concern' during this period, a figure that rose as high as 38% in the South Atlantic, 59% along the Pacific Coast, and 64% in the Gulf of Mexico (Coleman et al. 2004).

The growing popularity of marine recreational fishing presents a range of challenges for fisheries management agencies. Historically, most marine recreational fisheries have been open-access, with management directed towards limiting individual catches through conventional measures such as gear restrictions and bag and size limits (McPhee et al. 2002). In some cases, the absence of constraints on total recreational catch threatens the sustainability of fisheries, and new approaches to protect stocks have been developed, including spatial and temporal closures, licensing systems, catch quotas, and restrictions on the sex or reproductive condition of fish that may be kept (Jackson et al. 2005). The need for these approaches has been highlighted by improved understanding of the extent and impact of recreational fishing. In Australia, which has one of the highest levels of recreational fishing participation in the world (Henry and Lyle 2003), the number of recreational fishing surveys and associated scientific and socio-economic studies undertaken by state fisheries agencies, universities and private consultants proliferated during the late 1990s and 2000s. These included one of the first nationwide surveys of recreational fishing in 2001 (Henry and Lyle 2003). Techniques for estimating total catches, catch rates and stock size have been continually refined during this period, as scientists have employed improved approaches to quantifying the impacts of recreational fishing. The issues that confront management agencies—high participation rates, increasing recreational fishing efficiency, and declining stocks that are often shared with commercial operations, indicate that sustainable management is an increasing challenge in many jurisdictions.

The fishery for pink snapper *Pagrus auratus* in the inner gulfs of Shark Bay, a large marine embayment on the coast of Western Australia, provides a classic illustration of a 'tragedy of commons' scenario created by growing recreational fishing activity, overlayed on a history of a small and steady commercial fishery. The Shark Bay case-study also highlights the role that effective biological research and robust management intervention can play in assisting the recovery of a stock fished to the brink of collapse. Whilst the recovery of the inner gulf snapper stocks is continuing, it constitutes one of the few documented examples worldwide of the successful recovery of a marine recreational fishery through the promotion of sustainable, scientifically-based recreational harvest levels. The focus on biological outcomes and other factors that contributed to the successful restoration of inner Shark Bay's recreational snapper fishery are being recognised as essential elements in the reform of recreational fisheries management elsewhere.

Shark Bay is internationally recognized as a World Heritage Area, one of the few properties worldwide that meets all four criteria for inclusion on UNESCO's register of Natural Places (Christensen 2008; Shaw 2000). Each of these criteria relate in some way to the physical and biological processes that shape the Bay's unique subtropical marine environment. Comprised of open deeper waters in the north and the shallower waters of Denham Sound and Freycinet Estuary in Western Gulf and the Eastern Gulf which form the inner gulfs in the south, Shark Bay is characterized by highly elevated salinities that result from the combined effects of low rainfall, limited tidal exchange, persistent winds and high evaporation (Logan and Cebulski 1970). Snapper (*Pagrus auratus*), a large and long-lived member of the Sparidae (sea breams and porgies), found throughout the warm temperate and subtropical waters of the western Indo-Pacific (Paulin 1990) has very successfully adapted to the elevated salinity levels inside Shark Bay. Research conducted over more than 20 years has shown that four distinct stocks of snapper occur here: an oceanic stock, which inhabits the oceanic waters of the continental shelf outside of the Bay; a Denham Sound stock, partially isolated from the larger oceanic population; and reproductively-isolated stocks in the metahaline waters (around 1.5 times more saline than ocean waters) of Freycinet Estuary and the Eastern Gulf. Tagging and other studies have confirmed that little or no mixing occurs between the inner gulf stocks, or between the inner gulf stocks and the oceanic stock (Jackson 2007; Fig. 13.1).

Shark Bay has supported a number of important commercial fisheries since the arrival of Europeans and the commencement of pearling in the 1850s (Edwards 2000). Among these was a major fishery for snapper pioneered by line fishing vessels based at Geraldton and Fremantle to the south during the early 1900s (Cooper 1997). The snapper fishery remained small scale during its early decades, but with the modernization of Western Australia's commercial fisheries after the Second World War the fishery expanded, and following the introduction of baited steel traps into the fishery in 1958–1959 catches rose sharply (Marriott et al. 2012). Most of the snapper fishing was concentrated in the waters outside and at the entrances with the Bay (Cooper 1997). The first review of the commercial fishery, in 1959/1960, indicated that around 10 % of the total catch of approximately 600 t per year was taken in the inner gulfs by fishers out of Denham, the main settlement on Peron Peninsula (Bowen 1961). Trapping had been introduced in 1959 and by 1960 had become the main commercial fishing method for snapper. However, due to concerns about its impact on juveniles, the use of traps in shallower waters around the margins of the Bay was progressively prohibited until, by 1987, virtually all of Shark Bay inside Bernier, Dorre and Dirk Hartog Islands was closed to the use of traps.

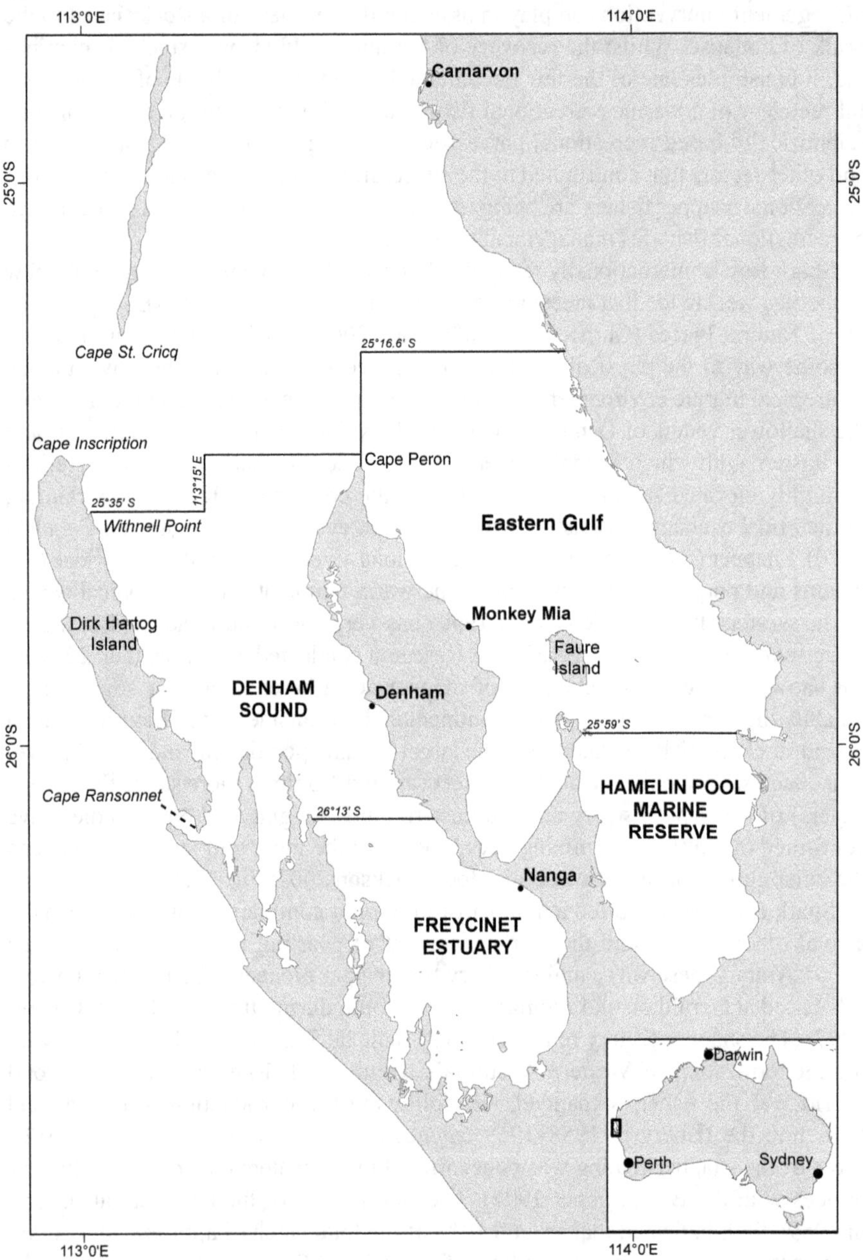

Fig. 13.1 Map of Shark Bay, showing Denham Sound, Freycinet Estuary, and the Eastern Gulf

Recreational fishers had been visiting Shark Bay since at least the 1920s. This began to change markedly during the 1960s when road access to Denham was improved and new caravan parks and camping grounds were developed.[1] Most visitors favoured Shark Bay's mild winter season, which presented a pleasant contrast to the cold and wet weather in the Western Australian capital, Perth, and the agricultural districts in the southern regions of the state. The peak tourism season coincided with the spawning season for pink snapper, when the fish congregated in large schools around key spots within the inner gulfs, placing them within easy reach of the small boats that fishers brought to the Bay on trailers or atop their 4WD vehicles. It was not long before a dedicated recreational fishery, targeting pink snapper and attracting hundreds of recreational fishermen on an annual pilgrimage, emerged. By the 1970s Shark Bay was among the most popular destinations for boat-based recreational fishermen in Western Australia (e.g. *Sunday Times* [Perth] 25 July 1971, p. 72; 9 July 1978, p. 109; 13 August 1978, p. 116; 5 August 1979, p. 110)

Around this time recreational fishers first began to observe declines in the size and number of snapper being caught. One early work on marine angling in Western Australia, Philip Bodeker's *The Sandgroper's Trail* (1971), suggested that amateurs regularly caught undersized juvenile snapper at Shark Bay and were responsible for declining catches in the inner gulfs (Bodeker 1971). More commonly, however, the blame for any perceived declines in the quality of the snapper fishing was directed towards commercial fishers. This became more widespread in the 1970s and early 1980s. Recreational fishers began to agitate for further restrictions on commercial trap fishing, joining more conventional commercial line fishers in objecting to the use of traps, and conveying their protests directly to Western Australia's Department of Fisheries ('the Department') (Department of Fisheries 1980/1981). It was assumed at this time that all fishers, amateur and professional alike, were exploiting the same single stock of snapper in Shark Bay. Under these conditions it was all too easy for amateurs to blame professionals for declining stocks, a well-established pattern of behavior that had led to various restrictions on the operations of the commercial sector during earlier decades in other parts of Western Australia (Christensen 2009). Indeed the acknowledgement that recreational fishing was the primary cause of the depletion of inner gulf snapper stocks came only later and somewhat grudgingly, many years after the separate breeding stocks in the inner gulfs had been identified, and only when these stocks had been fished to the brink of collapse.

Pushing Recreational Fishing to the Limit: The Growth of the Fishery

The first surveys of participation in recreational fishing in Western Australia present a picture of a highly popular pastime. A groundbreaking study in 1984 estimated that 469,000 people over the age of 13, 43 % of the state's population in that age

---

[1] The sealing of the Old Coast Road between Perth and Geraldton in 1969 and the opening of the Brand highway in 1976 made large stretches coastline accessible to recreational fishers.

group, had fished recreationally in the preceding year (PA Management Consultants 1984). In 1987 a second survey put the state's number of recreational fishers at 284,000 or 27% of the total eligible population (ABS 1987). Further surveys between 1996 and 2001 showed that recreational fishing was increasing in popularity as the state's population continued to grow, with 642,000 or 37% of the total population, going fishing at least once per year by 2001 (*Western Fisheries* March/April 1990). Increasing affluence, available leisure time, a climate conducive to outdoor activity and the proximity of most of the state's residents to the coast also underpinned high rates of private boat ownership. There were 25.0 registered boats per 1,000 of the state's population in 1990, 27.6 in 2000, and 30.0 in 2005 (DPI 2007). Recreational fishing was the bedrock of a sizeable industry that comprised of bait and tackle suppliers, gear manufacturers, boat builders, repairers and dealers, charter boat operators, and proprietors of caravan parks, camping grounds, and other tourist facilities. Newspaper columns and dedicated magazines and television programs fuelled the popularity of recreational fishing. A study in 1989/1990 estimated that the combined economic value of recreational fishing in Western Australia was $389 million and the employment impact was 5,900 full-time jobs. A second survey in 1998 put the total economic impact at $569 million and employment at 7,000 full-time jobs (Lindner and McLeod 1991; *Fisheries Management Paper* 136 2000). The popularity and economic significance of recreational fishing translated to considerable potential political influence that cut across established political divisions and, underpinned by strong and widely held public belief in the open-access nature of marine recreational fisheries, exerted a subtle but profound influence on the Western Australian Department of Fisheries.

In recognition of the social and economic importance of recreational fishing, and of the management challenges presented by high and growing participation rates in the context of strong population growth, the Western Australian Government appointed a Recreational Fishing Advisory Committee (RFAC) in 1989 to undertake a 2-year review of recreational fishing. The review aimed to develop a long-term policy direction for recreational fishing, and involved a comprehensive examination of the existing bag and size limits that acted as controls on recreational fishers. As a result of the review state-wide bag limits for 'reef fish' species that included pink snapper were reduced from ten to eight fish per angler per day.

This reduction was not based on any specific management objective or scientific rationale, but rather reflected the social and political principle that recreational fishers should not exploit common resources for anything other than their own enjoyment and the provision of fish for domestic consumption by a family unit. These same principles underpinned the definition of recreational fishing's purpose proposed by the RFAC in 1990:

> To aim to catch a feed for oneself and family and, for a variety of personal reasons, to enjoy the experience along the way. (Cusack and Cribb 1991)

The review recommended that the Fisheries Department urgently commence a dedicated program of research, compliance, management and education for recreational fisheries.

The review was promoted during the early 1990s as an example that other Australian states should follow, and the Department's efforts were lauded for promoting sustainable practises in the recreational sector (Kearney 1995). The reality, however, was that some visitors to Shark Bay had been pushing the concept of recreational fishing to the limit for many years beforehand. Since the 1970s it had been commonplace for amateurs to arrive for the snapper season equipped with large portable freezers, which were typically emptied of food and beverages over the course of a holiday and filled in turn with fillets of snapper that would be taken away at the conclusion of the visit. It was not uncommon for as much as 500 kg of fish fillets (approximately equal to 1.0–1.5 t of whole snapper) to be taken away from the Bay by a single angling party. Dubbed 'shamateurs' or the 'freezer brigade', these fishers were rumoured to make several fishing trips per day, and to sell or barter their catches on returning to Perth (Zekulich 1981). Their behaviour occasionally attracted the attention of newspapers, but for the most part it went unmonitored, so that the full impact of recreational fishing during the 1970s and early 1980s is unknown. But as snapper are a long-lived species, living to 40+ years in Western Australian waters (Norriss and Crisafulli 2010), the effects of high levels of fishing in the 1970s were still seen in the age structure of snapper populations in the 1990s, when the stock was coming under increasing pressure (Jackson et al. 2010).

The first survey of recreational fishing in Shark Bay was undertaken by the Department in 1983. This survey was part of a major review of the region's commercial snapper fisheries that subsequently led to the introduction of a limited-entry quota fishery in 1987. The research included tagging and genetic studies, the results of which indicated the existence of separate breeding stocks in the inner gulfs waters. The recreational survey combined aerial boats counts with creel surveys (angler surveys) undertaken at boat ramps during the peak winter season. Although many recreational boats were observed during the aerial counts, often more than 100 each day at the peak of the season, the creel survey showed that although snapper was the species most targeted by anglers, other demersal species accounted for the majority of their catches, and few fishers were found in possession of the daily bag limit for pink snapper.[2] Based on the results of this survey, the recreational snapper catch in 1983 was estimated to be around 7 t from 6,500 recreational fisher days in the Eastern Gulf, around 12 t from 3,500 days in Denham Sound, and around 17 t from 4,500 days in Freycinet Estuary (Jackson et al. 2002). The total recreational catch of snapper for the whole Shark Bay region including a minor contribution from oceanic waters outside the Bay was estimated at around 45 t (Jackson and Moran 2012).

Yet just as this picture emerged, two factors combined to fundamentally alter both the level and intensity of recreational fishing at Shark Bay. The first was a significant improvement in road access to Peron Peninsula. In 1986 the road that linked Denham to the North-West coastal highway, the main connection with the large population centre of Perth to the south, was sealed, or bituminised, for the first

---

[2] The results of this survey, conducted by the Western Australian Fisheries Department's Mike Moran, have not been published (See Fisheries Department 1980/1981).

time. A year later the sealed road was extended to the camping and resort settlement of Monkey Mia, a popular destination famous for daily visits by bottlenose dolphins on the shores of the Eastern Gulf. Tourism to Shark Bay increased significantly, with visitor numbers passing 100,000 annually by the mid-1990s (Marshall and Moore 2000). The majority of tourists visited during the winter season, and with the new road enabling more and larger boats to be brought into Shark Bay, the snapper fishery experienced a similar surge in participation. Secondly, this increase in recreational fishing effort coincided with the era when fish-finding technologies were becoming more affordable to the average recreational fisher. Fish sounders, used on commercial fishing vessels in Western Australia since the 1950s (Marriot et al. 2011), began to be more widely fitted to recreational vessels in the 1970s, and when colour sounders became available in the early 1980s the uptake amongst amateurs was rapid. The introduction of GPS devices was also rapid from the late 1980s onwards. By 1999, over 70% of recreational vessels inspected at Shark Bay were equipped with fish sounders, and over 50% with GPS devices (Sumner and Malseed 2002). As it does with commercial fishing, the use of this technology resulted in the increased efficiency of recreational vessels. Instead of spending hours locating the 'patches' around which snapper congregated, fishers could simply locate fish using their sounders and, once good ground had been found, record the location in their GPS so that the same area could be easily relocated on a subsequent visit. The increased effort combined with increased efficiency put severe pressure on the inner gulf snapper populations.

The impact of these changes in the recreational fishery came to a head during the 1995 season. In the Eastern Gulf a major snapper spawning site out from Monkey Mia, known previously only to a handful of local residents, was 'discovered' by a local charter boat (fishing tour) operator and, as a result, became widely known to visiting anglers who were in the habit of following charter boats at sea. Between May and July 1995 this cluster of aggregations, known colloquially as 'the Patch', was targeted relentlessly by a sizeable fleet of recreational fishing boats. An estimated 64 t of snapper were taken during this period. At the time, the Department of Fisheries suggested that the total catch in the Eastern Gulf that year might have been in the order of 100 t when recreational fishing out of other locations (such as Gladstone, on the eastern shores of Shark Bay, and the town of Carnarvon, on the Bay's northern shores) was taken into consideration. In 1996, an informal survey of recreational fishing in Shark Bay found low numbers of larger fish observed in recreational catches landed at Monkey Mia in contrast to the previous year, which indicated that the minimum length at that time (41 cm) did not offer adequate protection for the breeding stock at periods of high fishing (Sumner and Steckis 1999). In the Western Gulf, recreational snapper catches were continuing to rise, and concerns were also held there for the overall sustainability of the recreational snapper fishery.

Crisis and Response: Management and Science Address Overfishing

The situation in Shark Bay in the mid-1990s presented an unusual challenge to Western Australia's Department of Fisheries. The looming collapse of such an important recreational fishery, so soon after Shark Bay had been added to the World Heritage register, was challenging not only for the Department but for the Government in general. It was apparent that recreational fishing had been the principal cause of the crisis. By this time it was understood that the Bay's snapper constituted at least three separate breeding populations, with discrete populations in the Western and Eastern Gulf where recreational fishing was concentrated. Anecdotal evidence and the available survey data pointed to a marked increase in both total recreational effort and in the overall intensity of recreational fishing between the mid-1980s and mid-1990s. In 1998 the Director of Recfishwest, Western Australia's peak recreational fishing organisation, publicly acknowledged the role of the recreational sector in the depletion of the inner gulf snapper stocks, and endorsed the response that had been implemented by the Department's policy staff and research scientists (Prokop 1998). Yet some fishers continued to refuse to recognise that overfishing had occurred and strongly rejected the need for new or tighter restrictions on their fishing activities. The controversy that played out in 1997 and 1998 highlighted the importance of research to management—or in other words, the provision of scientific information that could provide a basis for the implementation of more effective fishery management strategies.

Even following the review of recreational fishing completed in 1990, most recreational fisheries in Western Australia, as elsewhere in Australia, were essentially un-managed from a stock sustainability perspective. The marine finfish stocks targeted by recreational fishers at the time had neither clear management objectives for the commercial or recreational sectors nor resource level biomass targets, and in most cases the information available on recreational catch and effort and the biology of targeted species was inadequate. The regulations that did apply, namely minimum size limits and bag limits, were based largely on 'social' rather than scientific criteria, a situation that was alluded to by the Department's 1990 definition of the objectives of recreational fishing. Furthermore, although recreational fishers were subject to some form of 'social-based' regulation, the notion that fish constituted a common-property resource and that private citizens had a 'right to fish' in marine waters remained untested principles of recreational fisheries management, even if the state's Fish Resources Management Act 1994 provided scope for declaring closed seasons and protecting certain species in addition to gear restrictions and bag and size limits. Social values, along with the absence of a proper management framework, therefore limited the range of management options that could be expected to achieve political support and popular compliance. However, the unprecedented situation that had developed at Shark Bay pointed to the need for new approaches to the management of recreational fishing.

In 1995, when the impact of recreational snapper fishing in the Eastern Gulf was becoming more apparent, the Denham Regional Recreational Fishing Advisory Committee (one of a network of Regional Recreational Fishing Advisory Committees, or RRFACs, established under the main RFAC after 1989) developed a proposal for a daily bag limit for pink snapper of four fish per day, introduce a 'slot limit' prohibiting retention of fish less than 45 cm and greater than 70 cm, and introduce an overall possession limit. But the proposal was not widely supported outside of the Denham RRFAC, and none of the suggestions were formally put forward to the Department. However, by the end of the 1996 season the indications that the Eastern Gulf stock was seriously depleted had become increasingly clear, presenting the Department with the need to take action (Jackson 2007). The response came in time for the start of the next season. Based on the results of the 1996 recreational survey (Sumner and Steckis 1999) and trawl surveys that indicated very low levels of juvenile snapper recruitment (Jackson and Moran 2012), the Department, in May 1997, took the unprecedented step of introducing a moratorium on fishing for snapper in the Eastern Gulf. It was the first time that a marine scalefish fishery had been closed to public access in Western Australia.

The reaction to the moratorium was furious. Some fishers objected on principle to the introduction of such a drastic restriction. Others, misunderstanding the consequences of the aggregating behavior of spawning snapper, rejected the suggestion that stocks in the Eastern Gulf were dangerously low. The local Shire Council and some business operators were particularly vocal in their opposition to the Department's decision. Tourism had grown in the late 1980s to surpass commercial fishing and pastoralism in importance to the local economy, and several Denham-based operators objected to the ban that coincided with the peak tourism period. Others, including the Shark Bay Shire Council, claimed that prawn and scallop trawlers operating in the deeper waters of Denham Sound were causing greater damage to snapper stocks than recreational fishers, and called for a ban on trawling to be introduced instead (Amalfi 1997). As the moratorium was reported in daily newspapers and fishing magazines many fishers took to telephoning the Fisheries Department or writing directly to the Fisheries Minister to register their protest. Two months later, the Minister relented in direct response to the outcry, and in July 1997 the moratorium was overturned (Marshall and Moore 2000). Fishing for snapper in the Eastern Gulf resumed but under more conservative regulations that included a slot limit of 50 to 70 cm, a daily bag limit of two fish per day, and a 'no fishing zone' covering four square nautical miles and the more than 20 spawning locations in the vicinity of the 'Patch' reported by anglers. Some indication of the depth of feelings that had been raised over the moratorium was revealed later in 1997 when buoys delineating the no-fishing zone were found washed ashore near Monkey Mia riddled with bullet holes (Carlish 2010). New regulations were also introduced in Denham Sound and the Freycinet Estuary to protect the Western Gulf stock, including an increase in the minimum legal length from 41 to 45 cm, the introduction of a maximum legal length of 70 cm that applied to two fish per day, and a reduced overall bag limit of four fish per day.

In 1997, the Department initiated a broad research programme aimed at improving understanding of snapper biology and population dynamics, and to provide estimates of the recreational catch and fishing effort in inner Shark Bay. The research included an evaluation of the daily egg production method (DEPM), a technique that had been employed successfully to estimate spawning biomass of snapper populations elsewhere in South Australia and New Zealand. A series of pilot DEPM surveys were completed in 1997, yielding valuable insights into the spatial and temporal variations in spawning patterns across the inner gulfs, and confirming the usefulness of the technique (Jackson et al. 2012). Other aspects of the Department's research included new investigations into stock structure using genetics in conjunction with otolith chemistry techniques and a tagging study to ascertain more about the movement of snapper within the inner gulfs (Norriss et al. 2012). Preliminary estimates from the pilot DEPM surveys of the size of the snapper spawning stock in the Eastern Gulf, and for Denham Sound and Freycinet Estuary in the Western Gulf, were initially presented to the local RRFAC in early 1998. These estimates suggested that the Eastern Gulf stock was depleted to a level estimated to be less than 5 % of the unexploited stock, with only around 10–20 t of spawning biomass remaining (Jackson 2007). Supported now with quantitative information on stock size that was unavailable in 1997, the Department reintroduced a moratorium on snapper fishing in the Eastern Gulf in June 1998.

The move precipitated a renewed controversy. Protests against the moratorium were again based on the contrary perception of snapper abundance that some recreational fishers formed on the basis of aggregating behavior during the spawning season, and on continued concern around the perceived impact of prawn trawlers that sustained the ingrained prejudices held by recreational fishers towards the commercial sector (Zekulich 1998). The science behind the moratorium was questioned by some. The President of the Shark Bay Shire, among the most vocal of critics, articulated a popular criticism of the DEPM survey when he claimed that it was 'flimsy, done at the wrong place and at the wrong time' (Baylen 1998). While the Department recognized some limitations in the sampling in the 1997 pilot surveys, the results provided a baseline of quantitative data that could not be ignored. Samples from Denham Sound and Freycinet Estuary were more positive by comparison, but still indicative of stocks under pressure (Jackson 2007). In 1998, the second year of DEPM surveys, results confirmed that spawning stocks remained low across the inner gulfs and in the Eastern Gulf in particular. The moratorium remained in place, and research activities continued.

During 1999 the findings of the first Gascoyne wide recreational fishing survey became available (Sumner et al. 2002). An estimated 12,000 + snapper, equivalent to around 38 t, had been retained by recreational fishers in the Western Gulf during the period April 1998 to May 1999; around 12 t had been taken from Denham Sound, and around 25 t from Freycinet Estuary (the Eastern Gulf was closed to snapper fishing at this time). Over 90 % of the snapper caught in Denham Sound were undersized and returned to the water, with a further 15,000 fish, 70 % of the total catch, returned in Freycinet Estuary (Sumner and Malseed 2002). By 2000, results from the genetics, otolith chemistry and tagging studies that commenced in

1998 were indicating that stock structure in the Western Gulf was more complex than previously thought, and that separate spawning populations existed in Denham Sound and Freycinet Estuary (Jackson 2007). Management arrangements were re-framed accordingly in 2000 to recognize the existence of three separate inner gulf stocks, in the Eastern Gulf, Denham Sound and Freycinet Estuary. The minimum size limit was increased to 50 cm and the bag limit reduced to two fish per day for both Denham Sound (accessed through Denham) and Freycinet Estuary (accessed through Nanga), with a slot limit of only one fish over 70 cm applying for Freycinet Estuary, and 6 week spawning closure introduced in the Freycinet Estuary from 15 August to 31 October. This latter move upset many of the Nanga 'regulars' as it coincided, not by accident, with their favourite visiting period, when the snapper in that part of the Bay were active and more catchable.

Back from the Brink: New Approaches to Recreational Fisheries Management

The implementation of a seasonal closure and the adoption of a slot limit for Freyci-net Estuary was a major turning point in the management of snapper in Shark Bay. A 'four-stock' management regime was now in place, comprising of a set of arrange-ments that applied to the commercial fishery that targeted the large oceanic stock in the Bay's outer waters and the adjoining continental shelf, and regulations that ap-plied to recreational fishers and pertained to each of the three inner gulf populations in Freycinet Estuary, Denham Sound, and the Eastern Gulf (Jackson 2007). Science and policy had promoted a historic shift from socially-based to scientifically-based management of recreational fishing impacts, with management arrangements now reflecting the fact that the Bay's unique ecology produced differences in the biological characteristics of snapper at a fine spatial scale. Yet the new arrangements announced in 2000 were introduced in the midst of an ongoing controversy that meant the future of this recreational fishery was far from assured. The unwillingness of some anglers and members of the local community to accept the role that recreational fishing had played in the stock depletion, or the scientific findings, and the often acrimonious public debate that ensued between 1997 and 2000, initially fostered considerable op-position to what were seen as overly harsh management measures aimed at contain-ing the total recreational impact on each of the inner gulf stocks. However, as the wider angling community and the recreational fishing media considered the problem, stronger support for long-term and effective management both within and outside the Denham community began to develop. As the new decade began, the Department of Fisheries remained confronted by the challenge of building sufficient political and public support for the tough approach that was necessary to ensure that Shark Bay's recreational snapper fishery was sustainably managed.

Further research revealed the complexity of this challenge. DEPM surveys early in the early 2000s confirmed that breeding stocks were recovering in the Eastern Gulf, were satisfactory in Denham Sound, but were declining in the Freycinet Estuary. Further recreational fishing surveys between May 2000 and April 2001, and May 2001 to April 2002, gave cause for concern. The estimated recreational snapper catch in Freycinet Estuary increased from around 15 t in 2000–2001 to more than 22 t in 2001–2002 (Sumner and Malseed 2002). While this represented a reduction from the total catch in 1998–1999, the new regulations had not achieved the reductions in snapper catch in the Freycinet area that were required. In Denham Sound the snapper catch fell from around 9 to 7 t over the same period. Additional evidence of the recovery of the Eastern Gulf stock was provided by figures that indicated that around 12,000 snapper, 84 % of which were less than 500 mm in length, had been released during the same 12-month period (Sumner and Malseed 2002). These results were a mixed bag for the Department. At the very time that the situation in Freycinet Estuary had become critical, an expectation was emerging that the Eastern Gulf would be reopened to recreational snapper fishing in the very near future.

As the pressure to re-open the Eastern Gulf built, the Department was determined to learn from the experiences of the late 1990s. In 2002, the Minister for Fisheries, seeking to avoid the large and boisterous public meetings that had been a hallmark of the consultation process in the late 1990s, acted on the Department's advice to establish a working group to examine management options and review research findings on the Shark Bay recreational snapper fishery. The Department was asked to undertake a review of research and management between 1997 and 2001. For this first time in this fishery, stock assessment models were developed to investigate the effects of different catch scenarios and management regimes on each of the three inner gulf populations (Stephenson and Jackson 2005). The Department proposed that the key management objective for the fishery should be to rebuild and maintain each of the inner gulf stocks to 40 % of the unexploited level. This type of management target was the international scientific benchmark for similarly long-lived species. To achieve this, the working group adopted an approach more commonly used in some of Western Australia's main commercial fisheries but which remained untried with a recreational fishery anywhere in Australia: the setting of a Total Allowable Catch, or TAC, for each snapper stock. In another first, the TACs were allocated on the basis of 75 % of the total catch in each area to recreational fishers and 25 % to the commercial fishers based in Denham who had a long history of taking small quantities of snapper throughout the Bay. The TACs were initially set at 10 t (equivalent to approx. 3,300 fish) in Denham Sound, 5 t (approx. 1,200 fish) in Freycinet Estuary, and 15 t (approx. 5,000 fish) in the Eastern Gulf (Jackson et al. 2005).

The adoption of TACs brought about further changes to the management arrangements. To avoid confusion the same bag and size limits were applied across all three areas with a slot limit of 50–70 cm determining what could be legally retained, and a bag limit of one fish per day. Seasonal closures to protect spawning snapper were used in the Eastern Gulf (1 April to 31 July), and in Freycinet Estuary (15 August to 30 September). Under these arrangements, and after almost 5 years since the moratorium was introduced, recreational fishing for snapper was resumed in the Eastern Gulf in March 2003. In the case of the Freycinet Estuary, where the

TAC represented only around 20–25% of the average recreational catches between 1998 and 2001, a third and particularly innovative approach was adopted in the form of a quota-tag system (Mitchell et al. 2008). From 2003 onwards, to be able to take snapper in the Freycinet Estuary, recreational fishers were required to pre-purchase management quota-tags from the Fisheries Department. The closed season and bag limit of one snapper still applied, but no snapper could be landed from Freycinet Estuary without a tag in place; each tag was inserted through the snapper's mouth and secured using a tamper-proof locking mechanism prior to landing. Individual fishers were limited to two tags each under the lottery system. Tags were released to the public from March 2003 at the Fisheries Department offices in Denham and Carnarvon, and at the Department's headquarters in Perth at a cost of $10 each.

In a recreational fishing setting where no licenses had ever been required for marine angling, the setting of the tag price at a near equivalent of the commercial beach price for the targeted species was almost as controversial as the introduction of the tags themselves. Perhaps inevitably, the public response to the new quota-tag system for Freycinet Estuary was not always positive. Interviews with recreational fishers in Shark Bay revealed that 97% of respondents rejected the new arrangement as a preferred policy option. Although 900 tags were allocated to the recreational sector in 2003, only 264 applicants entered the lottery, with 528 tags actually being issued; the commercial allocation of 300 tags were returned by Denham's commercial fishers in a conservation gesture to assist stock recovery. But acceptance grew in following years, as the Department continued to publicise the workings and objectives of the scheme, and equity issues were addressed through the introduction of a lottery system in place of the original in-person applications that were the basis of the first allocation. In 2004, the number of applicants for tags doubled to 526, with 880 of the 900 tags issued, and applications increased again to reach 627 in 2005, when all 900 tags were issued for the first time (Jackson et al. 2005). The combination of tight bag limits, spawning closures and a quota-tag system in Freycinet Estuary appeared to be succeeding. In 2005, at the conclusion of the first 3-year period since the introduction of the TACs, the Department indicated that the 40% management target had been achieved in both the Eastern Gulf and Denham Sound and while the Freycinet stock was still only around 25% it was showing signs of slow recovery. For the next 3-year management cycle (2006–2008) the TAC for Denham Sound was increased from 10 to 15 t, the closed season for the Eastern Gulf was reduced by 1 month to run from 1 May to 31 July, and the number of tags available to recreational fishers in Freycinet Estuary was increased from 900 to 1,050 (Jackson 2007).

The recovery of the inner gulf stocks masked a less conspicuous but nonetheless significant change in the attitudes of recreational fishers themselves. Although the Department had been at pains to keep the public informed of decision-making and to make the results of research widely available, the roots of this change can be traced back to the first collaborative initiatives between the Fisheries Department, the Shark Bay community and the recreational fishing community that emerged out of the heated controversies of the late 1990s. The tagging project involving the Department and local volunteer recreational fishers and directed at improving understanding of the movement of snapper in the inner gulfs, which commenced in 1998, was among the first such initiatives (Norriss et al. 2012). Local volunteers had also assisted with DEPM surveys by collecting samples of spawning fish from 1998 onwards (Jackson et al. 2012). This involvement of volunteers had spread from a core group of Denham-based fishers to include visiting recreational fishers at Nanga, who were particularly impacted by both the closed season and quota-tag system and had, at least initially, been among the most ardent critics of management reforms since 2000. Public engagement in the Department's activities reached a high point during the winter of 2004 with a new tagging study in the Eastern Gulf, where the objective was to use mark-recapture approach to estimate stock size. Some 70 recreational fishers, using 21 private boats, tagged 4,200 large snapper over a 10-day period under the supervision of the Department's research scientists (Anderson 2004; Jackson et al. 2005). Research outcomes and management decisions continued to be disseminated via media releases, interviews with fishing magazines, and at open public meetings, and with the same research and policy staff featuring year after year and the success of management reforms becoming apparent, public confidence in the Department began to build.

Between 2003 and 2006 recreational catches of snapper stabilised. The conservative catch controls remained in place. The TAC of 15 t remained in the Eastern Gulf and Denham Sound, and at 5 t in the Freycinet Estuary. Research was moving into more of a monitoring role with DEPM surveys undertaken less regularly in the Eastern Gulf and Denham Sound while recreational catches were monitored annually (Wise et al. 2012). As the stocks recovered and the conservative regulations remained in place, the proportion of snapper released by recreational fishers remained high, averaging around 90 % of all fish caught during the period 1998–2007. The working group met again in 2008, with no changes to the management arrangements for the period 2009–2011, and again in 2011. The most recent stock assessments indicated that that Eastern Gulf and Denham Sound stocks remain well above the management target level while the Freycinet Estuary stock continues to slowly rebuild. In late 2012, it was announced by the Minister for Fisheries that the daily bag limit for pink snapper at Shark Bay would increase from one to two in the year ahead.

Conclusion: Science, Policy, and Recreational Fisheries Management

In 2006, the Fisheries Department was awarded the Western Australian Premier's Award for Excellence in Public Sector Management, the state's leading award for Government agencies, in recognition of the restoration of Shark Bay's inner gulf snapper populations. The Award acknowledged the innovative work of the Department over the preceding decade across criteria that included community partnership and collaboration, innovation, achievement of sustainability goals and environmental protection, and 'contribution to the Western Australian lifestyle'

(Jackson 2007). Much had been learned from the Shark Bay experience. New research techniques had been introduced to the state and modified to suit local conditions. Research had been published in the international scientific literature that highlighted the adaptation of this iconic species to a unique marine environment. The susceptibility of highly localised fish populations to overfishing by recreational fishing had been proven. The various management measures trialled had yielded insights into the effectiveness of different policy options that assisted with emerging policy challenges in other key recreational fisheries, such as the snapper fishery in Cockburn Sound near Perth, where a closed season to protect spawning aggregations was introduced in 2000. Additionally, the successful engagement of stakeholders in research and decision-making processes had shown how science and policy could combine to produce sustainable management, and sustainable behaviour, in a popular recreational fishery. Some of the recreational volunteers who started out with very negative views of the Department and its intervention in Shark Bay have, through their increasing involvement with the research program, gained an understanding of the role that science has played and are now very supportive.

Since the mid-1990s Shark Bay's inner gulf snapper fishery has become one of the most intensively-studied and better-understood marine recreational fisheries in Australia. From a historical standpoint it provides an important case-study of the impact that recreational fishing can have on highly targeted stocks, showing that recreational fishers, by virtue of their greater numbers and their uptake of technologies developed for the commercial sector, can have an equal if not greater total catch than professional fishers in the same or similar fisheries. The Shark Bay case-study also demonstrates the complexity of the challenges associated with sustainably managing marine recreational fisheries. The effectiveness of traditional recreational management measures is increasingly being questioned. As more and more jurisdictions move towards implementing ecosystem-based management approaches, strategies to ensure sustainable harvests will be required for all sectors—commercial, recreational and artisanal alike. The Shark Bay case-study also highlights the difficulty of gaining support for management intervention in the absence of baseline biological information and in a political environment where the principles of precaution and risk are neither well understood nor widely accepted. Where recreational fishers lack the understanding of biological concepts and management principles that are more typically present amongst commercial fishers, it is critical that the results of research and the objectives of management are communicated adequately to the recreational fishing community. The benefits of positive and regular interactions between recreational fishers, scientists and managers are now widely recognized (Granek et al. 2008). Once change is enacted, ongoing research is needed to monitor the effects of the new arrangements over the longer term. The politics of recreational fishing can be a stormy sea to navigate, but the Shark Bay experience shows that combining science and policy with stakeholder engagement makes for a smoother voyage.

References

Aas O (ed) (2008) Global challenges in recreational fisheries. Blackwell, Oxford

ABS (1987) Recreational fishing in Western Australia, July 1987. Cat. No. 7602.5. Australian Bureau of Statistics, Canberra

Amalfi C (1997) Shark Bay angry at fishing ban. West Australian, 29 April 1997

Anderson C (2004) Preserving the patch: plotting a future for Shark Bay snapper. Western Fisheries, Autumn 2004:14–15

Baylen G (1998) Snapper row draws angry line. West Australian, 27 May 1998

Bodeker P (1971) The Sandgropers' trail: an angling safari from Perth to the Kimberley. A.H. and A.W. Read, Sydney

Bowen BK (1961) The Shark Bay fishery on snapper chrysophrys unicolor (Quoy and Gaimard), Western Australian Fisheries Department

Carlish B (2010) Pinks back from the brink. Western Fisheries, July 2010:6–12

Christensen J (2008) Shark Bay 1616–1991: the spread of science and the emergence of ecology in a world heritage area. Ph. D. Thesis, The University of Western Australia, Perth

Christensen J (2009) Recreational fishing and fisheries management: a HMAP Asia project paper. Asia Research Centre Working Papers, No. 157, School of Social Sciences and Humanities, Murdoch University, Perth

Coleman FC, Figueira WF et al (2004) The impact of United States recreational fisheries on marine fish populations. Science 305:1958–1960

Cooke SJ, Cowx IG (2006) Contrasting recreational and commercial fishing: searching for common issues to promote unified conservation of fisheries resources and aquatic environments. Biol Conserv 128:93–108

Cooper R (1997) Shark Bay legends. L.J. Cogan, Geraldton

Cusack R, Cribb A (1991) The future of recreational fishing: final report of the recreational fishing advisory committee Western Australia. Fisheries management paper no. 41, Fisheries Department of Western Australia, Perth

Department of Fisheries (1980/1981) Project number 36—Shark Bay snapper research programme. File 80/81, vols 1–2, Cons. 5602, State Records Office of Western Australia, Perth

DPI (2007) Draft Perth recreational boating facilities study. Department of Planning and Infrastructure, Government of Western Australia, Perth

Edwards H (2000) Shark Bay through four centuries 1616 to 2000. Shire of Shark Bay, Denham

Granek EF, Madin EM et al (2008) Engaging recreational fishers in management and conservation: global case studies. Conserv Biol 22(5):1125–1134

Henry GW, Lyle JM (2003) The national recreational and indigenous fishing survey. Fisheries and Research Development Corporation report 1999/158, NSW Fisheries, Cronulla

Jackson G (2007) Fisheries biology and management of pink snapper, *Pagrus auratus*, in the inner gulfs of Shark Bay, Western Australia. Ph.D. thesis, Murdoch University, Perth

Jackson G, Cheng YW, Wakefield CB (2012) An evaluation of the daily egg production method to estimate spawning biomass of snapper (*Pagrus auratus*) stocks in inner Shark Bay, Western Australia, following more than a decade of surveys 1997–2007. Fisheries Research 117–118: 22–34

Jackson G, Moran M (2012) Recovery of inner Shark Bay snapper (*Pagrus auratus*) stocks: relevant research and effective recreational fisheries management in a World Heritage context. Mar Freshw Res 63:1180–1190

Jackson G, Norriss JV et al (2010) Spatial variation in life history characteristics of snapper (*Pagrus auratus*) within Shark Bay, Western Australia. N Z J Mar Freshw Res 44(1):1–15

Jackson G, Sumner N et al (2005) Comparing conventional 'social-based' and alternative output-based management models for recreational finfish fisheries using Shark Bay pink snapper as a case study. Department of Fisheries Western Australia, Perth

Jackson GR, Lenanton R et al (2002) Research and management of snapper, *Pagrus auratus*, stocks in the inner sulfs of Shark Bay, Western Australia. In: Coleman APM (ed) 3rd world recreational fishing conference, 21–24 May 2002. Amateur Fishermen's Association NT and Fisheries Group, Department of Business, Industry and Resource Development, Darwin

Kearney RE (1995) Recreational fishing: what's the catch? In: Hancock DA (ed) Recreational fishing: what's the catch? Proceedings no. 12, Australian Society for Fish Biology, Canberra

Kearney RE (2002) Recreational fishing: value is in the eye of the beholder. In: Pitcher TJ, Hollingworth CE (eds) Recreational fisheries: ecological, economic and social evaluation. Blackwell Science, Oxford

Lewin WC, Arlinghaus R et al (2004) Documented and potential biological impacts of recreational fishing: insights for management and conservation. Rev Fish Sci 14(4):305–367

Lindner RK, McLeod PB (1991) An economic impact of recreational fishing in Western Australia. Fisheries Management Paper No. 38. Department of Fisheries, Western Australia

Logan BW, Cebulski DE (1970) Sedimentary environments of Shark Bay, Western Australia. Am Assoc Pet Geol Mem 13:1–37

Marriott R, Wise B, St John J (2011) Historical changes in fishing efficiency in the west coast demersal scalefish fishery, Western Australia. ICES J Mar Sci 68(1):76–86

Marriott R, Jackson G et al (2012) Biology and stock status of demersal indicator species in the Gascoyne coast bioregion. Fisheries research report no. 228, Department of Fisheries Western Australia, Perth

Marshall DG, Moore SA (2000) Tragedy of the commons and the neglect of science: planning and management in the Shark Bay world heritage area. Environ Plan Law J 17:126–137

McPhee DP, Leadbitter D, Skilleter GA (2002) Swallowing the bait: is recreational fishing in Australia ecologically sustainable? Pac Conserv Biol 8:40–51

Mitchell RWD, Baba O et al (2008) Comparing management of recreational *Pagrus* fisheries in Shark Bay (Australia) and Sagami Bay (Japan): conventional catch controls versus stock enhancement. Mar Policy 32:27–37

Norriss JV, Crisafulli B (2010) Longevity in Australian snapper *Pagrus auratus* (Sparidae). J Roy Soc West Aust 93:129–132

Norriss JV, Moran M, Jackson G (2012) Tagging studies reveal restricted movement of snapper (*Pagrus auratus*) within Shark Bay, supporting fine-scale fisheries management. Mar Freshw Res 63:1191–1199

PA Management Consultants (1984) National Survey of participation in recreational fishing. Report to the Australian recreational fishing confederation. PA Management Consultants, South Melbourne

Paulin CD (1990) *Pagrus auratus*, a new combination for the species known as "snapper" in Australasian waters (Pisces: Sparidae). N Z J Mar Freshw Res 24(2):259–265

Prokop FB (1998) Where have all the snapper gone? The plight of pink snapper in Shark Bay. Western Fisheries, Winter 1998:18–25

Shaw J (2000) Gascoyne region. Fisheries environmental management review no. 1. Department of Fisheries Western Australia, Perth

Stephenson P, Jackson G (2005) Managing depleted snapper stocks in inner Shark Bay, Western Australia. In: Kruse GH, Gallucci VF et al (eds) Fisheries assessment and management in data-limited situations. Alaska Sea Grant Program, Fairbanks

Sumner NR, Malseed BE (2002) Quantification of changes in recreational catch and effort on inner Shark Bay snapper species following implementation of responsive management changes. Department of Fisheries Western Australia, Perth

Sumner NR, Steckis RA (1999) Statistical analysis of Gascoyne region recreational fishing study July 1996. Fisheries research report no. 115, Department of Fisheries Western Australia, Perth

Sumner NR, Williamson PC, Malseed BE (2002) A 12-month survey of recreational fishing during 1998–99. Fisheries research report no. 139, Department of Fisheries Western Australia, Perth

Wise B, Telfer C et al (2012) Long-term on-site monitoring of boat-based recreational fishing in Shark Bay, Western Australia: providing advice for fisheries management in a world heritage area. Mar Freshw Res 63:1129–1141

Zekulich M (1981) Amateur fishermen plundering Shark Bay. West Australian, 15 August 1981

Zekulich M (1998) Shire wants trawling halted in Shark Bay. West Australian, 30 March 1998

Chapter 14
Conclusion: Learning from Asian and Indo-Pacific Fisheries history

Poul Holm

Abstract Until very recently very little had been published on the history of Asian fisheries. This sorry state of neglect changed at one stroke with the publication of Butcher's 2004 study of South East Asian fisheries. The present volume demonstrates the breadth and depth of progress in recent years. It is high time that historians from outside the region take note and reconsider some of our well-established patterns of thought in light of what we may learn from these new perspectives.

Keywords Asian fisheries history · Indo-Pacific fisheries history · HMAP Asia · Marine environmental history

Until very recently it seemed as if Asian and Indo-Pacific fisheries had no history. Certainly, if one looks in the established literature there is an abundance of works on the Atlantic but apart from some ethnographic studies very little on the history of Asian fisheries. A basic work like Cushing's *The Provident Sea* (1988) opens with Hornell's descriptions of early twentieth century Indian fishing and has some text on the development of modern Japanese fisheries and the West Australian fishery for rock lobster. Cushing was conscious that 'much has happened elsewhere' but clearly had little information to hand. This sorry state of neglect changed at one stroke with the publication of Butcher's study of South East Asian fisheries, and quite rightly several contributors to the present volume identify this book as a path breaking work (Butcher 2004). Other more specialised works like David Luke Howell's study of nineteenth-century Japanese fisheries, Chen's study of twentieth-century Taiwanese fisheries, and Muscolino's history of fishing wars in Imperial and modern China have contributed to make Asian fisheries history an established academic field (Chen 2009; Muscolino 2009; Howell 1995). The present volume demonstrates the breadth and depth of progress in recent years. It is high time that historians from outside the region take note and reconsider some of our well-established patterns of thought in light of what we may learn from these new perspectives.

Concepts like diversity, communities, and colonialism are key to many contributions to this volume. In Chap. 2, Joseph Christensen attributes the late rise of Asian

---

P. Holm (✉)
Trinity Long Room Hub, Trinity College Dublin,
College Green, Dublin 2, Ireland
e-mail: holmp@tcd.ie

J. Christensen, M. Tull (eds.), *Historical Perspectives of Fisheries Exploitation in the Indo-Pacific*, MARE Publication Series 12, DOI 10.1007/978-94-017-8727-7_14,

fisheries history to the complexity and richness of ecosystems and human cultures of the Indo-Pacific region, which defy a periodization according to the pattern historians often use for the Atlantic fisheries. Building on Butcher's work in relation to South East Asia, he observes that the Indo-Pacific region did not have one or two overwhelmingly important commercial fisheries, such as cod and herring in the Atlantic, around which a history may be structured. This point is well taken but looking at the richness of the present volume I wonder if that concentration on major commercial fisheries has not sometimes worked to the detriment of Western fisheries history by focusing our attention too narrowly. Asian fisheries historians demonstrate a keen awareness of species diversity and regional patterns. A recalibration of Western fisheries history with more attention to ecological specificities would certainly enrich the analysis and might provide us with some surprising insights. The diversity of Asian coastal fishing communities depended on rich and highly diverse local resource bases well into the twentieth century and many still do to this day. Most, but not all, Atlantic fishing communities had lost that local dependency by the late nineteenth century as they pursued the riches of the large commercial fisheries of the open sea. But rather than perceive of this development as an almost automatic effect of serial depletion, we may become more sensitive to the costs to local communities by learning from Asian fisheries history.

Several studies in this volume make clear the crucial importance of the Asian colonial experience for problems of availability of capital, the role of the state and the slow development of scientific support for the fisheries. Problems of colonialism and post colonialism are just as relevant for the Atlantic experience, and yet they remain understudied. Historians of early modern North American fisheries do of course take account of the colonial setting but we have still no good studies of European colonialism and the fisheries, and there is a glaring lack of studies of twentieth-century post colonialism and Atlantic fisheries, not least with regard to African and Latin American fisheries.

Atlantic historians will need to collaborate with Asian historians as we focus our attention on the globalization of fisheries in the second half of the twentieth century. The seas of the world have become one ocean in economic terms. Fisheries for tuna and sea cucumber have driven Asian vessels into the Atlantic just as earlier Atlantic whalers made the Pacific their home grounds. The global trade in marine foods is a vast and understudied field as indeed is the development of illegal and unregulated high-seas fisheries. The rise of aquaculture, not least in South East Asia, is a phenomenon of worldwide importance both as regards food security and changing urban life styles and has come at tremendous environmental cost but still lacks comprehensive historical treatment. The agenda for future collaboration between Western and Eastern historians is vast and rapidly expanding.

If there is much work to be done, it may be fair to ask if we are prepared to undertake the challenge? Without wanting to be too optimistic, I think the answer is that we know what needs to be done and the solution is largely up to ourselves. In the last 20 years, the discipline of history has changed dramatically and for the better. The discipline used to be mired in nationalistic agendas, constrained by language, and largely ignorant of what was written in other countries if not to speak of other continents. If these problems are still with us, the good news is that new approaches such as 'world history' and 'environmental history' are changing the discipline for good. We now have journals of global and environmental history and historians are talking to and increasingly publishing with scholars from other disciplines. Most importantly, there has been a 'historical turn' of the natural sciences. In the last two or three decades we—citizens and politicians in rich and poor countries—have come to recognize that our planet is small and vulnerable. This recognition necessitates a historical perspective on modern existence. One of the most important themes of this revitalization of history is the relationship between humankind and nature. Collaboration across disciplines is recognized as essential to address problems of global environmental change, and history provides that insight into the long term, which is badly needed to provide a sense of proportion and understanding of when, how and why humans change perceptions and behaviour in our interaction with surroundings (Holm et al. 2013).

Marine environmental history is an example of this change, and the capacity building, which took place in the early years of the HMAP project, may offer some valuable experience for the future. HMAP was established as an alliance of people from many disciplines with a mission to do what had until then been unthinkable: to bring history, archaeology, biology, statistics and all other relevant disciplines together for a historical dimension to our understanding of human impact and dependence on the sea. No discipline trains students for the kind of 'environmental literacy' of the many disciplines, which may inform such a quest for new knowledge. As the very first step, we had summer schools to train a new generation of marine environmental historians and historical ecologists from around the globe to practice interdisciplinary collaboration. Some contributors to this volume benefitted from this experience and it has since sustained the Oceans Past conferences and the many outputs (more than 200 peer-reviewed papers) of the HMAP project. International summer schools are vital for future academic development of our understanding of the sea as they bring together students from across disciplines to learn how multidisciplinary perspective may enrich their own disciplinary insights.

Another lesson, which I would draw from HMAP, is the key role of individuals and institutional commitment. Right from its establishment in 2000, HMAP was designed to be a global meeting ground for marine environmental history. In the first years, however, the Steering Group consisted of Western academics and in our search for Eastern partners we were defeated by our own ignorance, and tenuous leads to possible interested parties in Asia came to nought. It was only when Professor Malcolm Tull and the Asia Research Centre of Murdoch University took up the challenge that sustained research began on the questions raised by HMAP. Their commitment to build capacity for the study of the marine environmental history of the Asian and Indo-Pacific regions was crucial.

In future years, marine environmental history is likely to grow and benefit not only from a dialogue with the natural sciences as has been the case in the past 15 years or so, but increasingly also from dialogues within the humanities. History is not the only discipline to have been 'environed'. Environmental humanities is a broad concept, which encompasses the exciting developments happening also in disciplines such as literature and media studies, art, philosophy and educational studies. As that dialogue across disciplines evolves, historians may find themselves no longer at one end of a spectrum of disciplines but perhaps as bridge-builders between analytical sciences and interpretative and creative practices of the arts. We all stand to benefit by bringing diverse forms of knowledge into play.

References

Butcher JG (2004) The closing of the frontier: a history of the marine fisheries of Southeast Asia, c.1850–2000. ISEAS Publications, Singapore

Chen HT (2009) Taiwanese distant-water fisheries in Southeast Asia, 1936–1977. International Maritime Economic History Association, St. John's, Newfoundland

Holm P, Goodsite M, Cloetingh S, Vanheusden B, Yusoff K, Agnoletti M, Bedřich M, Lang D, Leemans R, Oerstroem Moeller J, Pardo Buendia M, Pohl W, Sors A, Zondervan R (2013) Collaboration between the natural, social and human sciences in global change studies. Environ Sci Policy 28:25–35

Howell DL (1995) Capitalism from within: economy, society, and the state in a Japanese fishery. University of California Press, Berkley

Muscolino M (2009) Fishing wars and environmental change in late imperial and modern China. Harvard East Asian Monographs, Harvard University Press, Cambridge

Index

J. Christensen, M. Tull (eds.), *Historical Perspectives of Fisheries Exploitation in the Indo-Pacific*, MARE Publication Series 12, DOI 10.1007/978-94-017-8727-7,
© Springer Science+Business Media Dordrecht 2014